Springer Series in Computational Mathematics

35

T0155892

Peter Deuflhard

Newton Methods
for Nonlinear Problems

Affine Invariance and Adaptive Algorithms

With 49 Figures

 Springer

Peter Deuflhard
Zuse Institute Berlin (ZIB)
Takustr. 7
14195 Berlin, Germany
and
Freie Universität Berlin
Dept. of Mathematics and Computer Science
deuflhard@zib.de

Mathematics Subject Classification (2000): 65-01, 65-02, 65F10, 65F20,
65H10, 65H20, 65J15, 65L10, 65L60, 65N30, 65N55, 65P30

ISSN 0179-3632
ISBN 978-3-540-21099-7 (hardcover)   e-ISBN 978-3-642-23899-4
ISBN 978-3-642-23898-7 (softcover)
DOI 10.1007/978-3-642-23899-4
Springer Heidelberg Dordrecht London New York

Library of Congress Control Number: 2011937965

Cover design: deblik, Berlin

Printed on acid-free paper

Springer is part of Springer Science+Business Media (www.springer.com)

Preface

In 1970, my former academic teacher Roland Bulirsch gave an exercise to his students, which indicated the fascinating invariance of the *ordinary* Newton method under general affine transformation. To my surprise, however, nearly all *global* Newton algorithms used damping or continuation strategies based on residual norms, which evidently lacked affine invariance. Even worse, nearly all convergence theorems appeared to be phrased in not affine invariant terms, among them the classical Newton-Kantorovich and Newton-Mysovskikh theorem. In fact, in those days it was common understanding among numerical analysts that convergence theorems were only expected to give qualitative insight, but not too much of quantitative advice for application, apart from toy problems.

This situation left me deeply unsatisfied, from the point of view of both mathematical aesthetics and algorithm design. Indeed, since my first academic steps, my scientific guideline has been and still is that 'good' mathematical theory should have a palpable influence on the construction of algorithms, while 'good' algorithms should be as firmly as possible backed by a transparently underlying mathematical theory. Only on such a basis, algorithms will be efficient enough to cope with the enormous difficulties of real life problems.

In 1972, I started to work along this line by constructing global Newton algorithms with affine invariant damping strategies [59]. Early companions on this road were Hans-Georg Bock, Gerhard Heindl, and Tetsuro Yamamoto. Since then, the tree of affine invariance has grown lustily, spreading out in many branches of Newton-type methods. So the plan of a comprehensive treatise on the subject arose naturally. Florian Potra, Ekkehard Sachs, and Andreas Griewank gave highly valuable detailed advice. Around 1992, a manuscript on the subject with a comparable working title had already swollen to 300 pages and been distributed among quite a number of colleagues who used it in their lectures or as a basis for their research. Clearly, these colleagues put screws on me to 'finish' that manuscript.

However, shortly after, new relevant aspects came up. In 1993, my former coworker Andreas Hohmann introduced *affine contravariance* in his PhD thesis [120] as a further coherent concept, especially useful in the context of inexact Newton methods with GMRES as inner iterative solver. From then on, the former 'affine invariance' had to be renamed, more precisely, as *affine covariance*. Once the door had been opened, two more concepts arose: in 1996, myself and Martin Weiser formulated *affine conjugacy* for convex optimization [84]; a few years later, I found *affine similarity* to be important for steady state problems in dynamical systems. As a consequence, I decided to rewrite the whole manuscript from scratch, with these four affine invariance concepts representing the columns of a structural matrix, whose rows are the various Newton and Gauss-Newton methods. A presentation of details of the contents is postponed to the next section.

This book has two faces: the first one is that of a *textbook* addressing itself to graduate students of mathematics and computational sciences, the second one is that of a *research monograph* addressing itself to numerical analysts and computational scientists working on the subject.

As a *textbook*, selected chapters may be useful in classes on Numerical Analysis, Nonlinear Optimization, Numerical ODEs, or Numerical PDEs. The presentation is striving for structural simplicity, but not at the expense of precision. It contains a lot of theorems and proofs, from affine invariant versions of the classical Newton-Kantorovich and Newton-Mysovskikh theorem (with proofs simpler than the traditional ones) up to new convergence theorems that are the basis for advanced algorithms in large scale scientific computing. I confess that I did not work out all details of all proofs, if they were folklore or if their structure appeared repeatedly. More elaboration on this aspect would have unduly blown up the volume without adding enough value for the construction of algorithms. However, I definitely made sure that each section is self-contained to a reasonable extent. At the end of each chapter, exercises are included. Web addresses for related software are given.

As a *research monograph*, the presentation (a) quite often goes into the depth covering a large amount of otherwise unpublished material, (b) is open in many directions of possible future research, some of which are explicitly indicated in the text. Even though the experienced reader will have no difficulties in identifying further open topics, let me mention a few of them: There is no complete coverage of all possible combinations of local and global, exact and inexact Newton or Gauss-Newton methods in connection with continuation methods—let alone of all their affine invariant realizations; in other words, the above structural matrix is far from being full. Moreover, apart from convex optimization and constrained nonlinear least squares problems, general optimization and optimal control is left out. Also not included are recent results on interior point methods as well as inverse problems in L^2, even though affine invariance has just started to play a role in these fields.

Generally speaking, finite dimensional problems and techniques dominate the material presented here—however, with the declared intent that the finite dimensional presentation should filter out promising paths into the infinite dimensional part of the mathematical world. This intent is exemplified in several sections, such as

- Section 6.2 on ODE initial value problems, where stiff problems are analyzed via a simplified Newton iteration in function space—replacing the Picard iteration, which appears to be suitable only for nonstiff problems,
- Section 7.4.2 on ODE boundary value problems, where an adaptive multi-level collocation method is worked out on the basis of an inexact Newton method in function space,
- Section 8.1 on asymptotic mesh independence, where finite and infinite dimensional Newton sequences are synoptically compared, and
- Section 8.3 on elliptic PDE boundary value problems, where inexact Newton multilevel finite element methods are presented in detail.

The *algorithmic paradigm*, given in Section 1.2.3 and used all over the whole book, will certainly be useful in a much wider context, far beyond Newton methods.

Unfortunately, after having finished this book, I will probably lose all my scientific friends, since I missed to quote exactly that part of their work that should have been quoted by all means. I cannot but apologize in advance, hoping that some of them will maintain their friendship nevertheless. In fact, as the literature on Newton methods is virtually unlimited, I decided to not even attempt to screen or pretend to have screened all the relevant literature, but to restrict the references essentially to those books and papers that are either intimately tied to affine invariance or have otherwise been taken as direct input for the presentation herein. Even with this restriction the list is still quite long.

At this point it is my pleasure to thank all those coworkers at ZIB, who have particularly helped me with the preparation of this book. My first thanks go to Rainer Roitzsch, without whose high motivation and deep TeX knowledge this book could never have appeared. My immediate next thanks go to Erlinda Körnig and Sigrid Wacker for their always friendly cooperation over the long time that the manuscript has grown. Moreover, I am grateful to Ulrich Nowak, Andreas Hohmann, Martin Weiser, and Anton Schiela for their intensive computational assistance and invaluable help in improving the quality of the manuscript.

Nearly last, but certainly not least, I wish to thank Harry Yserentant, Christian Lubich, Matthias Heinkenschloss, and a number of anonymous reviewers for valuable comments on a former draft. My final thanks go to Martin Peters from Springer for his enduring support.

Berlin, February 2004

Peter Deuflhard

Preface to Second Printing

The enjoyably fast acceptance of this monograph has made a second printing necessary. Compared to the first one, only minor corrections and citation updates have been made.

Berlin, November 2005

Peter Deuflhard

Table of Contents

Outline of Contents

This book is divided into eight chapters, a reference list, a software list, and an index. After an elementary introduction in Chapter 1, it splits into two parts: Part I, Chapter 2 to Chapter 5, on finite dimensional Newton methods for *algebraic equations*, and Part II, Chapter 6 to Chapter 8, on extensions to ordinary and partial *differential equations*. Exercises are added at the end of each chapter.

Chapter 1. This introductory chapter starts from the historical root, Newton's method for scalar equations (Section 1.1). The method can be derived either *algebraically*, which leads to *local* Newton methods only (presented in Chapter 2), or *geometrically*, which leads to *global* Newton methods via the concept of the Newton path (see Chapter 3).

The next Section 1.2 contains the *key to the basic understanding of this monograph*. First, four affine invariance classes are worked out, which represent the four basic strands of this treatise:

- *affine covariance*, which leads to *error* norm controlled algorithms,
- *affine contravariance*, which leads to *residual* norm controlled algorithms,
- *affine conjugacy*, which leads to *energy* norm controlled algorithms, and
- *affine similarity*, which may lead to *time* step controlled algorithms.

Second, the affine invariant local estimation of affine invariant Lipschitz constants is set as the central *paradigm* for the construction of adaptive Newton algorithms.

In Section 1.3, we give a roadmap of the large variety of Newton-type methods—essentially fixing terms to be used throughout the book such as ordinary and simplified Newton method, Newton-like methods, inexact Newton methods, quasi-Newton methods, Gauss-Newton methods, quasilinearization, or inexact Newton multilevel methods. In Section 1.4, we briefly collect details about iterative linear solvers to be used as inner iterations within finite dimensional inexact Newton algorithms; each affine invariance class is linked with a special class of inner iterations. In view of function space oriented inexact Newton algorithms, we also revisit linear multigrid methods. Throughout this section, we emphasize the role of adaptive error control.

PART I. The following Chapters 2 to 5 deal with *finite dimensional* Newton methods for algebraic equations.

Chapter 2. This chapter deals with *local* Newton methods for the numerical solution of systems of nonlinear equations with finite, possibly large dimension. The term 'local' refers to the situation that 'sufficiently good' initial guesses of the solution are assumed to be at hand. Special attention is paid to the issue of how to recognize, whether a given initial guess x^0 is 'sufficiently good'. Different affine invariant formulations give different answers to this question, in theoretical terms as well as by virtue of the algorithmic paradigm of Section 1.2.3. Problems of this structure are called 'mildly nonlinear'; their computational complexity can be bounded a-priori in units of the computational complexity of the corresponding linearized system.

As it turns out, different affine invariant Lipschitz conditions, which have been introduced in Section 1.2.2, lead to different characterizations of local convergence domains in terms of error oriented norms, residual norms, or energy norms, which, in turn, give rise to corresponding variants of Newton algorithms. We give three different, strictly affine invariant convergence analyses for the cases of affine covariant (error oriented) Newton methods (Section 2.1), affine contravariant (residual based) Newton methods (Section 2.2), and affine conjugate Newton methods for convex optimization (Section 2.3). Details are worked out for ordinary Newton algorithms, simplified Newton algorithms, and inexact Newton algorithms—synoptically for each of the three affine invariance classes. Moreover, affine covariance is naturally associated with Broyden's 'good' quasi-Newton method, whereas affine contravariance corresponds to Broyden's 'bad' quasi-Newton method.

Affine invariant *globalization*, which means global extension of the convergence domains of local Newton methods in the affine invariant frame, is possible along several lines:

- global Newton methods with damping strategy—see Chapter 3,
- parameter continuation methods—see Chapter 5,
- pseudo-transient continuation methods—see Section 6.4.

Chapter 3. This chapter deals with *global* Newton methods for systems of nonlinear equations with finite, possibly large dimension. The term 'global' refers to the situation that here, in contrast to the preceding chapter, 'sufficiently good' initial guesses of the solution are no longer assumed. Problems of this structure are called 'highly nonlinear'; their computational complexity depends on topological details of Newton paths associated with the nonlinear mapping and can typically not be bounded a-priori.

In Section 3.1 we survey globalization concepts such as

- steepest descent methods,
- trust region methods,
- the Levenberg-Marquardt method, and
- the Newton method with damping strategy.

In Section 3.1.4, a rather general geometric approach is taken: the idea is to derive a globalization concept without a pre-occupation to any iterative method, just starting from the requirement of affine covariance as a 'first principle'. Surprisingly, this general approach leads to a topological derivation of Newton's method with damping strategy via Newton paths.

In order to accept or reject a new iterate, *monotonicity tests* are applied. We study different such tests, according to different affine invariance requirements:

- the most popular *residual* monotonicity test, which is related to affine contravariance (Section 3.2),
- the error oriented so-called *natural* monotonicity test, which is related to affine covariance (Section 3.3), and
- the convex functional test as the natural requirement in convex optimization, which reflects affine conjugacy (Section 3.4).

For each of these three affine invariance classes, *adaptive trust region strategies* are designed in view of an efficient choice of damping factors in Newton's method. They are all based on the *paradigm* of Section 1.2.3. On a theoretical basis, details of algorithmic realization in combination with either *direct* or *iterative* linear solvers are worked out. As it turns out, an efficient determination of the steplength factor in global inexact Newton methods is intimately linked with the accuracy matching for affine invariant combinations of inner and outer iteration.

Chapter 4. This chapter deals with both *local* and *global Gauss-Newton* methods for *nonlinear least squares* problems in finite dimension—a method, which attacks the solution of the nonlinear least squares problem by solving a sequence of linear least squares problems. Affine invariance of both theory and algorithms will once again play a role, here restricted to *affine contravariance* and *affine covariance*. The theoretical treatment requires considerably more sophistication than in the simpler case of Newton methods for nonlinear equations.

In order to lay some basis, unconstrained and equality constrained *linear* least squares problems are first discussed in Section 4.1, introducing the useful calculus of generalized inverses. In Section 4.2, an affine contravariant convergence analysis of Gauss-Newton methods is given and worked out in the direction of *residual* based algorithms. Local convergence turns out to be only guaranteed for 'small residual' problems, which can be characterized in theoretical and algorithmic terms. Local and global convergence analysis as well as adaptive trust region strategies rely on some *projected residual* monotonicity test. Both *unconstrained* and *separable* nonlinear least squares problems are treated.

In the following Section 4.3, local convergence of *error* oriented Gauss-Newton methods is studied in affine covariant terms; again, Gauss-Newton methods are seen to exhibit guaranteed convergence only for a restricted problem class, named 'adequate' nonlinear least squares problems, since they are seen to be adequate in terms of the underlying statistical problem formulation. The globalization of these methods is done via the construction of two topological paths: the local and the global Gauss-Newton path. In the special case of nonlinear equations, the two paths coincide to one path, the Newton path. On this theoretical basis, adaptive trust region strategies (including rank strategies) combined with a natural extension of the *natural* monotonicity test are presented in detail for *unconstrained*, for *separable*, and—in contrast to the residual based approach—also for nonlinearly *constrained* nonlinear least squares problems. Finally, in Section 4.4, we study *underdetermined* nonlinear systems. In this case, a *geodetic Gauss-Newton path* exists generically and can be exploited to construct a quasi-Gauss-Newton algorithm and a corresponding adaptive trust region method.

Chapter 5. This chapter discusses the numerical solution of parameter dependent systems of nonlinear equations, which is the basis for parameter studies in systems analysis and systems design as well as for the globalization of local Newton methods. The key concept behind the approach is the (possible) existence of a *homotopy path* with respect to the selected parameter. In order to follow such a path, we here advocate *discrete continuation methods*, which consist of two essential parts:

- a *prediction* method, which, from given points on the homotopy path, produces some 'new' point assumed to be 'sufficiently close' to the homotopy path,

- an iterative *correction* method, which, from a given starting point close to, but not on the homotopy path, supplies some point on the homotopy path.

For the prediction step, *classical* or *tangent continuation* are the canonical choices. Needless to say that, for the iterative correction steps, we here concentrate on local Newton and (underdetermined) Gauss-Newton methods. Since the homotopy path is a mathematical object in the domain space of the nonlinear mapping, we only present the *affine covariant* approach.

In Section 5.1, we derive an adaptive *Newton continuation* algorithm with the ordinary Newton method as correction; this algorithm terminates locally in the presence of critical points including turning points. In order to follow the path beyond turning points, a *quasi-Gauss-Newton continuation* algorithm is worked out in Section 5.2, based on the preceding Section 4.4. This algorithm still terminates in the neighborhood of any higher order critical point. In order to overcome such points as well, we exemplify a scheme to construct *augmented systems*, whose solutions are just selected critical points of higher order—see Section 5.3. This scheme is an appropriate combination of Lyapunov-Schmidt reduction and topological universal unfolding. Details of numerical realization are only worked out for the computation of diagrams including simple bifurcation points.

PART II. The following Chapters 6 to 8 deal predominantly with *infinite dimensional*, i.e., function space oriented Newton methods. The selected topics are stiff initial value problems for ordinary differential equations (ODEs) and boundary value problems for ordinary and partial differential equations (PDEs).

Chapter 6. This chapter deals with *stiff* initial value problems for ODEs. The discretization of such problems is known to involve the solution of non-linear systems per each discretization step—in one way or the other.

In Section 6.1, the contractivity theory for linear ODEs is revisited in terms of *affine similarity*. Based on an affine similar convergence theory for a simplified Newton method in *function space*, a *nonlinear contractivity* theory for stiff ODE problems is derived in Section 6.2, which is quite different from the theory given in usual textbooks on the topic. The key idea is to replace the Picard iteration in function space, known as a tool to show uniqueness in nonstiff initial value problems, by a simplified Newton iteration in function space to characterize stiff initial value problems. From this point of view, *linearly implicit* one-step methods appear as direct realizations of the simplified Newton iteration in function space. In Section 6.3, exactly the same theoretical characterization is shown to apply also to *implicit* one-step methods, which require the solution of a nonlinear system by some finite dimensional Newton-type method at each discretization step.

Finally, in a deliberately longer Section 6.4, we discuss *pseudo-transient continuation* algorithms, whereby steady state problems are solved via stiff integration. This type of algorithm is particularly useful, when the Jacobian matrix is singular due to hidden dynamical invariants (such as mass conservation). The (nearly) affine similar theoretical characterization permits the derivation of an *adaptive (pseudo-)time step strategy* and an accuracy matching strategy for a residual based inexact variant of the algorithm.

Chapter 7. In this chapter, we consider nonlinear two-point boundary value problems for ODEs. The presentation and notation is closely related to Chapter 8 in the textbook [71]. Algorithms for the solution of such problems can be grouped into two approaches: *initial value* methods such as multiple shooting and *global discretization* methods such as collocation. Historically, affine covariant Newton methods have first been applied to this problem class—with significant success.

In Section 7.1, the realization of Newton and discrete continuation methods within the standard multiple shooting approach is elaborated. Gauss-Newton methods for parameter identification in ODEs are discussed in Section 7.2, also based on multiple shooting. For periodic orbit computation, Section 7.3 presents Gauss-Newton methods, both in the shooting approach (Sections 7.3.1 and 7.3.2) and in a Fourier collocation approach, also called Urabe or harmonic balance method (Section 7.3.3).

In Section 7.4 we concentrate on *polynomial* collocation methods, which have reached a rather mature status including affine covariant Newton methods. In Section 7.4.1, the possible discrepancy between discrete and continuous solutions is studied including the possible occurrence of so-called 'ghost solutions' in the nonlinear case. On this basis, the realization of *quasilinearization* is seen to be preferable in combination with collocation. The following Section 7.4.2 is then devoted to the key issue that quasilinearization can be interpreted as an *inexact Newton method in function space*: the approximation errors in the infinite dimensional setting just replace the inner iteration errors arising in the finite dimensional setting. With this insight, an adaptive multilevel control of the collocation errors can be realized to yield an adaptive inexact Newton method in function space—which is the bridge to adaptive Newton multilevel methods for PDEs (compare Section 8.3).

Chapter 8. This chapter deals with Newton methods for boundary value problems in nonlinear PDEs. There are two principal approaches: (a) finite dimensional Newton methods applied to a given system of already discretized PDEs, also called *discrete Newton methods*, and (b) function space oriented Newton methods applied to the continuous PDEs, at best in the form of *inexact Newton multilevel methods*.

Before we discuss the two principal approaches in detail, we present an affine covariant analysis of *asymptotic mesh independence* that connects the finite dimensional and the infinite dimensional Newton methods, see Section 8.1. In Section 8.2, we assume the standard situation in industrial technology software, where the grid generation module is strictly separated from the solution module. Consequently, nonlinear PDEs arise there as discrete systems of nonlinear equations with fixed finite, but usually high dimension and large sparse ill-conditioned Jacobian matrix. This is the domain of applicability of finite dimensional inexact Newton methods. More advanced, but often less favored in the huge industrial software environments, are *function space* oriented inexact Newton methods, which additionally include the adaptive manipulation of discretization meshes within a multilevel or multigrid solution process. This situation is treated in Section 8.3 and compared there with *finite dimensional* inexact Newton techniques.

1 Introduction

This chapter is an elementary introduction into the general theme of this book. We start from the historical root, Newton's method for scalar equations (Section 1.1): the method can be derived either *algebraically*, which leads to *local* Newton methods only (see Chapter 2), or *geometrically*, which leads to *global* Newton methods via the topological Newton path (see Chapter 3).

Section 1.2 contains the *key to the basic understanding of this monograph*. First, four affine invariance classes are worked out, which represent the four basic strands of this treatise:

- *affine covariance*, which leads to *error* norm controlled algorithms,
- *affine contravariance*, which leads to *residual* norm controlled algorithms,
- *affine conjugacy*, which leads to *energy* norm controlled algorithms, and
- *affine similarity*, which may lead to *time* step controlled algorithms.

Second, the affine invariant local estimation of affine invariant Lipschitz constants is set as the central *paradigm* for the construction of adaptive Newton algorithms.

In Section 1.3, we fix terms for various Newton-type methods to be named throughout the book: ordinary and simplified Newton method, Newton-like methods, inexact Newton methods, quasi-Newton methods, quasilinearization, and inexact Newton multilevel methods.

In Section 1.4, details are given for the iterative linear solvers GMRES, PCG, CGNE, and GBIT to an extent necessary to match them with finite dimensional inexact Newton algorithms. In view of function space oriented inexact Newton algorithms, we also revisit multiplicative, additive, and cascadic multigrid methods emphasizing the role of adaptive error control therein.

1.1 Newton-Raphson Method for Scalar Equations

Assume we have to solve the scalar equation

$$f(x) = 0$$

with an appropriate guess x^0 of the unknown solution x^* at hand.

Algebraic approach. We use the *perturbation*

$$\Delta x = x^* - x^0$$

for Taylor's expansion

$$0 = f(x^0 + \Delta x) = f(x^0) + f'(x^0)\Delta x + O(|\Delta x|^2).$$

Upon dropping terms of order higher than linear in the perturbation, we arrive at the approximate equation

$$f'(x^0)\Delta x \approx -f(x^0),$$

which, assuming $f'(x^0) \neq 0$, leads to the precise equation

$$x^1 - x^0 = \Delta x^0 = -\frac{f(x^0)}{f'(x^0)}$$

for a first correction of the starting guess. From this, an *iterative* scheme is constructed by repetition

$$x^{k+1} = \Phi(x^k) = x^k - \frac{f(x^k)}{f'(x^k)}, \quad k = 0, 1, \dots .$$

If we study the contraction mapping Φ in terms of a *contraction factor* Θ, we arrive at

$$\Theta = \max_{x \in I} \Phi'(x) = \max_{x \in I} \frac{f(x)f''(x)}{(f'(x))^2}$$

with I an appropriate interval containing x^*. From this, we have at least *linear* convergence

$$|x^{k+1} - x^*| \leq \Theta|x^k - x^*|$$

in a neighborhood of x^*, where $\Theta < 1$. In passing we note that this contraction factor Θ remains unchanged, if we rescale the equation according to

$$\alpha f(\beta y) = 0, \quad \alpha\beta \neq 0, \quad x = \beta y.$$

An extension of this kind of observation to rather general nonlinear problems will lead to fruitful theoretical and algorithmical consequences below. For starting guesses x^0 'sufficiently close' to x^* even *quadratic* convergence of the iterates can be shown in the sense that

$$|x^{k+1} - x^*| \leq C|x^k - x^*|^2, \quad k = 0, 1, 2 \dots .$$

The algebraic derivation in terms of the linear perturbation treatment carries over to rather general nonlinear problems up to operator equations such as boundary value problems for ordinary or partial differential equations.

Geometric approach. Looking at the *graph* of $f(x)$—as depicted in Figure 1.1—any root can be interpreted as the intersection of this graph with the real axis. Since this intersection cannot be constructed other than by tedious sampling of f, the graph of $f(x)$ is replaced by its *tangent* $p(x)$ in x_0 and the first iterate x_1 is defined as the intersection of the tangent with the real axis. Upon repeating this geometric process, the close-by solution point x^* can be constructed up to any desired accuracy. By geometric insight, the iterative process will converge *globally* for *convex* (or concave) f—which includes the case of arbitrarily 'bad' initial guesses as well! At first glance, this geometric derivation seems to be restricted to the scalar case, since the graph of $f(x)$ is a typically one-dimensional concept. A careful examination of the subject in more than one dimension, however, naturally leads to a topological path called *Newton path*—see Section 3.1.4 below.

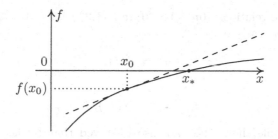

Fig. 1.1. Geometric interpretation: Newton's method for a scalar equation.

HISTORICAL NOTE. Strictly speaking, Newton's method could as well be named as Newton-Raphson-Simpson method—as elaborated in recent articles by N. Kollerstrom [134] or T.J. Ypma [203]. According to these careful historical studies, the following facts seem to be agreed upon among the experts:

- In the year 1600, Francois Vieta (1540–1603) had (first?) designed a *perturbation* technique for the solution of the scalar polynomial equations, which supplied one decimal place of the unknown solution per step via the explicit calculation of successive polynomials of the successive perturbations. It seems that this method had also been detected independently by al-Kāshī and simplified around 1647 by Oughtred.
- Isaac Newton (1643–1727) got to know Vieta's method in 1664. Up to 1669 he had improved it by *linearizing* these successive polynomials. As an example, he discussed the numerical solution of the cubic polynomial

$$f(x) := x^3 - 2x - 5 = 0 \,.$$

Newton first noted that the integer part of the root is 2 setting $x_0 = 2$. Next, by means of $x = 2 + p$, he obtained the polynomial equation

$$p^3 + 6p^2 + 10p - 1 = 0.$$

Herein he neglected terms higher than first order and thus put $p \approx 0.1$. He inserted $p = 0.1 + q$ and constructed the polynomial equation

$$q^3 + 6.3q^2 + 11.23q + 0.061 = 0.$$

Again he neglected terms higher than linear and found $q \approx -0.0054$. Continuation of the process one more step led him to $r \approx 0.00004853$ and therefore to the third iterate

$$x_3 = x_0 + p + q + r = 2.09455147.$$

Note that the relations $10p - 1 = 0$ and $11.23q + 0.061 = 0$ given above correspond precisely to

$$p = x_1 - x_0 = -f(x_0)/f'(x_0)$$

and to

$$q = x_2 - x_1 = -f(x_1)/f'(x_1).$$

As the example shows, he had also observed that by keeping all decimal places of the corrections, the number of accurate places would *double* per each step—i.e., *quadratic convergence*. In 1687 (Philosophiae Naturalis Principia Mathematica), the first nonpolynomial equation showed up: it is the well-known equation from astronomy

$$x - e\sin(x) = M$$

between the *mean anomaly M* and the *eccentric anomaly x*. Here Newton used his already developed polynomial techniques via the series expansion of *sin* and *cos*. However, no hint on the derivative concept is incorporated!

- In 1690, Joseph Raphson (1648–1715) managed to avoid the tedious computation of the successive polynomials, playing the computational scheme back to the original polynomial; in this now fully *iterative* scheme, he also kept all decimal places of the corrections. He had the feeling that his method differed from Newton's method at least by its derivation.

- In 1740, Thomas Simpson (1710–1761) actually introduced derivatives ('fluxiones') in his book 'Essays on Several Curious and Useful Subjects in Speculative and Mix'd Mathematicks, Illustrated by a Variety of Examples'. He wrote down the true *iteration* for one (nonpolynomial) equation and for a system of two equations in two unknowns thus making the correct extension to *systems* for the first time. His notation is already quite close to our present one (which seems to go back to J. Fourier).

Throughout this book, we will use the name 'Newton-Raphson method' only for scalar equations. For general equations we will use the name 'Newton method'—even though the name 'Newton-Simpson method' would be more appropriate in view of the just described historical background.

1.2 Newton's Method for General Nonlinear Problems

In contrast to the preceding section, we now approach the general case. Assume we have to solve a nonlinear operator equation

$$F(x) = 0 \,,$$

wherein $F : D \subset X \to Y$ for Banach spaces X, Y endowed with norms $\| \cdot \|_X$ and $\| \cdot \|_Y$. Let F be at least once continuously differentiable. Suppose we have a starting guess x^0 of the unknown solutions x^* at hand. Then *successive linearization* leads to the general Newton method

$$F'(x^k)\Delta x^k = -F(x^k), \ \ x^{k+1} = x^k + \Delta x^k, \ \ \ k = 0, 1, \ldots . \tag{1.1}$$

Obviously, this method attacks the solution of a nonlinear problem by solving a *sequence of linear problems of the same kind.*

1.2.1 Classical convergence theorems revisited

A necessary assumption for the solvability of the above linear problems is that the derivatives $F'(x)$ are *invertible* for all occurring arguments. For this reason, standard convergence theorems typically require a-priori that the inverse $F'(x)^{-1}$ exists and is bounded

$$\|F'(x)^{-1}\|_{Y \to X} \le \beta < \infty, \ x \in D, \tag{1.2}$$

where$\| \cdot \|_{Y \to X}$ denotes an *operator norm*. From a computational point of view, such a theoretical quantity β defined over the domain D seems to be hard to get, apart from rather simple examples. Sampling of *local* estimates like

$$\|F'(x^0)^{-1}\|_{Y \to X} \le \beta_0 \tag{1.3}$$

seems to be preferable, but is still quite expensive. Moreover, a well-known rule in Numerical Analysis states that the actual computation of inverses should be avoided. Rather, such a condition should be monitored implicitly in the course of solving linear systems with specific right hand sides.

In order to study the convergence properties of the above Newton iteration, some *second derivative* information is needed, as already stated in the scalar equation case (Section 1.1 above). The classical standard form to include this information is via a *Lipschitz condition* of the type

$$\|F'(x) - F'(\bar{x})\|_{X \to Y} \leq \gamma \|x - \bar{x}\|_X , \quad x, \bar{x} \in D . \qquad (1.4)$$

With this additional assumption, the *operator perturbation lemma* (sometimes also called Banach perturbation lemma) proves the existence of some upper bound β such that

$$\|F'(x)^{-1}\|_{Y \to X} \leq \beta \leq \frac{\beta_0}{1 - \beta_0 \gamma \|x - x^0\|_X}$$

for

$$\|x - x^0\|_X < \frac{1}{\beta_0 \gamma} , \quad x \in D .$$

The proof is left as Exercise 1.1. Classical convergence theorems for Newton's method use certain combinations of these assumptions.

Newton-Kantorovich theorem. This first classical convergence theorem for Newton's method in abstract spaces (see [127, 163]) requires assumptions (1.3) and (1.4) to show *existence* and *uniqueness* of a solution x^* as well as quadratic convergence of the Newton iterates within a neighborhood characterized by a so-called *Kantorovich quantity*

$$h_0 := \|\Delta x^0\|_X \, \beta_0 \gamma < \tfrac{1}{2}$$

and a corresponding convergence ball around x^0 with radius $\rho_0 \sim 1/\beta_0 \gamma$. This theorem is also the standard tool to prove the classical *implicit function theorem*—compare Exercise 1.2.

Newton-Mysovskikh theorem. This second classical convergence theorem (see [155, 163]) requires assumptions (1.2) and (1.4) to show *uniqueness* (not existence!) and quadratic convergence within a neighborhood characterized by the slightly different quantity

$$h_0 := \|\Delta x^0\|_X \, \beta \gamma < 2$$

and a corresponding convergence ball around x^0 with radius $\rho \sim 1/\beta \gamma$.

Both theorems seem to require the actual computation of the Lipschitz constant γ. However, such a quantity is certainly hard if not hopeless to compute in realistic nonlinear problems. Moreover, even computational local estimates of β and γ are typically far off any use in practical applications. That is why, for quite a time, people believed that convergence results are of theoretical interest only, but not of any value for the actual implementation of Newton algorithms. An illustrating simple example is given as Exercise 2.3.

This undesirable gap between convergence analysis and algorithm construction has been the motivation for the present book. As will become apparent, the key to closing this gap is supplied by *affine invariance in both convergence theory and algorithmic realization.*

1.2.2 Affine invariance and Lipschitz conditions

In order to make the essential point clear enough, it is sufficient to regard simply systems of nonlinear equations, which means that $X = Y = \mathbb{R}^n$ for fixed dimension $n > 1$ and the same norm in X and Y. Recall Newton's method in the form

$$F'(x^k)\Delta x^k = -F(x^k), \quad x^{k+1} = x^k + \Delta x^k \quad k = 0, 1, \ldots .$$

Scaling. In sufficiently complex problems, scaling or re-gauging of variables (say, from km to $miles$) needs to be carefully considered. Formally speaking, with preselected nonsingular *diagonal* scaling matrices D_L, D_R for left and right scaling, we may write

$$(D_L F'(x^k) D_R)(D_R^{-1}\Delta x^k) = -D_L F(x^k)$$

for the scaled linear system. Despite its formal equivalence with (1.1), all standard norms used in Newton algorithms must now be replaced by *scaled norms* such that (dropping the iteration index k)

$$\|\Delta x\|, \|F\|, \|F + F'(x)\Delta x\| \longrightarrow \|D_R^{-1}\Delta x\|, \|D_L F\|, \|D_L(F + F'(x)\Delta x)\|.$$

With the change of norms comes a change of the criteria for the acceptance or rejection of new iterates. The effect of scaling on the iterative performance of Newton-type methods is a sheet lightning of the more general effects caused by affine invariance, which are the topic of this book.

Affine transformation. Let $A, B \in \mathbb{R}^{n \times n}$ be *arbitrary nonsingular* matrices and study the affine transformations of the nonlinear system as

$$G(y) = AF(By) = 0, \quad x = By.$$

Then Newton's method applied to $G(y)$ reads

$$G'(y^k)\Delta y^k = -G(y^k), \quad y^{k+1} = y^k + \Delta y^k \quad k = 0, 1, \ldots .$$

With the relation

$$G'(y^k) = AF'(x^k)B$$

and starting guess $y^0 = B^{-1}x^0$ we immediately obtain

$$x^k = By^k, \quad k = 0, 1, \ldots .$$

Obviously, the iterates are invariant under transformation of the image space (by A)—an invariance property described by *affine covariance*. Moreover, they are transformed just as the whole original space (by B)—a property denoted by *affine contravariance*.

It is only natural to require that the above affine invariance properties are inherited by any theoretical characterization. As it turns out, the inheritance of the full invariance property is impossible. That is why we restrict our study to four special invariance classes.

Affine covariance. In this setting, we keep the domain space of F fixed ($B = I$) and look at the *whole class of problems*

$$G(x) = AF(x) = 0$$

that is generated by the class $\mathrm{GL}(n)$ of nonsingular matrices A. The Newton iterates are the same all over the whole class of nonlinear problems. For this reason, an affine covariant theory about their convergence must be possible. Upon revisiting the above theoretical assumptions (1.2), (1.3), and (1.4) we now obtain

$$\|G'(x)^{-1}\| \le \beta(A)\,, \ \ \|G'(x^0)^{-1}\| \le \beta_0(A)\,, \ \ \|G'(x) - G'(\bar{x})\| \le \gamma(A)\|x - \bar{x}\|\,.$$

Application of the classical convergence theorems then yields convergence balls with radius, say

$$\rho(A) \sim 1/\beta(A)\gamma(A)\,.$$

Compared with $\beta(I)$, $\gamma(I)$ we obtain (assuming best possible theoretical bounds)

$$\beta(A) \le \beta(I)\|A^{-1}\|\,, \ \gamma(A) \le \gamma(I)\|A\|$$

and therefore

$$\beta(A)\gamma(A) \le \beta(I)\gamma(I)\,\mathrm{cond}(A)\,. \tag{1.5}$$

For $n > 1$ we have $\mathrm{cond}(A) \ge 1$, even unbounded for $A \in \mathrm{GL}(n)$. Obviously, by a mean choice of A we can make the classical convergence balls shrink to nearly zero!

Fortunately, careful examination of the proof of the Newton-Kantorovich theorem shows that assumptions (1.3) and (1.4) can be telescoped to the requirement

$$\left\|F'(x^0)^{-1}\big(F'(x) - F'(\bar{x})\big)\right\| \le \omega_0 \|x - \bar{x}\|\,, \ \ \ x, \bar{x}, x^0 \in D\,. \tag{1.6}$$

The thus defined Lipschitz constant ω_0 is affine covariant, since

$$\begin{aligned}
G'(x^0)^{-1}\big(G'(x) - G'(\bar{x})\big) &= \big(AF'(x^0)\big)^{-1}A\big(F'(x) - F'(\bar{x})\big) \\
&= F'(x^0)^{-1}\big(F'(x) - F'(\bar{x})\big)
\end{aligned}$$

so that both sides of (1.6) are independent of A. This definition of ω_0 (assumed best possible) still has the disadvantage of containing an operator norm on the left side—which, however, is unavoidable, because the operator perturbation lemma is required in the proof. Examination of the Newton-Mysovskikh theorem shows that assumptions (1.2) and (1.4) can also be telescoped to an affine covariant Lipschitz condition, which this time only contains vector norms (and directional derivatives):

$$\left\|F'(x)^{-1}\big(F'(\bar{x}) - F'(x)\big)(\bar{x} - x)\right\| \le \omega\|\bar{x} - x\|^2\,, \ \ \ x, \bar{x} \in D\,. \tag{1.7}$$

This assumption allows a clean affine covariant theory about the local quadratic convergence of the Newton iterates including local uniqueness of the solution x^*—see Section 2.1 below. Moreover, this type of theorem will be the stem from which a variety of computationally useful convergence theorems branch off.

Summarizing, any affine covariant convergence theorems will lead to results in terms of *iterates* $\{x^k\}$, *correction norms* $\|\Delta x^k\|$ or *error norms* $\|x^k - x^*\|$.

BIBLIOGRAPHICAL NOTE. For quite a while, *affine covariance* held only in very few convergence theorems for local Newton methods, among which are Theorem 6. (1.XVIII) in the book of Kantorovich/Akhilov [127] from 1959, part of the theoretical results by J.E. Dennis [52, 53], or an interesting early paper by H.B. Keller [129] from 1970 (under the weak assumption of just Hölder continuity of $F'(x)$). None of these authors, however, seems to have been fully aware of the importance of this invariance property, since all of them neglected this aspect in their later work.

A systematic approach toward affine covariance, then simply called affine invariance, has been started in 1972 by the author in his dissertation [59], published two years later in [60]. His initial motivation had been to overcome severe difficulties in the actual application of Newton's method within multiple shooting—compare Section 7.1 below. In 1979, this approach has been transferred to convergence theory in a paper by P. Deuflhard and G. Heindl [76]. Following the latter paper, T. Yamamoto has preserved affine covariance in his subtle convergence estimates for Newton's method—see, e.g., his starting paper [202] and work thereafter. Around that time H.G. Bock [29, 31, 32] also joined the affine invariance crew and slightly improved the theoretical characterization from [76]. The first affine covariant convergence proof for inexact Newton methods is due to T.J. Ypma [203].

Affine contravariance. This setting is dual to the preceding one: we keep the image space of F fixed ($A = I$) and consider the *whole class of problems*

$$G(y) = F(By)\,, \quad x = By\,, \ B \in \mathrm{GL}(n)$$

that is generated by the class $\mathrm{GL}(n)$ of nonsingular matrices B. Consequently, a common convergence theory for the whole problem class will not lead to statements about the Newton iterates $\{y^k\}$, but only about the *residuals* $\{F(x^k)\}$, which are independent of any choice of B. Once more, the classical conditions (1.2) and (1.4) can be telescoped, this time in image space terms only:

$$\left\|\left(F'(\bar x) - F'(x)\right)(\bar x - x)\right\| \le \omega \|F'(x)(\bar x - x)\|^2\,. \tag{1.8}$$

Observe that both sides are independent of B, since, for example

$$G'(y)(\bar y - y) = F'(x)B(\bar y - y) = F'(x)(\bar x - x)\,.$$

A Newton-Mysovskikh type theorem on the basis of such a Lipschitz condition will lead to convergence results in terms of *residual norms* $\|F(x^k)\|$.

BIBLIOGRAPHICAL NOTE. The door to *affine contravariance* in the Lipschitz condition has been opened by A. Hohmann in his dissertation [120] , wherein he exploited it for the construction of a residual based inexact Newton method within an adaptive collocation method for ODE boundary value problems—compare Section 7.4 below.

At first glance, the above dual affine invariance classes seem to be the only ones that might be observed in actual computation. At second glance, however, certain couplings between the linear transformations A and B may arise, which are discussed next.

Affine conjugacy. Assume that we have to solve the *minimization problem*

$$f(x) = \min, \; f : D \subset \mathbb{R}^n \to \mathbb{R}$$

for a functional f, which is *convex* in a neighborhood D of the minimum point x^*. Then this problem is equivalent to solving the nonlinear equations

$$F(x) = \operatorname{grad} f(x) = f'(x)^T = 0, \; x \in D.$$

For such a gradient mapping F the Jacobian $F'(x) = f''(x)$ is *symmetric* and certainly *positive semi-definite*. Moreover, assume that $F'(x)$ is *strictly positive definite* so that $F'(x)^{1/2}$ can be defined. This also implies that f is *strictly convex*. Upon transforming the minimization problem to

$$g(y) = f(By) = \min, \; x = By,$$

we arrive at the transformed equations

$$G(y) = B^T F(By) = 0$$

and the transformed Jacobian

$$G'(y) = B^T F'(x)B, \; x = By.$$

The Jacobian transformation is *conjugate*, which motivates the name of this special affine invariance. Due to Sylvester's theorem (compare [151]), it conserves the index of inertia, so that all G' are symmetric and strictly positive definite. Affine conjugate theoretical terms are, of course, functional values $f(x)$ and, in addition, so–called *local energy products*

$$(u, v) = u^T F'(x)v, \; u, v, x \in D.$$

Just note that energy products are invariant under this kind of affine transformation, since

$$u, v, x \to \bar{u} = Bu, \bar{v} = Bv, x = By$$

implies

$$u^T G'(y)v = \bar{u}^T F'(x)\bar{v}.$$

Local energy products induce *local energy norms*

$$\|F'(x)^{1/2}u\|^2 = (u, u) = u^T F'(x)u, \quad u, x \in D.$$

In this framework, telescoping the theoretical assumptions (1.2) and (1.4) leads to an affine conjugate Lipschitz condition

$$\|F'(x)^{-1/2}\big(F'(\bar{x}) - F'(x)\big)(\bar{x} - x)\| \le \omega \|F'(x)^{1/2}(\bar{x} - x)\|^2. \tag{1.9}$$

Affine conjugate convergence theorems will lead to results in terms of *functional values* $f(x)$ and *energy norms of corrections* $\|F'(z)^{1/2}\Delta x^k\|$ or *errors* $\|F'(z)^{1/2}(x^k - x^*)\|$.

BIBLIOGRAPHICAL NOTE. The concept of *affine conjugacy* dates back to P. Deuflhard and M. Weiser, who, in 1997, defined and exploited it for the construction of an adaptive Newton multilevel FEM for nonlinear elliptic PDEs—see [84, 85] and Section 8.3.

Affine similarity. This invariance principle is more or less common in the differential equation community—apart perhaps from the name given here. Consider the case that the solution of the nonlinear system $F(x) = 0$ can be interpreted as *steady state* or *equilibrium point* of the *dynamical system*

$$\dot{x} = F(x). \tag{1.10}$$

Arbitrary affine transformation

$$A\dot{x} = AF(x) = 0$$

here affects both the domain and the image space of F in the same way— of course, differentiability with respect to time differs. The corresponding problem class to be studied is then

$$G(y) = AF(A^{-1}y) = 0, \quad y = Ax,$$

which gives rise to the Jacobian transformation

$$G'(y) = AF'(x)A^{-1}.$$

This *similarity* transformation (which motivates the name affine similarity) is known to leave the *Jacobian eigenvalues* λ invariant. Note that a theoretical characterization of *stability* of the equilibrium point involves their real parts $\Re(\lambda)$. In fact, an upper bound of these real parts, called the *one-sided* Lipschitz constant, will serve as a substitute of the Lipschitz constant of F, which is known to restrict the analysis to nonstiff differential equations. As an affine similar representative, we may formally pick the (possibly complex) *Jordan canonical form* J, known to consist of elementary Jordan blocks for each separate eigenvalue. Let the Jacobian at any selected point $\hat{x}$ be decomposed such that

$$F'(\hat{x}) = T(\hat{x})J(\hat{x})T(\hat{x})^{-1} = TJT^{-1},$$

which implies

$$G'(\hat{y}) = AF'(x)A^{-1} = (AT)J(AT)^{-1}.$$

Consequently, any theoretical results phrased in terms of the *canonical norm*

$$|\cdot| := \|T^{-1}\cdot\|$$

will meet the requirement of affine similarity. We must, however, remain aware of the fact that numerical Jordan decomposition may be *ill-conditioned*, whenever eigenvalue clusters arise—a property, which is reflected in the size of $\mathrm{cond}(T)$. With this precaution, an affine similar approach will be helpful in the analysis of *stiff* initial value problems for ODE's (see Chapter 6).

In contrast to the other invariance classes, note that here not only Newton's iteration exhibits the correct affine similar pattern, but also any *fixed point iteration* of the type

$$x^{k+1} = x^k + \alpha_k F(x^k),$$

assuming the parameters α_k are chosen by some affine similar criterion. Hence, any linear combination of Newton and fixed point iteration may be considered as well: this leads to an iteration of the type

$$\left(I - \tau F'(x^k)\right)(x^{k+1} - x^k) = \tau F(x^k),$$

which is nothing else than a *linearly implicit Euler discretization* of the above ordinary differential equation (1.10) with timestep τ to be adapted. As worked out in Section 6.4, such a *pseudo-transient continuation* method can be safely applied only, if the equilibrium point is dynamically *stable*—a condition anyway expected from geometrical insight. As a 'first choice', we then arrive at the following Lipschitz condition

$$|\left(F'(\bar{x}) - F'(x)\right)u| \le \omega|\bar{x} - x||u|.$$

Unfortunately, the canonical norm is computationally not easily available and at the same time may suffer from ill-conditioning—reflected in the size of $\mathrm{cond}(T)$. Therefore, upon keeping in mind that in affine similar problems domain and image space of F have the same transformation behavior, we are led to realize a 'second best' choice: we may switch from the canonical norm $|\cdot|$ to the standard norm $\|\cdot\|$ thus obtaining a Lipschitz condition of the structure

$$\|\left(F'(\bar{x}) - F'(x)\right)u\| \le \omega\|\bar{x} - x\|\cdot\|u\|.$$

However, in this way we lose the affine similarity property in the definition of ω, which means we have to apply careful scaling at least. In passing, we note that here the classical Lipschitz condition (1.4) arises directly from affine invariance considerations; however, a bounded inverse assumption like (1.2) is not needed in this context, but replaced by other conditions.

Scaling invariance. Scaling as discussed at the beginning of this section is a special affine transformation. In general, we will want to realize a scaling invariant algorithm, i.e. an algorithm that is invariant under the choice of units in the given problem. Closer examination shows that the four different affine invariance classes must be treated differently.

In an *affine covariant* setting, the formal assumption $B = I$ will certainly cover any fixed scaling transformation of the type $B = D$ so that 'dimensionless' variables

$$y = D^{-1}x, \quad D = \mathrm{diag}(\alpha_1,\ldots,\alpha_n), \quad \alpha_i > 0$$

are used at least inside the codes (internal scaling). For example, with components $x = (x_1,\ldots,x_n)$, *relative* scaling could mean any a-priori choice like

$$\alpha_i = |x_i^0|, \quad \text{if } |x_i^0| \neq 0$$

or an iterative adaptation like

$$\alpha_i^{k+1} = \max\{|x_i^k|,\ |x_i^{k+1}|\}.$$

Whenever these choices guarantee $\alpha_i > 0$, then scaling invariance is assured: to see this, just re-scale the components of x according to

$$x_i \longrightarrow \hat{x}_i = \beta_i x_i,$$

which implies

$$\alpha_i \longrightarrow \hat{\alpha}_i = \beta_i \alpha_i$$

and leaves

$$y_i = \frac{\hat{x}_i}{\hat{\alpha}_i} = \frac{x_i}{\alpha_i}$$

unchanged. In reality, however, *absolute threshold values* $\alpha_{\min} > 0$ have to be imposed in the form, say

$$\bar{\alpha}_i = \max\{\alpha_i, \alpha_{\min}\}$$

to avoid overflow for values close to zero. By construction, such threshold values spoil the nice scaling invariance property, unless they are defined for dimensionless components of the variable y.

In an *affine contravariant* setting, scaling should be applied in the image space of F, which means for the residual components

$$F \to G = D^{-1}F$$

with appropriately chosen diagonal matrix D.

For *affine similarity*, simultaneous scaling should be applied in both domain and image space

$$x, \, F \to y = D^{-1}x, \, G = D^{-1}F.$$

Finally, the *affine conjugate* energy products can be verified to be scaling invariant already by construction.

Further affine invariance classes. The four affine invariance classes mentioned so far actually represent the dominant classes of interest. Beyond these, certain combinations of these classes play a role in problems with appropriate substructures, each of which gives rise to one of the 'grand four'. As an example take optimization with equality constraints, which may require affine covariance or contravariance in the constraints, but affine conjugacy in the functional—see, e.g., the recent discussion [193] by S. Volkwein and M. Weiser.

1.2.3 The algorithmic paradigm

The key question treated in this book is how theoretical results from convergence analysis can be exploited for the construction of *adaptive* Newton algorithms. The key answer to this question is to realize *affine invariant computational estimates* of *affine invariant Lipschitz constants* that are cheaply available in the course of the algorithms. The realization is done as follows:

We identify some *theoretical local Lipschitz constant* ω defined over a nonempty domain D such that

$$\omega = \sup_{x,y,z \in D} g(x,y,z) \qquad (1.11)$$

in terms of some scalar expression $g(x, y, z)$ that will only contain affine invariant terms. For ease of writing, we will mostly just write

$$g(x, y, z) \le \omega \text{ for all } x, y, z \in D,$$

even though we mean the best possible estimates (1.11) to characterize non-linearity by virtue of Lipschitz constants. Once such a g has been selected, we exploit it by defining some corresponding *computational local estimate* according to

$$[\omega] = g(\hat{x}, \hat{y}, \hat{z}) \text{ for specific } \hat{x}, \hat{y}, \hat{z} \in D.$$

By construction, $[\omega]$ and ω share the same affine invariance property and satisfy the relation

$$[\omega] \le \omega.$$

Illustrating example. For the affine covariant Lipschitz condition (1.6) we have

$$\omega_0 = \sup_{x,y \in D} g(x,y,x^0) = \frac{\| F'(x^0)^{-1} \left(F'(x) - F'(y) \right) \|}{\| x - y \|}. \tag{1.12}$$

As a local affine covariant estimate, we may choose

$$[\omega_0] = g(x^1, x^0, x^0) = \frac{\| F'(x^0)^{-1} \left(F'(x^1) - F'(x^0) \right) \|}{\| x^1 - x^0 \|} \tag{1.13}$$

in terms of the anyway computed Newton iterates x^0, x^1. In actual implementation, we will apply estimates different from (1.12) and (1.13), but preferable in the algorithmic context. The art in this kind of approach is to find out, among many possible theoretical characterizations, those ones that give rise to 'cheap and suitable' computational estimates and, in turn, lead to the construction of efficient algorithms.

There remains some gap $\omega - [\omega] \geq 0$, which can be reduced by appropriate reduction of the domain D. As will turn out, efficient adaptive Newton algorithms can be constructed, if $[\omega]$ catches at least one leading binary digit of ω—for details see the various *bit counting lemmas* scattered all over the book.

Remark 1.1   If the paradigm were realized *without* a strict observation of affine invariance of Lipschitz constants and estimates, then undesirable geometrical distortion effects (like those described in detail in (1.5)) would lead to totally unrealistic estimates and thus could not be expected to be a useful basis for any efficient algorithm.

BIBLIOGRAPHICAL NOTE. The general *paradigm* described here was, in an intuitive sense, already employed by P. Deuflhard in his 1975 paper on adaptive damping for Newton's method [63]. In 1979, the author formalized the whole approach introducing the notation [·] for computational estimates and exploited it for the construction of adaptive continuation methods [61]. Early on, H.G. Bock also took up the paradigm in his work on multiple shooting techniques for parameter identification and optimal control problems [29, 31, 32].

1.3 A Roadmap of Newton-type Methods

There is a large variety of Newton-type methods, which will be discussed in the book and therefore named and briefly sketched here.

Ordinary Newton method. For general nonlinear problems, the classical ordinary Newton method reads

$$F'(x^k)\Delta x^k = -F(x^k), \ x^{k+1} = x^k + \Delta x^k, \qquad k = 0, 1, \ldots . \qquad (1.14)$$

For $F : D \subset \mathbb{R}^n \to \mathbb{R}^n$ a Jacobian (n, n)-matrix is required. Sufficiently accurate Jacobian approximations can be computed by symbolic differentiation or by numerical differencing—see, for example, the automatic differentiation due to A. Griewank [112].

The above form of the linear system deliberately reflects the actual sequence of computation: first, compute the Newton corrections Δx^k, then improve the iterates x^k to obtain x^{k+1}—to avoid possible cancellation of significant digits, which might occur, if we solve for the new iterates x^{k+1} directly.

Simplified Newton method. This variant of Newton's method is characterized by keeping the initial derivative throughout the whole iteration:

$$F'(x^0)\overline{\Delta x}^k = -F(x^k), \ x^{k+1} = x^k + \overline{\Delta x}^k, \ k = 0, 1, \ldots .$$

Compared to the ordinary Newton method, computational cost per iteration is saved—at the possible expense of increasing the number of iterations and possibly decreasing the convergence domain of the thus defined iteration.

Newton-like methods. This type of Newton method is characterized by the fact that, in finite dimension, the Jacobian matrices are either replaced by some fixed 'close by' Jacobian $F'(z)$ with $z \neq x^0$, or by some approximation so that

$$M(x^k)\delta x^k = -F(x^k), \quad x^{k+1} = x^k + \delta x^k, \qquad k = 0, 1, \ldots .$$

As an example, deliberate 'sparsing' of a large Jacobian, which means dropping of 'weak couplings', will permit the use of a *direct sparse solver* for the Newton-like corrections and therefore possibly help to reduce the work per iteration; if really only weak couplings are dropped, then the total iteration pattern will not deteriorate significantly.

Exact Newton methods. Any of the finite dimensional Newton-type methods requires the numerical solution of the linear equations

$$F'(x^k)\Delta x^k = -F(x^k).$$

Whenever *direct* elimination methods are applicable, we speak of *exact* Newton methods. However, naive application of direct elimination methods may cause serious trouble, if scaling issues are ignored.

BIBLIOGRAPHICAL NOTE. There are numerous excellent books on the numerical solution of linear systems—see, e.g., the classic by G.H. Golub and C.F. van Loan [107]. Programs for direct elimination in *full* or *sparse mode* can be found in the packages LAPACK [5], SPARSPAK [100], or [27]. As a rule, these codes leave the scaling issue to the user—for good reasons, since the user will typically know the specifications behind the problem that define the necessary scaling.

Local versus global Newton methods. Local Newton methods require 'sufficiently good' initial guesses. Global Newton methods are able to compensate for bad initial guesses by virtue of damping or adaptive trust region strategies. Exact global Newton *codes* for the solution of nonlinear equations are named NLEQ plus a characterizing suffix. We give details about

- NLEQ-RES for the residual based approach,
- NLEQ-ERR for the error oriented approach, or
- NLEQ-OPT for convex optimization.

Inexact Newton methods. For extremely large scale nonlinear problems the arising linear systems for the Newton corrections can no longer be solved directly ('exactly'), but must be solved *iteratively* ('inexactly')—which gives the name *inexact* Newton methods. The whole scheme then consists of an *inner iteration* (at Newton step k)

$$
\begin{aligned}
F'(x^k)\,\delta x_i^k &= -F(x^k) + r_i^k\,, \quad k = 0, 1, \dots, \\
x_i^{k+1} &= x^k + \delta x_i^k\,, \qquad\qquad i = 0, 1, \dots, i_{\max}^k
\end{aligned}
\tag{1.15}
$$

in terms of *residuals* r_i^k and an *outer iteration* where, given x^0, the iterates are defined as

$$
x^{k+1} = x_i^{k+1} \quad \text{for} \quad i = i_{\max}^k, \; k = 0, 1, \dots \,.
$$

Compared with the exact Newton corrections in (1.14), *errors* $\delta x_i^k - \Delta x^k$ arise. Throughout the book, we will mostly drop the inner iteration index i for ease of notation.

In an *adaptive* inexact Newton method, the accuracy of the inner iteration should be matched to the outer iteration, preferably such that the Newton convergence pattern is essentially unperturbed—which means an appropriate control of $i_{\max}$ above. Criteria for the choice of the *truncation index* $i_{\max}$ depend on affine invariance, as will be worked out in detail. With this aspect in mind, inexact Newton methods are sometimes also called *truncated Newton methods*.

Inexact global Newton *codes* for the solution of large scale nonlinear equations are named GIANT plus a suffix characterizing the combination with an inner iterative solver. The name GIANT stands for Global Inexact Affine invariant Newton Techniques. We will work out details for

- GIANT-GMRES for the residual based approach,
- GIANT-CGNE and GIANT-GBIT for the error oriented approach, or
- GIANT-PCG for convex optimization.

As for the applied iterative solvers, see Section 1.4 below.

Preconditioning. A compromise between direct and iterative solution of the arising linear Newton correction equations is obtained by direct elimination of 'similar' linear systems, which can be used in a wider sense than just scaling as mentioned above. For its characterization we write

$$C_L F'(x^k) C_R C_R^{-1} \delta x_i^k = -C_L \left(F(x^k) - r_i^k \right) , \qquad i = 0, 1, \ldots i_{\max} \qquad (1.16)$$

or, equivalently, also

$$C_L F'(x^k) C_R C_R^{-1} \left(\delta x_i^k - \Delta x_i^k \right) = C_L r_i^k , \qquad i = 0, 1, \ldots i_{\max} .$$

Consequently, within the algorithms any residual or error norms need to be replaced by their preconditioned counterparts

$$\| r_i^k \| , \| \delta x_i^k - \Delta x_i^k \| \quad \longrightarrow \quad \| C_L r_i^k \| , \| C_R^{-1} \left(\delta x_i^k - \Delta x_i^k \right) \| .$$

Matrix-free Newton methods. Linear iterative solvers within inexact Newton methods only require the evaluation of Jacobian matrix vector products so that numerical difference approximations

$$F'(x)v \doteq \frac{F(x + \delta v) - F(x)}{\delta}$$

can be conveniently realized. Note, however, that the quality of such directional difference approximations will heavily depend on the choice of the relative deviation parameter δ and the mantissa length of the used arithmetic. A numerically stable realization will use *automatic differentiation* as suggested by A. Griewank [112].

Secant method. For *scalar* equations, say $f(x) = 0$, this type of method is derived from Newton's method by substituting the tangent by the secant

$$f'(x^k + \delta x_k) \longrightarrow \frac{f(x^k + \delta x_k) - f(x^k)}{\delta x_k} = j_{k+1}$$

and computing the correction as

$$\delta x_{k+1} = -\frac{f(x^{k+1})}{j_{k+1}} , \quad x^{k+1} = x^k + \delta x_k .$$

The thus constructed secant method is known to converge locally *superlinearly*.

Quasi-Newton methods. This class of methods extends the secant idea to systems of equations. In this case only a so-called *secant condition*

$$J\delta x_k = F(x^{k+1}) - F(x^k) \tag{1.17}$$

can be imposed, wherein J represents some Jacobian approximation to be specified. The above condition does not determine a unique J, but a whole class of matrices. If we recur to the previous quasi-Newton step as

$$J_k \delta x_k = -F(x^k),$$

we may select special *Jacobian rank-1 updates* as

$$J_{k+1} = J_k + \frac{F(x^{k+1})z^T}{z^T \delta x_k}, \qquad z \in \mathbb{R}^n, \; z \neq 0,$$

where the vector z is arbitrary, in principle. As will be shown below in detail, the specification of z is intimately linked with affine invariance. Once z has been specified, the next quasi-Newton step

$$J_{k+1} \delta x_{k+1} = -F(x^{k+1})$$

is determined. In the best case, *superlinear* local convergence can be shown to hold again. A specification to *linear* systems is the algorithm GBIT described in Section 1.4.4 below.

Gauss-Newton methods. This type of method applies to nonlinear least squares problems, whether unconstrained or constrained. The method requires the nonlinear least squares problems to be statistically well-posed, characterized either as 'small residual' (Section 4.2.1) or as 'adequate' problems (Section 4.3.2). For this problem class, *local* Gauss-Newton methods are appropriate, when 'sufficiently good' initial guesses are at hand, while *global* Gauss-Newton methods are used, when only 'bad initial guesses' are available. In the statistics community Gauss-Newton methods are also called *scoring methods*.

Quasilinearization. Infinite dimensional Newton methods for operator equations are also called *Newton methods in function space* or quasilinearization. The latter name stems from the fact that the nonlinear operator equation is solved via a sequence of corresponding linearized operator equations. Of course, the linearized equations for the Newton corrections can only be solved *approximately*. Consequently, *inexact* Newton methods supply the correct theoretical frame, within which now the 'truncation errors' represent *approximation errors*, typically *discretization errors*.

Inexact Newton multilevel methods. We reserve this term for those multilevel schemes, wherein the arising infinite dimensional linear Newton systems are approximately solved by some linear multilevel or multigrid method; in such a setting, Newton methods act in function space. The highest degree of sophistication of an inexact Newton multilevel method would be an *adaptive* Newton multilevel method, where the approximation errors are controlled within an abstract framework of inexact Newton methods.

Multilevel Newton methods. Unfortunately, the literature is often not unambiguous in the choice of names. In particular, the name 'Newton multigrid method' is often given to schemes, wherein a finite dimensional Newton multigrid method is applied on each level—see, e.g., the classical textbook [113] by W. Hackbusch or the more recent treatment [135] by R. Kornhuber, who uses advanced functional analytic tools. In order to avoid confusion, such a scheme will here be named 'multilevel Newton method'.

Nonlinear multigrid methods. For the sake of clarity, it may be worth mentioning that 'nonlinear multigrid methods' are not Newton methods, but fixed point iteration methods, and therefore not treated within the scope of this book.

BIBLIOGRAPHICAL NOTE. The classic among the textbooks for the numerical solution of finite dimensional systems of nonlinear equations has been the 1970 book of J.M. Ortega and W.C. Rheinboldt [163]. It has certainly set the state of the art for quite a long time. The monograph [177] by W.C. Rheinboldt guides into related more recent research areas. The popular textbook [132] by C.T. Kelley offers a nice introduction into finite dimensional inexact Newton methods—see also references therein. The technique of 'preconditioning' is usually attributed to O. Axelsson—see his textbook [11] and references therein. Multigrid Newton methods are worked out in detail in the meanwhile classic text of W. Hackbusch [113]; a detailed convergence analysis of such methods for certain smooth as well as a class of non-smooth problems has been recently given by R. Kornhuber [135].

1.4 Adaptive Inner Solvers for Inexact Newton Methods

As stated in Section 1.3 above, *inexact* Newton methods require the linear systems for the Newton corrections to be solved *iteratively*. Different affine invariance concepts naturally go with different concepts for the iterative solution. In particular, recall that

- *residual* norms go with *affine contravariance*,
- *error* norms go with *affine covariance*,

- *energy* norms go with *affine conjugacy*.

For the purpose of this section, let the inexact Newton system (1.15) be written as

$$Ay_i = b - r_i, \qquad i = 0, 1, \dots i_{\max}$$

in terms of iterative approximations y_i for the solution y and iterative residuals r_i. In order to control the number $i_{\max}$ of iterations, several *termination criteria* may be realized:

- Terminate the iteration as soon as the *residual norm* $\|r_i\|$ is small enough.
- Terminate the iteration as soon as the iterative *error norm* $\|y - y_i\|$ is small enough.
- If the matrix A is symmetric positive definite, terminate the iteration as soon as the *energy norm* $\|A^{1/2}(y - y_i)\|$ of the error is small enough.

In what follows, we briefly sketch some of the classical iterative linear solvers with particular emphasis on appropriate termination criteria for use within inexact Newton algorithms. We will restrict our attention to those iterative solvers, which minimize or, at least, reduce

- the residual norm (GMRES, Section 1.4.1),
- the energy norm of the error (PCG, Section 1.4.2), and
- the error norm (CGNE, Section 1.4.3, and GBIT, Section 1.4.4).

We include the less known solver GBIT, since it is a *quasi-Newton method* specialized to the solution of *linear* systems.

Preconditioning. This related issue deals with the iterative solution of systems of the kind

$$C_L A C_R C_R^{-1} y_i = C_L(b - r_i), \qquad i = 0, 1, \dots i_{\max}, \qquad (1.18)$$

where left preconditioner C_L and right preconditioner C_R arise. A proper choice of preconditioner will exploit information from the problem class under consideration and often crucially affect the convergence speed of the iterative solver.

Bi-CGSTAB. Beyond the iterative algorithms selected here, there are numerous further ones of undoubted merits. An example is the iterative solver Bi-CG and its stabilized variant Bi-CGSTAB due to H.A. van der Vorst [189]. This solver might actually be related to affine similarity as treated above in Section 1.2; as a consequence, this code would be a natural candidate within an inexact pseudo–continuation method (see Section 6.4.2). However, this combination of inner and outer iteration would require a rather inconvenient norm (Jordan canonical norm). That is why we do not incorporate this candidate here. However, further work along this line might be promising.

BIBLIOGRAPHICAL NOTE. A good survey on many aspects of the iterative solution of linear equation systems can be found in the textbook [181] by Y. Saad. Preconditioning techniques are described, e.g., in the textbook [11] by O. Axelsson.

Multilevel discretization. For the adaptive realization of inexact Newton methods in function space, discretizations on successively finer levels play the role of the inner iteration. That is why we additionally treat linear multigrid methods in Section 1.4.5 below. Skipping any technical details here, multilevel methods permit an adaptive control of discretization errors on each level—for example, see Section 7.3.3 on Fourier-Galerkin methods for periodic orbit computation, Section 7.4.2 on polynomial collocation methods for ODE boundary value problems, and Section 8.3 on adaptive multigrid methods for elliptic PDEs.

1.4.1 Residual norm minimization: GMRES

A class of iterative methods aims at the successive reduction of the residual norms $\|r_i\|$ for increasing index i. Outstanding candidates among these are those solvers that even *minimize the residual norms* over some Krylov subspace—such as GMRES and CGNR. Since algorithm GMRES requires less matrix/vector multiplies per step, we focus our attention on it here.

Algorithm GMRES. Given an initial approximation $y_0 \approx y$ compute the initial residual $r_0 = b - Ay_0$. Set $\beta = \|r_0\|_2$, $v_1 = r_0/\beta$, $V_1 = v_1$. For $i = 1, 2, \ldots, i_{\max}$:

I. Orthogononalization:
$$\hat{v}_{i+1} = Av_i - V_i h_i$$
$$\text{where } h_i = V_i^T Av_i$$

II. Normalization:
$$v_{i+1} = \hat{v}_{i+1}/\|\hat{v}_{i+1}\|_2$$

III. Update:
$$V_{i+1} = (V_i \ v_{i+1})$$

$$H_i = \begin{pmatrix} H_{i-1} & h_i \\ 0 & \|\hat{v}_{i+1}\|_2 \end{pmatrix}$$

H_i is an $(i+1, i)$-Hessenberg matrix (for $i = 1$ drop the left block column)

IV. Least squares problem for z_i : $\|\beta e_1 - H_i z\| = \min$

V. Approximate solution $(i = i_{\max})$: $y_i = V_i z_i + y_0$

Array storage. Up to iteration step i, the above implementation requires to store $i + 2$ vectors of length n.

Computational amount. In each iteration step i, there is one matrix/vector multiply needed. Up to step i, the Euclidean inner products sum up to $\sim i^2 n$ flops.

As already stated, this algorithm *minimizes* the residual norms over the Krylov subspace

$$\mathcal{K}_i(r_0, A) = \text{span}\ \{r_0, \dots, A^{i-1} r_0\}\,.$$

By construction, the inner residuals will *decrease monotonically*

$$\|r_{i+1}\|_2 \leq \|r_i\|_2\,.$$

Therefore, a reasonable *inner termination criterion* will check whether the final residual $\|r_i\|_2$ is 'small enough'. Moreover, starting with arbitrary initial guess y_0 and initial residual $r_0 \neq 0$, we have the *orthogonality relation* (in terms of the Euclidean inner product $\langle\,\cdot\,,\cdot\,\rangle$)

$$\langle r_i, r_i - r_0 \rangle = 0\,,$$

which directly implies that

$$\|r_0\|_2^2 = \|r_0 - r_i\|_2^2 + \|r_i\|_2^2 \tag{1.19}$$

throughout the inner iteration. If we define

$$\eta_i = \frac{\|r_i\|_2}{\|r_0\|_2}\,,$$

then we will generically have $\eta_i < 1$ for $i > 0$ and

$$\eta_{i+1} < \eta_i\,, \quad \text{if} \quad \eta_i \neq 0\,.$$

This implies that, after a number of iterations, any adaptive truncation criterion

$$\eta_i \leq \bar{\eta}$$

for a prescribed threshold value $\bar{\eta} < 1$ can be met. In passing we note that then (1.19) can be rewritten as

$$\|r_0 - r_i\|_2^2 = (1 - \eta_i^2)\|r_0\|_2^2\,. \tag{1.20}$$

These detailed results are applied in Sections 2.2.4 and 3.2.3.

Preconditioning. Finally, if (1.18) is applied, then the Euclidean norms of the *preconditioned residuals* $\bar{r}^k = C_L r^k$ are iteratively *minimized* in GMRES, whereas C_R only affects the rate of convergence. Therefore, if strict residual minimization is aimed at, then only *right* preconditioning should be implemented, which means $C_L = I$.

BIBLIOGRAPHICAL NOTE. This iterative method has been designed as a rather popular code by Y. Saad and M.H. Schultz [182] in 1986; an earlier often overlooked derivation has been given by G.I. Marchuk and Y.A. Kuznetsov [146] already in 1968.

1.4.2 Energy norm minimization: PCG

With A *symmetric positive definite*, we are able to define the *energy product* $(\cdot, \cdot)$ and its induced *energy norm* $\| \cdot \|_A$ by

$$(u, v) = \langle u, Av \rangle, \quad \|u\|_A^2 = (u, u)$$

in terms of the Euclidean inner product $\langle \cdot, \cdot \rangle$. Let $B \approx A^{-1}$ denote some preconditioning matrix, assumed to be also *symmetric positive definite*. Usually the numerical realization of $z = Bc$ is much simpler and faster than the solution of $Ay = b$. Formally speaking, we may specify some $C_L = C_R^T = B^{1/2}$. This specification does not affect the energy norms, but definitely the speed of convergence of the iteration. Any *preconditioned conjugate gradient* (PCG) method reads:

Algorithm PCG. For given approximation $y_0 \approx y$ compute the initial residual $r_0 = b - Ay_0$ and the preconditioned residual $\bar{r}_0 = Br_0$. Set $p_0 = \bar{r}_0$ $\sigma_0 = \langle r_0, \bar{r}_0 \rangle = \|r_0\|_B^2$.
For $i = 0, 1, \ldots, i_{\max}$:

$$
\begin{aligned}
\alpha_i &= \frac{\|p_i\|_A^2}{\sigma_i} \\
y_{i+1} &= y_i + \frac{1}{\alpha_i} p_i \\
\gamma_i^2 &= \frac{\sigma_i}{\alpha_i} \quad \text{(energy error contribution} \quad \|y_{i+1} - y_i\|_A^2) \\
r_{i+1} &= r_i - \frac{1}{\alpha_i} Ap_i, \quad \bar{r}_{i+1} = Br_{i+1} \\
\sigma_{i+1} &= \|r_{i+1}\|_B^2, \quad \beta_{i+1} = \frac{\sigma_{i+1}}{\sigma_i} \\
p_{i+1} &= \bar{r}_{i+1} + \beta_{i+1} p_i.
\end{aligned}
$$

Array storage. Up to iteration step i, ignoring any preconditioners, the above implementation requires to store only 4 vectors of length n.

Computational amount. In each iteration step i, there is one matrix/vector multiply needed. Up to step i, the Euclidean inner products sum up to $\sim 5in$ flops.

This iteration successively *minimizes* the energy error norm $\|y - y_i\|_A$ within the associated Krylov subspace

$$\mathcal{K}_i(r_0, A) = \mathrm{span}\{r_0, \dots, A^{i-1}r_0\}.$$

By construction, we have the orthogonality relations (also called *Galerkin conditions*)

$$(y_i - y_0, y_{i+m} - y_i)_A = 0, \quad m = 1, \dots,$$

which imply the orthogonal decompositions (with $m = 1$)

$$\|y_{i+1} - y_0\|_A^2 = \|y_{i+1} - y_i\|_A^2 + \|y_i - y_0\|_A^2 \tag{1.21}$$

and (with $m = n - i$ and $y_n = y$)

$$\|y - y_0\|_A^2 = \|y - y_i\|_A^2 + \|y_i - y_0\|_A^2 . \tag{1.22}$$

From (1.21) we easily derive that

$$\|y_i - y_0\|_A^2 = \sum_{j=0}^{i-1} \|y_{j+1} - y_j\|_A^2 = \sum_{j=0}^{i-1} \gamma_j^2 . \tag{1.23}$$

Together with (1.22), we then obtain

$$\epsilon_i = \|y - y_i\|_A^2 = \sum_{j=i}^{n-1} \gamma_j^2 . \tag{1.24}$$

Estimation of PCG error. Any adaptive *affine conjugate* inexact Newton algorithm will require a reasonable estimate for the errors ϵ_i to be able to exploit the theoretical convergence results. Note that the monotonicity

$$\|y_i - y_0\|_A \le \|y_{i+1} - y_0\|_A \le \cdots \le \|y - y_0\|_A$$

can be derived from (1.21)—a saturation effect easily observable in actual computation. There are two basic methods to estimate the error.

(I) Assume we have a computable upper bound

$$\bar{\epsilon}_0 \ge \|y - y_0\|_A^2 ,$$

such as the ones suggested by G.H. Golub and G. Meurant [104] or by B. Fischer [92]. Then, with (1.22) and (1.23), we obtain the computable upper bound

$$\epsilon_i \le \bar{\epsilon}_0 - \sum_{j=0}^{i-1} \gamma_j^2 = [\epsilon_i] .$$

(II) As an alternative (see [68]), we may exploit the structure of (1.24) via the lower bound

$$[\epsilon_i] = \sum_{j=i}^{i+m} \gamma_j^2 \le \epsilon_i \tag{1.25}$$

for some sufficiently large index $m > 0$—which means to continue the iteration further just for the purpose of getting some error estimate. In the case of 'fast' convergence (usually for 'good' preconditioners only), few terms in the sum will suffice. Typically, we use this technique, since it does not require any choice of an upper bound $\bar{\epsilon}_0$.

Both techniques inherit the *monotonicity* $[\epsilon_{i+1}] \leq [\epsilon_i]$ from $\epsilon_{i+1} \leq \epsilon_i$. Hence, generically, after a number of iterations, any adaptive truncation criterion

$$[\epsilon_i] \leq \bar{\epsilon}$$

for a prescribed threshold value $\bar{\epsilon}$ can be met. In the inexact Newton-PCG algorithms to be worked out below we will use the *relative* energy error norms defined, for $i > 0$, as

$$\delta_i = \frac{\|y - y_i\|_A}{\|y_i\|_A} \approx \frac{\sqrt{[\epsilon_i]}}{\|y_i\|_A}.$$

Whenever $y_0 = 0$, then (1.21) implies the monotone increase $\|y_{i+1}\|_A \geq \|y_i\|_A$ and therefore the monotone decrease $\delta_{i+1} \leq \delta_i$. This guarantees that any relative error criterion

$$\delta_i \leq \bar{\delta}$$

can be met. Moreover, in this case we have the relation

$$\|y_i\|_A^2 = (1 + \delta_i^2)\|y\|_A^2. \tag{1.26}$$

For $y_0 \neq 0$, the above monotonicities and (1.26) no longer hold. This option, however, is not used in the inexact Newton-PCG algorithms to be derived in Sections 2.3.3 and 3.4.3 below.

1.4.3 Error norm minimization: CGNE

Another class of iterative solvers aims at the successive reduction of the (possibly scaled) Euclidean error norms $\|y - y_i\|$ for $i = 0, 1, \ldots$. Among these the ones that *minimize the error norms* over some Krylov subspace play a special role. For nonsymmetric Jacobian matrices the outstanding candidate with this feature seems to be CGNE. For its economic implementation, we recommend Craig's variant (see, e.g., [181]), which reads:

Algorithm CGNE. Given an initial approximation y_0, compute the initial residual $r_0 = b - Ay_0$ and set $p_0 = 0, \beta_0 = 0, \sigma_0 = \|r_0\|^2$.
For $i = 1, 2, \ldots, i_{\max}$:

$$
\begin{aligned}
p_i &= A^T r_{i-1} + \beta_{i-1} p_{i-1} \\
\alpha_i &= \sigma_{i-1}/\|p_i\|^2 \\
\gamma_{i-1}^2 &= \alpha_i \sigma_{i-1} \qquad \text{(Euclidean error contribution } \|y_i - y_{i-1}\|^2) \\
y_i &= y_{i-1} + \alpha_i p_i \\
r_i &= r_{i-1} - \alpha_i A p_i \\
\sigma_i &= \|r_i\|^2 \\
\beta_i &= \sigma_i/\sigma_{i-1}
\end{aligned}
$$

Array storage. Up to iteration step i, the above implementation requires to store only 3 vectors of length n.

Computational amount. In each iteration step i, there are *two* matrix/vector multiplies needed. Up to step i, the Euclidean inner products sum up to $\sim 5in$ flops.

This iteration successively *minimizes* the Euclidean norms $\|y - y_i\|$ within the Krylov subspace

$$
\mathcal{K}_i(A^T r_0, A^T A) = \text{span}\{A^T r_0, \ldots, (A^T A)^{i-1} A^T r_0\}.
$$

By construction, we have the orthogonality relations (also: *Galerkin conditions*)

$$
(y_i - y_0, y_{i+m} - y_i) = 0 , \quad m = 1, \ldots ,
$$

which imply the orthogonal decomposition (with $m = 1$)

$$
\|y_{i+1} - y_0\|^2 = \|y_{i+1} - y_i\|^2 + \|y_i - y_0\|^2 \tag{1.27}
$$

and (with $m = n - i$ and $y_n = y$)

$$
\|y - y_0\|^2 = \|y - y_i\|^2 + \|y_i - y_0\|^2 . \tag{1.28}
$$

From (1.27) we easily derive that

$$
\|y_i - y_0\|^2 = \sum_{j=0}^{i-1} \|y_{j+1} - y_j\|^2 = \sum_{j=0}^{i-1} \gamma_j^2 . \tag{1.29}
$$

Together with (1.28), we then obtain

$$
\epsilon_i = \|y - y_i\|^2 = \sum_{j=i}^{n-1} \gamma_j^2 . \tag{1.30}
$$

Estimation of CGNE error. Any adaptive *affine covariant* inexact Newton algorithm will require a reasonable estimate for the errors ϵ_i to be able to exploit the theoretical convergence results. Again, the saturation effect from the monotonicity

$$\|y_i - y_0\| \leq \|y_{i+1} - y_0\| \leq \cdots \leq \|y - y_0\|$$

can be derived from (1.27).

There are two basic methods to estimate the error within CGNE.

(I) Assume we have a computable upper bound

$$\bar{\epsilon}_0 \geq \|y - y_0\|^2$$

in the spirit of those suggested by G.H. Golub and G. Meurant [104] or by B. Fischer [92]. From this, with (1.28) and (1.29), we obtain the computable upper bound

$$\epsilon_i \leq \bar{\epsilon}_0 - \sum_{j=0}^{i-1} \gamma_j^2 = [\epsilon_i].$$

(II) As an alternative, transferring an idea from [68], we may exploit the structure of (1.30) to look at the lower bound

$$[\epsilon_i] = \sum_{j=i}^{i+m} \gamma_j^2 \leq \epsilon_i \tag{1.31}$$

for some sufficiently large index $m > 0$—which means to continue the iteration further just for the purpose of getting some error estimate. In the case of 'fast' convergence (for 'sufficiently good' preconditioner, see below), only few terms in the sum will be needed. Typically, we use this second technique.

Both techniques inherit the *monotonicity* $[\epsilon_{i+1}] \leq [\epsilon_i]$ from $\epsilon_{i+1} \leq \epsilon_i$. After a number of iterations, any adaptive truncation criterion

$$[\epsilon_i] \leq \bar{\epsilon}$$

for a prescribed threshold value $\bar{\epsilon}$ can generically be met. In the inexact Newton-ERR algorithms to be worked out below we will use the *relative* error norms defined, for $i > 0$, as

$$\delta_i = \frac{\|y - y_i\|}{\|y_i\|} \approx \frac{\sqrt{[\epsilon_i]}}{\|y_i\|}. \tag{1.32}$$

Whenever $y_0 = 0$, then (1.27) implies the monotonicities $\|y_{i+1}\| \geq \|y_i\|$ and $\delta_{i+1} \leq \delta_i$. The latter one guarantees that the relative error criterion

$$\delta_i \leq \bar{\delta} \tag{1.33}$$

can be eventually met. For this initial value we also have the relation

$$\|y_i\|^2 = (1 + \delta_i^2)\|y\|^2 \,. \tag{1.34}$$

These detailed results enter into the presentation of local inexact Newton-ERR methods in Section 2.1.5.

For $y_0 \neq 0$, the above monotonicities as well as the relation (1.34) no longer hold. This situation occurs in the global inexact Newton-ERR method to be derived in Section 3.3.4. Since the $\|y_i\|$ eventually approach $\|y\|$, we nevertheless require the relative truncation criterion (1.33).

Preconditioning. Finally, if (1.18) is applied, then the norms of the iterative *preconditioned errors* $C_R^{-1}(y - y_i)$ are minimized. Therefore, if *strict* unscaled error minimization is aimed at, then only *left* preconditioning should be realized. In addition, if 'good' preconditioners C_R or C_L are available (resulting in 'fast' convergence), then the simplification

$$\|C_R^{-1}(y - y_i)\| \approx \|C_R^{-1}(y_{i+1} - y_i)\|$$

will be sufficient.

Remark 1.2 Numerical experiments with large discretized PDEs in Section 8.2.1 document a poor behavior of CGNE, which seems to stem from a rather sensitive dependence on the choice of preconditioner. Generally speaking, a preconditioner, say B, is expected to reduce the condition number $\kappa(J)$ to some $\kappa(BJ) \ll \kappa(J)$. However, as the algorithm CGNE works on the normal equations, the characterizing dependence is on $\kappa^2(BJ) \gg \kappa(BJ)$. In contrast to this behavior, the preconditioned GMRES just depends on $\kappa(BJ)$ as the characterizing quantity.

1.4.4 Error norm reduction: GBIT

The quasi-Newton methods already mentioned in Section 1.3 can be specified to apply to *linear* systems as well. Following the original paper by P. Deuflhard, R. Freund, and A. Walter [74], a special *affine covariant* rank-1 update can be chosen, which turns out to be Broyden's 'good' update [40]. Especially for linear systems, an optimal line search is possible, which then gives the algorithm GBIT (abbreviation for Good Broyden ITerative solver for linear systems).

Preconditioner improvement. The main idea behind this algorithm is to improve any (given) initial preconditioner $B_0 \sim A$, or $H_0 \sim A^{-1}$, respectively, successively to $B_i \sim A, H_i \sim A^{-1}$. Let $E_i = I - A^{-1}B_i$ denote the preconditioning error, then each iterative step can be shown to realize some new preconditioner such that

$$\|E_{i+1}\|_2 \le \|E_i\|_2 , \quad i = 0, 1, \ldots .$$

Error reduction. In [74], this algorithm has been proven to converge under the sufficient assumption

$$\|E_0\|_2 < \tfrac{1}{3} , \tag{1.35}$$

in the sense that

$$\|y_{i+1} - y\| < \|y_i - y\| . \tag{1.36}$$

Moreover, asymptotic *superlinear* convergence can even be shown. Numerical experience shows that an assumption weaker than (1.35) might do, but there is no theoretical justification of such a statement yet.

The actual implementation of the algorithm does not store the improved preconditioners H_i explicitly, but exploits the Sherman-Morrison formula to obtain a cheap recursion. The following implementation is a recent slight improvement over the algorithm GB suggested in [74]—essentially replacing the iterative stepsize $t_i = \tau_i$ therein by some modification (see below and Exercise 1.4 for $\tau_{\max} = 1$).

Algorithm GBIT. Given an initial guess y_0, an initial preconditioner $H_0 \sim A^{-1}$, and some inner product $\langle u, v \rangle$.
Initialization:

$$
\begin{aligned}
r_0 &= b - Ay_0 \\
\Delta_0 &= H_0 r_0 \\
\sigma_0 &= \langle \Delta_0, \Delta_0 \rangle
\end{aligned}
$$

Iteration loop $i = 0, 1, \ldots, i_{\max}$:

$$
\begin{aligned}
q_i &= A\Delta_i \\
\zeta_0 &= H_0 q_i
\end{aligned}
$$

Update loop $m = 0, \ldots, i - 1$ (for $i \ge 1$):

$$\zeta_{m+1} = \zeta_m + \frac{\langle \Delta_m, \zeta_m \rangle}{\sigma_m} \left(\Delta_{m+1} - (1 - t_m)\Delta_m \right)$$

$$
\begin{aligned}
z_i &= \zeta_i \\
\gamma_i &= \langle \Delta_i, z_i \rangle \\
\tau_i &= \sigma_i / \gamma_i
\end{aligned}
$$

if $\tau_i < \tau_{\min}:$ **restart**

$$t_i = \tau_i$$

$$
\begin{aligned}
\text{if} \quad t_i \quad &> \quad \tau_{\max} : \quad t_i = 1 \\
y_{i+1} \quad &= \quad y_i + t_i \Delta_i \\
(r_{i+1} \quad &= \quad r_i - t_i q_i) \\
\Delta_{i+1} \quad &= \quad (1 - t_i + \tau_i)\Delta_i - \tau_i z_i \\
\sigma_{i+1} \quad &= \quad \langle \Delta_{i+1}, \Delta_{i+1} \rangle \\
\epsilon_i \quad &= \quad \tfrac{1}{2}\sqrt{\sigma_{i-1} + 2\sigma_i + \sigma_{i+1}}
\end{aligned}
$$

$$
\text{if} \quad \epsilon_i \quad \le \quad \rho\|y_{i+1}\| \cdot \text{ERRTOL}: \quad \textbf{solution found}
$$

The parameters $\tau_{\min}, \tau_{\max}$ are set internally such that $0 < \tau_{\min} \ll 1, \tau_{\max} \ge 1$, the safety factor $\rho < 1$ and the error tolerance ERRTOL are user prescribed values.

Array storage. Up to iteration step i, the above recursive implementation requires to store the $i + 3$ vectors

$$
\Delta_0, \ldots, \Delta_i, q, z \equiv \zeta .
$$

of length n.

Computational amount. In each iteration step i, the computational work is dominated by one matrix/vector multiply, one solution of a preconditioned system ($\zeta_0 = H_0 q$), and $\sim 2i \cdot n$ flops for the Euclidean inner product. Up to step i, this sums up to i preconditioned systems and $\sim i^2 n$ flops.

Inner product. Apart from the Euclidean inner product $\langle u, v \rangle = u^T v$ any scaled version such as $\langle u, v \rangle = (D^{-1}u)^T D^{-1} v$, with D a diagonal scaling matrix, will be applicable—and even preferable. For special problems like discrete PDE boundary value problems certain discrete L^2-products and norms are recommended.

Error estimation and termination criterion. By construction, we obtain the relation

$$
y_i - y = \Delta_i - E_i \Delta_i ,
$$

which, under the above preconditioning assumption, certainly implies the estimation property

$$
\left(1 - \frac{\|E_i \Delta_i\|}{\|\Delta_i\|}\right) \|\Delta_i\| \le \|y_i - y\| \le \left(1 + \frac{\|E_i \Delta_i\|}{\|\Delta_i\|}\right) \|\Delta_i\| .
$$

Hence, the true error can be roughly estimated as

$$
\|y_i - y\| \approx \|\Delta_i\| = \sqrt{\sigma_i} .
$$

In order to suppress possible outliers, an average of the kind

$$
\epsilon_i = \tfrac{1}{2}\sqrt{\sigma_{i-1} + 2\sigma_i + \sigma_{i+1}} \tag{1.37}
$$

is typically applied for $i > 1$. This estimator leads us to the relative termination criterion

$$\epsilon_i \le \rho \|y_{i+1}\| \cdot \text{ERRTOL} ,$$

as stated above.

Note that the above error estimator cannot be shown to inherit the monotonicity property (1.36). Consequently, this algorithm seems to be less efficient than CGNE within the frame of Newton-ERR algorithms. Surprisingly, this expectation is not at all in agreement with numerical experiments—see, for instance, Section 8.2.1.

Remark 1.3   On top of GBIT we also applied ideas of D.M. Gay and R.B. Schnabel [97] about successive orthogonalization of update vectors to construct some projected 'good' Broyden method for linear systems. The corresponding algorithm PGBIT turned out to require only slightly less iterations, but significantly more array storage and computational amount—and is therefore omitted here.

1.4.5 Linear multigrid methods

In Newton methods for *operator equations* the corresponding iterates and solutions live in appropriate *infinite* dimensional function spaces. For example, in steady state partial differential equations (PDEs), the solutions live in some Sobolev space—like H^α—depending on the prescribed boundary conditions. It is an important mathematical paradigm that any such infinite dimensional space should not just be represented by a single finite dimensional space of possibly high dimension, but by a *sequence of finite dimensional subspaces with increasing dimension*.

Consequently, any infinite dimensional Newton method will be realized via a *sequence* of finite dimensional linearized systems

$$Ay_j = b + r_j, \qquad j = 0, 1, \ldots j_{\max} ,$$

where the residuals r_j represent *approximation errors*, mostly *discretization errors*. Each of the subsystems is again solved iteratively, which gives rise to the question of accuracy matching of discretization versus iteration. This is the regime of linear *multigrid* or *multilevel* methods—see, e.g., the textbook of W. Hackbusch [113] and Chapter 8 below on Newton multilevel methods.

Adaptivity. In quite a number of application problems rather localized phenomena occur. In this case, uniform grids are by no means optimal, which, in turn, also means that the classical multigrid methods on uniform grids could not be regarded as optimal. For this reason, multigrid methods on *adaptive* grids have been developed quite early, probably first by R.E. Bank

[18] in his code PLTMG and later in the family UG of parallel codes [22, 21] by
G. Wittum, P. Bastian, and their groups.

Independent of the classical multigrid methods, a multilevel method based
on conjugate gradient iteration with some *hierarchical basis* (HB) *preconditioning* had been suggested for elliptic PDEs by H. Yserentant [204]. An
adaptive 2D version of the new method had been first designed and implemented by P. Deuflhard, P. Leinen, and H. Yserentant [78] in the code
KASKADE. A more mature version including also 3D has been worked out by
F. Bornemann, B. Erdmann, and R. Kornhuber [36]. The present version of
KASKADE [23] contains the original HB-preconditioner for 2D and the more
recent BPX-preconditioner due to J. Xu [200, 39] for 3D.

Additive versus multiplicative multigrid methods. In the interpretation of multigrid methods as abstract Schwarz methods as given by J. Xu
[201], which the author prefers to adopt, the classical multigrid methods
are now called *multiplicative multigrid methods*, whereas the HB- or BPX-
preconditioned conjugate gradient methods are called *additive multigrid
methods*. In general, any difference in speed between additive or multiplicative multigrid methods is only marginal, since the bulk of computing time is
anyway spent in the evaluation of the stiffness matrix elements and the right
hand side elements. For the orientation of the reader: UG is nearly exclusively
multiplicative, PLTMG is predominantly multiplicative with some additive options, KASKADE is predominantly additive with some multiplicative code for
special PDE eigenvalue problems.

Cascadic multigrid methods. These rather recent multigrid methods can
be understood as a confluence of additive and multiplicative multigrid methods. From the additive point of view, cascadic multigrid methods are characterized by the simplest possible preconditioner: either no or just a diagonal
preconditioner is applied; as a distinguishing feature, coarser levels are vis-
ited more often than finer levels—to serve as preconditioning substitutes.
From the multiplicative side, cascadic multigrid methods may be understood
as multigrid methods with an increased number of smoothing iterations on
coarser levels, but without any coarse grid corrections. A first algorithm of
this type, the *cascadic conjugate gradient method* (algorithm CCG) had been
proposed by the author in [68]. First rather restrictive convergence results
were due to V. Shaidurov [185]. The general cascadic multigrid method class
with arbitrary inner iterations beyond conjugate gradient methods has been
presented by F. Bornemann and P. Deuflhard [35].

Just to avoid mixing terms: *cascadic* multigrid methods are different from
the code KASKADE, which predominantly realizes *additive* multigrid methods.

Local error estimators. Any efficient implementation of *adaptive* multigrid methods (additive, multiplicative, cascadic) must be based on cheap *local*

error estimators or, at least, *local error indicators*. In the best case, these are derived from theoretical *a-posteriori error estimates*. These estimates will be local only, if local (right hand side) perturbations in the given problem remain local—i.e., if the Greens' function of the PDE problem exhibits local behavior. As a consequence of this elementary insight, *adaptive* multigrid methods will be essentially applicable to linear or nonlinear elliptic problems (among the stationary PDE problems). A comparative assessment of the different available local error estimators has been given by F. Bornemann, B. Erdmann, and R. Kornhuber in [37]. In connection with any error estimator, the local extrapolation method due to I. Babuška and W.C. Rheinboldt [14] can be applied. The art of refinement is quite established in 2D (see the 'red' and 'green' refinements due to R.E. Bank et al. [20]) and still under further improvement in 3D.

Summarizing, adaptive multilevel methods for linear PDEs play a dominant role in the frame of adaptive Newton multilevel methods for nonlinear PDEs—see, e.g., Section 8.3.

Exercises

Exercise 1.1   Given a nonlinear C^1-mapping $F : X \to Y$ over some domain $D \subset X$ for Banach spaces X, Y, each endowed with some norm $\|\cdot\|$. Assume a Lipschitz condition of the form

$$\|F'(x) - F'(y)\| \leq \gamma\|x - y\|, \ x, y \in D.$$

Let the derivative at some point x^0 have a bounded inverse with

$$\|F'(x^0)^{-1}\| \leq \beta_0.$$

Show that then, for all arguments $x \in D$ in some open ball $S(x^0, \rho)$ with $\rho = \dfrac{1}{\beta_0\gamma}$, there exists a bounded derivative inverse with

$$\|F'(x)^{-1}\| \leq \frac{\beta_0}{1 - \beta_0\gamma\|x - x^0\|}.$$

Exercise 1.2   Usual proofs of the implicit function theorem apply the Newton-Kantorovich theorem—compare Section 1.2. Revisit this kind of proof in any available textbook in view of affine covariance. In particular, replace condition (1.3) for a locally bounded inverse and Lipschitz condition (1.4) by some affine covariant Lipschitz condition like (1.6), which defines some local affine covariant Lipschitz constant ω_0. Formulate the thus obtained affine covariant implicit function theorem. Characterize the class of problems, for which $\omega_0 = \infty$.

Exercise 1.3 Consider the scalar monomial equation

$$f(x) = x^m - a = 0 \,.$$

We want to study the convergence properties of Newton's method. For this purpose consider the general corresponding contraction term

$$\Theta(x) = \frac{f f''}{f'^2} \,.$$

Verify this expression in general and calculate it for the specific case. What kind of convergence occurs for $m \neq 1$? How could the Newton method be 'repaired' such that quadratic convergence still occurs? Why is this, in general, not a good idea?

Hint: Study the convergence properties under small perturbations.

Exercise 1.4 Consider the linear iterative solver GBIT described in Section 1.4.4. In the notation introduced there, let the iterative error be written as $e_i = y - y_i$.

a) Verify the recursive relation

$$e_{i+1} = (1 - t_i)e_i + t_i \overline{E}_i e_i \,,$$

where

$$\overline{E}_i = -(I - E_i)^{-1} E_i \,.$$

b) Show that, under the assumption $\|\overline{E}_i\| < 1$ on a 'sufficiently good' preconditioner, any stepsize choice $0 < t_i \leq 1$ will lead to convergence, i.e.,

$$\|e_{i+1}\| < \|e_i\|, \quad \text{if } e_i \neq 0 \,.$$

c) Verify that $\|E_{i+1}\|_2 \leq \|E_i\|_2$ holds, so that $\|E_0\|_2 < \frac{1}{2}$ implies $\|\overline{E}_i\|_2 < 1$ for all indices $i = 0, 1, \dots$.

d) Compare the algorithm GBIT for the two steplength strategies $t_i = \tau_i$ and $t_i = \min(1, \tau_i)$ at your favorite linear system with nonsymmetric matrix A.

Part I

ALGEBRAIC EQUATIONS

2 Systems of Equations: Local Newton Methods

This chapter deals with the numerical solution of systems of nonlinear equations with finite, possibly large dimension n. The term *local* Newton methods refers to the situation that—only throughout this chapter—'sufficiently good' initial guesses of the solution are assumed to be at hand. Special attention is paid to the issue of how to recognize—in a computationally cheap way—whether a given initial guess x^0 is 'sufficiently good'. As it turns out, different affine invariant Lipschitz conditions, which have been introduced in Section 1.2.2, lead to different characterizations of local convergence domains in terms of error oriented norms, residual norms, or energy norms and convex functionals, which, in turn, give rise to corresponding variants of Newton algorithms.

We give three different, strictly affine invariant convergence analyses for the cases of affine covariant (error oriented) Newton methods (Section 2.1), affine contravariant (residual based) Newton methods (Section 2.2), and affine conjugate Newton methods for convex optimization (Section 2.3). Details are worked out for ordinary Newton algorithms, simplified Newton algorithms, and inexact Newton algorithms synoptically for the three affine invariance classes. Moreover, affine covariance appears as associated with Broyden's 'good' quasi-Newton method, whereas affine contravariance corresponds to Broyden's 'bad' quasi-Newton method.

2.1 Error Oriented Algorithms

A convergence analysis for any error oriented algorithm of Newton type will start from *affine covariant* Lipschitz conditions of the kind (1.7) and lead to results in the space of the iterates only. The behavior of the residuals will be ignored. For actual computation, scaling of any arising norms of Newton corrections is tacitly assumed.

2.1.1 Ordinary Newton method

Consider the ordinary Newton method in the notation

$$F'(x^k)\Delta x^k = -F(x^k), \quad x^{k+1} = x^k + \Delta x^k, \quad k = 0, 1, \ldots . \tag{2.1}$$

Convergence analysis. Because of its fundamental importance, we begin with an affine covariant version of the classical 'Newton-Kantorovich theorem'. Only at this early stage we state the theorem in Banach spaces—well aware of the fact that a Banach space formulation is not directly applicable to numerical methods: in the numerical solution of nonlinear operator equations both function and derivative *approximations* must be taken into account. As a consequence, *inexact* Newton methods in Banach spaces are the correct theoretical frame to study convergence of algorithms—to be treated below in Sections 7.4, 8.1, and 8.3.

Theorem 2.1 *Let $F : D \to Y$ be a continuously Fréchet differentiable mapping with $D \subseteq X$ open and convex. For a starting point $x^0 \in D$ let $F'(x^0)$ be invertible. Assume that*

$$\|F'(x^0)^{-1}F(x^0)\| \le \alpha,$$

$$\left\| F'(x^0)^{-1}\left(F'(y) - F'(x)\right) \right\| \le \overline{\omega}_0\|y - x\| \quad x, y \in D, \tag{2.2}$$

$$h_0 := \alpha\overline{\omega}_0 \le \tfrac{1}{2}, \tag{2.3}$$

$$\overline{S}(x^0, \rho_-) \subset D, \qquad \rho_- := \left(1 - \sqrt{1 - 2h_0}\right) / \overline{\omega}_0.$$

Then the sequence $\{x^k\}$ obtained from the ordinary Newton iteration is well-defined, remains in $\overline{S}(x^0, \rho_-)$, and converges to some x^* with $F(x^*) = 0$. For $h_0 < \tfrac{1}{2}$, the convergence is quadratic.

Proof. Rather than giving the classical 1948 proof [126] of L.V. Kantorovich, we here sketch an alternative affine covariant proof, which dates back to T. Yamamoto [202] in 1985.

The proof is by induction starting with $k = 0$. At iterate x^k, let the Fréchet derivative $F'(x^k)$ be invertible. Hence we may require the affine covariant Lipschitz condition

$$\left\| F'(x^k)^{-1}\left(F'(y) - F'(x)\right) \right\| \le \overline{\omega}_k\|y - x\|$$

and define an associated first majorant

$$\overline{\omega}_k\|\Delta x^k\| \le h_k.$$

As a preparation to show that with $F'(x^k)$ also the Fréchet derivative $F'(x^{k+1})$ is invertible, we define the operators

$$B_{k+1} := F'(x^k)^{-1}F'(x^{k+1})$$

and the associated second majorant

$$\|B_{k+1}^{-1}\| \leq \beta_{k+1}.$$

Consequently, for $k > 0$ we have the upper bound

$$\overline{\omega}_k \leq \beta_k \overline{\omega}_{k-1}.$$

By means of the operator perturbation lemma, we easily obtain

$$\beta_{k+1} = 1/(1 - h_k). \tag{2.4}$$

Next, in order to exploit the above Lipschitz condition, we apply standard analytical techniques to obtain

$$\|x^{k+1} - x^k\| = \left\| F'(x^k)^{-1} \int_{s=0}^{1} \left[F'(x^{k-1} + s\Delta x^{k-1}) - F'(x^{k-1}) \right] \Delta x^{k-1} ds \right\|,$$

which implies

$$\overline{\omega}_k \|x^{k+1} - x^k\| \leq \tfrac{1}{2}\beta_k^2 h_{k-1}^2 =: h_k. \tag{2.5}$$

Combination of the two relations (2.5) and (2.4) then yields the single recursive equation

$$h_k = \frac{\tfrac{1}{2}h_{k-1}^2}{(1 - h_{k-1})^2}.$$

Herein contraction occurs, if

$$\frac{\tfrac{1}{2}h_0}{(1 - h_0)^2} < 1,$$

which directly leads to $h_0 < \tfrac{1}{2}$. Under this assumption, the convergence is quadratic.

Things are more complicated for the limiting case $h_0 = \tfrac{1}{2}$, which requires extra consideration. In this case, we obtain $h_k = \tfrac{1}{2}$, $k = 1, 2, \ldots$, which implies

$$\beta_k = 2, \qquad \overline{\omega}_k \leq 2^k \overline{\omega}_0.$$

Insertion into the majorant inequality (2.5) then leads to

$$\lim_{k \to \infty} \overline{\omega}_k \|x^{k+1} - x^k\| \leq \lim_{k \to \infty} 2^k \overline{\omega}_0 \|x^{k+1} - x^k\| \leq \tfrac{1}{2},$$

which verifies that

$$\lim_{k \to \infty} \|x^{k+1} - x^k\| \leq \lim_{k \to \infty} \frac{1}{2^{k-1}\overline{\omega}_0} = 0.$$

In the latter case, the convergence is linear. $\qquad\qquad\square$

Remark 2.1 If we define $t^{**} = 1 + \sqrt{1 - 2h_0}$, $\rho_+ = t^{**}/\overline{\omega}_0$, and assume that $\bar{S}(x^0, \rho_+) \subset D$, the solution x^* can be shown to be unique in $S(x^0, \rho_+)$. The corresponding proof is omitted here.

BIBLIOGRAPHICAL NOTE. The name 'Newton-Kantorovich theorem' has been coined, since historically L.V. Kantorovich was probably the first to prove convergence for Newton's method in Banach spaces. In 1939, he actually showed *linear* convergence (see [125]), but not earlier than 1948 he published his famous proof of *quadratic* convergence (see [126]). Even though this early theorem has already been phrased in affine covariant terms, nearly all (with few exceptions) of his later published versions lack this desirable property (see, e.g., the book by L.V. Kantorovich and G. Akhilov [127]). In 1949, I. Mysovskikh [155] presented an alternative meanwhile classical convergence theorem, which today is called 'Newton-Mysovskikh theorem'. That theorem was not affine invariant in any sense; the following Theorem 2.2 is an affine covariant version of it. In 1970, an interesting theorem for local convergence of Newton's method, already in affine covariant formulation, has been proved by H.B. Keller in [129], under the relaxed assumption of Hölder continuity of $F'(x)$—see Exercise 2.4. Since then a huge literature concerning different aspects of the classical theorems has unfolded, typically in not affine invariant form—compare, e.g., the monograph of F.A. Potra and V. Pták [171].

Not earlier than 1979, affine invariance as a subject of its own right within convergence analysis has been emphasized by P. Deuflhard and G. Heindl in [76]; this paper included an affine covariant (then called affine invariant) rephrasing of the classical Newton-Kantorovich and Newton-Mysovskikh theorem and permitted a new local convergence theorem for Gauss-Newton methods for nonlinear least squares problems—see Section 4.3.1. Also around that time H.G. Bock [29, 31, 32] adopted affine invariance and slightly weakened the Lipschitz condition in the affine covariant Newton-Mysovskikh theorem that had been given in [76]. Following the affine invariance message of [76], T. Yamamoto has introduced affine covariance into his subtle convergence estimates for Newton's method—see, e.g., his starting paper [202] and work thereafter. Later on, the earlier convergence theorem due to L.B. Rall [174], proved under the assumptions

$$\|F'(x^*)^{-1}\| \le \beta_* , \ \|F'(x) - F'(y)\| \le \gamma \|x - y\|$$

has been put into an affine covariant form by G. Bader [15]. For the improved variant of Rall's theorem due to W.C. Rheinboldt [176] see Exercise 2.5.

Throughout the subsequent convergence analysis for local Newton-type methods, we will mostly study extensions of the Newton-Mysovskikh theorem [155], which have turned out to be an extremely useful basis for the construction of algorithms. The subsequent Theorem 2.3 is the 'refined Newton-Mysovskikh theorem' due to P. Deuflhard and F.A. Potra [82], which has no classical predecessor, since it relies on affine covariance in its proof.

Next, we present an affine covariant Newton-Mysovskikh theorem. In what follows, we will return to the case of finite dimensional nonlinear equations, i.e. to $F : D \subset \mathbb{R}^n \to \mathbb{R}^n$.

Theorem 2.2 *Let $F : D \to \mathbb{R}^n$ be a continuously differentiable mapping with $D \subset \mathbb{R}^n$ convex. Suppose that $F'(x)$ is invertible for each $x \in D$. Assume that the following affine covariant Lipschitz condition holds:*

$$\left\| F'(z)^{-1} \big(F'(y) - F'(x) \big)(y - x) \right\| \le \omega \|y - x\|^2$$

for collinear x, y, $z \in D$. For the initial guess x^0 assume that

$$h_0 := \omega \|\Delta x^0\| < 2 \tag{2.6}$$

and that $\bar{S}(x^0, \rho) \subset D$ for $\rho = \dfrac{\|\Delta x^0\|}{1 - \frac{1}{2}h_0}$.

Then the sequence $\{x^k\}$ of ordinary Newton iterates remains in $S(x^0, \rho)$ and converges to a solution $x^* \in \bar{S}(x^0, \rho)$. Moreover, the following error estimates hold

$$\|x^{k+1} - x^k\| \le \tfrac{1}{2}\omega\|x^k - x^{k-1}\|^2 , \tag{2.7}$$

$$\|x^k - x^*\| \le \frac{\|x^k - x^{k+1}\|}{1 - \frac{1}{2}\omega\|x^k - x^{k+1}\|} . \tag{2.8}$$

Proof. First, the ordinary Newton iteration is used for k and $k - 1$:

$$\|\Delta x^k\| = \left\| F'(x^k)^{-1} \left[F(x^k) - \big(F(x^{k-1}) + F'(x^{k-1})\Delta x^{k-1} \big) \right] \right\| .$$

Application of the above Lipschitz condition yields

$$\|\Delta x^k\| \le \tfrac{1}{2}\omega\|\Delta x^{k-1}\|^2 ,$$

which is (2.7). For the purpose of repeated induction, introduce the following notation:

$$h_k := \omega \left\| \Delta x^k \right\| .$$

Multiplication of (2.7) by ω then leads to

$$h_k \le \tfrac{1}{2}h_{k-1}^2 .$$

Contraction of the $\{h_k\}$ is obtained, if $h_0 < 2$ is assumed, which is just (2.6). From this, as in the proofs of the preceding theorems, we have $h_k < h_{k-1} < h_0 < 2$ so that there exists

$$\lim_{k \to \infty} h_k = 0 .$$

A straightforward induction argument shows that

$$\|\Delta x^l\| \le \left(\tfrac{1}{2}h_k\right)^{l-k} \|\Delta x^k\| \text{ for } l \ge k .$$

Hence

$$\|x^{l+1} - x^k\| \leq \|\Delta x^l\| + \cdots + \|\Delta x^k\| \leq \|\Delta x^k\| \sum_{j=0}^{\infty} \left(\tfrac{1}{2}h_k\right)^j = \frac{\|\Delta x^k\|}{1 - \tfrac{1}{2}h_k}.$$

The special case $k = 0$ implies that all Newton iterates remain in $S(x^0, \rho)$. Moreover, the results above show that $\{x^k\}$ is a Cauchy sequence, so it converges to some $x^* \in S(x^0, \rho)$. Taking the limit $l \to \infty$ on the previous estimate yields (2.8). Finally, with $\omega < \infty$ from (2.6) we have that x^* is a solution point. $\qquad\square$

The following theorem has been named 'refined Newton-Mysovskikh theorem' in [82].

Theorem 2.3 *Let $F : D \to \mathbb{R}^n$ be a continuously differentiable mapping with $D \subset \mathbb{R}^n$ open and convex. Suppose that $F'(x)$ is invertible for each $x \in D$. Assume that the following affine covariant Lipschitz condition holds*

$$\left\|F'(x)^{-1}\left(F'(y) - F'(x)\right)(y - x)\right\| \leq \omega\|y - x\|^2$$

for x, y, $\in D$. Let $F(x) = 0$ have a solution x^*.
For the initial guess x^0 assume that $\bar{S}(x^*, \|x^0 - x^*\|) \subset D$ and that

$$\omega\|x^0 - x^*\| < 2. \tag{2.9}$$

Then the ordinary Newton iterates defined by (2.1) remain in the open ball $S(x^*, \|x^0 - x^*\|)$ and converge to x^* at an estimated rate

$$\|x^{k+1} - x^*\| \leq \tfrac{1}{2}\omega\|x^k - x^*\|^2. \tag{2.10}$$

Moreover, the solution x^* is unique in the open ball $S(x^*, 2/\omega)$.

Proof. We define $e_k := x^k - x^*$ and proceed for $\lambda \in [0, 1]$ as follows:

$$\begin{aligned}
\|x^k + \lambda\Delta x^k - x^*\| &= \|e_k - \lambda F'(x^k)^{-1}\left(F(x^k) - F(x^*)\right)\| \\
&= \|F'(x^k)^{-1}\left(\lambda\left(F(x^*) - F(x^k)\right) + F'(x^k)e_k\right)\| \\
&= \|(1 - \lambda)e_k + \lambda F'(x^k)^{-1}\int_{s=0}^{1}\left(F'(x^k + se_k) - F'(x^k)\right)e_k ds\| \\
&\leq (1 - \lambda)\|e_k\| + \tfrac{1}{2}\omega\|e_k\|^2.
\end{aligned}$$

For the purpose of repeated induction assume that $\omega\|e_k\| \leq \omega\|e_0\| < 2$ so that $x^k \in D$ is guaranteed. Then the above estimate can be continued to supply

$$\|x^k + \lambda\Delta x^k - x^*\| < (1 - \lambda)\|e_k\| + \lambda\|e_k\| = \|e_k\| \leq \|e_0\|.$$

From this, any statement $x^k + \lambda\Delta x^k \notin S(x^*, \|x^0 - x^*\|)$ would lead to a contradiction. Hence, $x^{k+1} \in D$ and

$$\|e_{k+1}\| \le \tfrac{1}{2}\omega\|e_k\|^2\,,$$

which is just (2.10). In order to prove uniqueness in $S(x^*, 2/\omega)$, let $x^0 := x^{**}$ for some $x^{**} \ne x^*$ with $F(x^{**}) = 0$, which implies $x^1 = x^{**}$ as well. Insertion into (2.10) finally yields the contradiction

$$\|x^{**} - x^*\| \le \omega/2 \, \|x^{**} - x^*\|^2 < \|x^{**} - x^*\|\,.$$

This completes the proof.   □

In view of actual computation, we may combine the results of Theorem 2.2 and 2.3: if we require $h_k \le 1$, then contraction towards x^* shows up, since

$$\frac{\|x^{k+1} - x^*\|}{\|x^k - x^*\|} \le \tfrac{1}{2}\omega\|x^k - x^*\| \le \frac{\tfrac{1}{2}h_k}{1 - \tfrac{1}{2}h_k} \le 1\,.$$

Convergence monitor. We are now ready to exploit both convergence theorems for actual implementation of Newton's method. First, we define the contraction factors

$$\Theta_k := \frac{\|\Delta x^{k+1}\|}{\|\Delta x^k\|}\,,$$

which in terms of the unknown theoretical quantities h_k are known to satisfy

$$\Theta_k = \frac{h_{k+1}}{h_k} \le \tfrac{1}{2}h_k < 1\,. \tag{2.11}$$

Whenever $\Theta_k \ge 1$, then the ordinary Newton iteration is classified as 'not convergent'.

Computational Kantorovich estimates. Obviously, the assumption $h_0 \le 1$ implies $\Theta_0 \le 1/2$. We define the computationally available *a-posteriori* estimates

$$[h_k]_1 = 2\Theta_k \le h_k\,, \quad k = 0, 1, \dots$$

and, recalling $h_{k+1} = \Theta_k h_k$ and shifting the index $k + 1 \to k$, also corresponding *a-priori* estimates

$$[h_k] := \Theta_{k-1}[h_{k-1}]_1 = 2\Theta_{k-1}^2 \le h_k\,, \quad k = 1, 2, \dots\,.$$

Bit counting lemma. The relative accuracy of these estimates is considered in the following lemma, the type of which will appear repeatedly in different context.

Lemma 2.4 *Assume that the just introduced Kantorovich estimates $[h_k]$ satisfy the relative accuracy requirement*

$$0 \le \frac{h_k - [h_k]}{[h_k]} \le \sigma < 1\,, \quad k = 0, 1, \dots\,.$$

Then

$$\Theta_{k+1} \le (1 + \sigma)\Theta_k^2\,, \quad k = 0, 1, \dots\,.$$

Proof. We collect the above relations to obtain

$$\Theta_{k+1} \le \tfrac{1}{2} h_{k+1} \le \tfrac{1}{2}(1+\sigma)[h_{k+1}] = (1+\sigma)\Theta_k^2 .$$

$\square$

Restricted convergence monitor. With $\sigma \to 1$ we then end up with

$$\Theta_k \le 2\Theta_{k-1}^2 , \qquad k = 0, 1, \ldots ,$$

which leads us to the requirement

$$\Theta_k \le \tfrac{1}{2} , \quad k = 0, 1, \ldots , \tag{2.12}$$

a convergence criterion more restrictive than (2.11) above. Otherwise we diagnose *divergence* of the ordinary Newton iteration.

Termination criterion. A desirable criterion to terminate the iteration would be

$$\|x^k - x^*\| \le \mathrm{XTOL} , \tag{2.13}$$

with XTOL a user prescribed error tolerance. In view of (2.8) and with $h_k \to [h_k] = 2\Theta_{k-1}^2$ we will replace this condition by its cheaply computable substitute

$$\frac{\|\Delta x^k\|}{1 - \Theta_{k-1}^2} \le \mathrm{XTOL} . \tag{2.14}$$

Note that XTOL can be chosen quite relaxed here, since $x^{k+1} = x^k + \Delta x^k$ is cheaply available with an accuracy of $O(\mathrm{XTOL}^2)$.

2.1.2 Simplified Newton method

Consider the simplified Newton iteration as introduced above:

$$F'(x^0)\overline{\Delta x}^k = -F(x^k) , \; x^{k+1} = x^k + \overline{\Delta x}^k , \; k = 0, 1, \ldots . \tag{2.15}$$

Convergence analysis. We study the influence of the fixed initial Jacobian on the convergence behavior. The theorems to be derived are slight improvements of well-known theorems of J.M. Ortega and W.C. Rheinboldt—see [163].

Theorem 2.5 *Let $F : D \to \mathbb{R}^n$ be a continuously differentiable mapping with $D \subset \mathbb{R}^n$ open and convex. Let $x^0 \in D$ denote a given starting point so that $F'(x^0)$ is invertible. Assume the affine covariant Lipschitz condition*

$$\|F'(x^0)^{-1}\big(F'(x) - F'(x^0)\big)\| \le \omega_0 \|x - x^0\| \tag{2.16}$$

for all $x \in D$. Let

$$h_0 := \omega_0 \|\overline{\Delta x}^0\| \leq \tfrac{1}{2} \tag{2.17}$$

and define

$$t^* = 1 - \sqrt{1 - 2h_0}\,, \quad \rho = \frac{t^*}{\omega_0}\,.$$

Moreover, assume that $\bar{S}(x^0, \rho) \subset D$. Then the simplified Newton iterates (2.15) remain in $\bar{S}(x^0, \rho)$ and converge to some x^* with $F(x^*) = 0$. The convergence rate can be estimated by

$$\frac{\|x^{k+1} - x^k\|}{\|x^k - x^{k-1}\|} \leq \tfrac{1}{2}(t_k + t_{k-1})\,, \quad k = 1, 2, \dots \tag{2.18}$$

and

$$\|x^k - x^*\| \leq \frac{t^* - t_k}{\omega_0}\,, \quad k = 0, 1, \dots \tag{2.19}$$

with $t_0 = 0$ and

$$t_{k+1} = h_0 + \tfrac{1}{2}t_k^2\,, \quad k = 0, 1, \dots\,.$$

Proof. We follow the line of the proofs in [163] and use (2.16) to obtain

$$\|x^{k+1} - x^k\| \leq \tfrac{1}{2}\omega_0\|x^k - x^{k-1}\|\left(\|x^{k-1} - x^0\| + \|x^k - x^0\|\right). \tag{2.20}$$

The result is slightly more complicated than for the ordinary Newton iteration. We therefore turn to a slightly more sophisticated proof technique by introducing the *majorants*

$$\omega_0\|x^{k+1} - x^k\| \leq h_k\,, \quad \omega_0\|x^k - x^0\| \leq t_k$$

with initial values $t_0 = 0$, $h_0 \leq \tfrac{1}{2}$. Because of

$$\|x^{k+1} - x^0\| \leq \|x^k - x^0\| + \|x^{k+1} - x^k\|$$

and

$$\omega_0\|x^{k+1} - x^k\| \leq \tfrac{1}{2}h_{k-1}(t_k + t_{k-1}) =: h_k$$

we select the two majorant equations

$$t_{k+1} = t_k + h_k\,, \quad h_k = \tfrac{1}{2}h_{k-1}(t_k + t_{k-1})\,,$$

which can be combined to a single equation of the form

$$t_{k+1} - t_k = (t_k - t_{k-1})\left(t_{k-1} + \tfrac{1}{2}(t_k - t_{k-1})\right) = \tfrac{1}{2}(t_k^2 - t_{k-1}^2)\,.$$

Rearrangement of this equation leads to

$$t_{k+1} - \tfrac{1}{2}t_k^2 = t_k - \tfrac{1}{2}t_{k-1}^2\,.$$

Since here the right hand side is just an index shift (downward) of the left hand side, we can apply the so-called *Ortega trick* to obtain

$$t_{k+1} - \tfrac{1}{2}t_k^2 = t_1 - \tfrac{1}{2}t_0^2 = h_0 \,,$$

which may be rewritten as the simplified Newton iteration

$$t_{k+1} - t_k = -\frac{g(t_k)}{g'(t_0)} = g(t_k)$$

for the scalar equation

$$g(t) = h_0 - t + \tfrac{1}{2}t^2 = 0 \,.$$

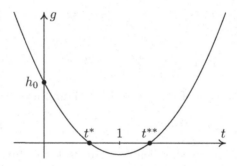

Fig. 2.1. Ortega trick: simplified Newton iteration.

As can be seen from Figure 2.1, the iteration starting at $t = 0$ will converge to the root t^*, which exists, if the above quadratic equation has two real roots. This implies the necessary condition $h_0 \leq 1/2$, which has been imposed above. Also from Figure 2.1 we immediately see that $g(t_{k+1}) < g(t_k)$, which is equivalent to $h_{k+1} < h_k$. Moreover with

$$t_k \leq t^* = 1 - \sqrt{1 - 2h_0} \,,$$

we immediately have

$$x^k \in \bar{S}(x^0, \rho) \subset D \,.$$

Hence, for the solution x^* we also get $x^* \in \bar{S}(x^0, \rho)$. As for the convergence rates, just observe that

$$\omega_0 \|x^k - x^*\| \leq \sum_{i=k}^{\infty} h_i = t^* - t_k$$

and use (2.20) to verify the remaining statements of the theorem.   □

Convergence monitor. From Theorem 2.5 we derive that

$$\Theta_k = \frac{\|\overline{\Delta x}^{k+1}\|}{\|\overline{\Delta x}^{k}\|} \leq \frac{h_{k+1}}{h_k} = \tfrac{1}{2}(t_{k+1} + t_k).$$

With $t_0 = 0, t_1 = h_0$, the condition $h_0 \leq 1/2$ induces the condition

$$\Theta_0 = \frac{\|\overline{\Delta x}^{1}\|}{\|\overline{\Delta x}^{0}\|} \leq \tfrac{1}{2}h_0 \leq \tfrac{1}{4}, \qquad (2.21)$$

which characterizes the local convergence domain of the simplified Newton method. In comparison with $\Theta_0 < 1$ for the ordinary Newton method, where a new Jacobian is used at each step, this is a clear reduction. The above result also shows that the convergence rate may slow down to

$$\Theta_k < t^* = 1 - \sqrt{1 - 2h_0}.$$

We may replace the theoretical quantity t^* by its computationally available bounds

$$[t^*] = 1 - \sqrt{1 - 4\Theta_0} \leq 1 - \sqrt{1 - 2h_0} = t^* \leq 1.$$

Then *divergence* of the simplified Newton iteration will be defined to occur when $\Theta_k \geq [t^*]$.

Termination criterion. From (2.19) we may derive the upper bound

$$\|x^k - x^*\| \leq \frac{t^* - t_k}{\omega_0}.$$

This line is just a different form of the repeated triangle inequality used in the proof so that

$$\|x^k - x^*\| \leq \sum_{j=k}^{\infty} \|\overline{\Delta x}^{j}\|.$$

This gives rise to the upper bound

$$\|x^k - x^*\| \leq \|\overline{\Delta x}^{k}\| (1 + \Theta_k + \Theta_{k+1}\Theta_k + \ldots) \leq \frac{\|\overline{\Delta x}^{k}\|}{1 - t^*}.$$

Upon insertion of the estimate $[t^*] \leq t^*$ from above, we are led to the *approximate* termination criterion

$$\frac{\|\overline{\Delta x}^{k}\|}{\sqrt{1 - 4\Theta_0}} \leq \text{XTOL},$$

where XTOL is the user prescribed final error tolerance. Of course, the application of such a criterion will require to start with some $\Theta_0 < \tfrac{1}{4}$.

2.1.3 Newton-like methods

Consider a rather general Newton-like iteration of the form

$$M(x^k)\delta x^k = -F(x^k), \quad x^{k+1} = x^k + \delta x^k, \quad k = 0, 1, \ldots. \quad (2.22)$$

Convergence analysis. From the basic construction idea, such an iteration will converge, if $M(x)$ is a 'sufficiently accurate' approximation of $F'(x)$. The question will be how to measure the approximation quality and to quantify the vague term 'sufficiently accurate'.

Theorem 2.6 *Let $F : D \to \mathbb{R}^n$ be a continuously differentiable mapping with $D \subset \mathbb{R}^n$ open and convex. Let M denote an approximation of F'. Assume that one can find a starting point $x^0 \in D$ with $M(x^0)$ invertible and constants $\alpha, \overline{\omega}_0, \delta_0, \delta_1, \delta_2 \geq 0$ such that for all $x, y \in D$*

$$\left\| M(x^0)^{-1} F(x^0) \right\| \leq \alpha,$$

$$\left\| M(x^0)^{-1} \big(F'(y) - F'(x) \big) \right\| \leq \overline{\omega}_0 \|y - x\|,$$

$$\left\| M(x^0)^{-1} \big(F'(x) - M(x) \big) \right\| \leq \delta_0 + \delta_1 \|x - x^0\|,$$

$$\left\| M(x^0)^{-1} \big(M(x) - M(x^0) \big) \right\| \leq \delta_2 \|x - x^0\|,$$

$$\delta_0 < 1, \quad \sigma := \max(\overline{\omega}_0, \delta_1 + \delta_2), \quad h := \frac{2\alpha\sigma}{(1 - \delta_0)^2} \leq 1, \quad (2.23)$$

$$\overline{S}(x^0, \rho) \subset D \quad \text{with} \quad \rho := \frac{2\alpha}{1 - \delta_0} \Big/ \Big(1 + \sqrt{1 - h} \Big).$$

Then the sequence $\{x^k\}$ generated from the Newton-like iteration (2.22) is well-defined, remains in $\overline{S}(x^0, \rho)$ and converges to a solution point x^* with $F(x^*) = 0$. With the notation

$$\overline{h} := \frac{\overline{\omega}_0}{\sigma} h, \quad \rho_\pm = \frac{2\alpha}{1 - \delta_0} \Big/ \big(1 \mp \sqrt{1 - \overline{h}} \,\big)$$

the solution $x^* \in \overline{S}(x^0, \rho_-)$ is unique in

$$\overline{S}(x^0, \rho) \cup \big(D \cap S(x^0, \rho_+) \big).$$

Proof. For the usual induction proof, the following majorants are convenient

$$\begin{aligned} \overline{\omega}_0 \|\delta x^k\| &\leq h_k, & h_0 &:= \alpha\overline{\omega}_0, \\ \overline{\omega}_0 \|x^k - x^0\| &\leq t_k, & t_0 &:= 0, \end{aligned}$$

together with

$$t_{k+1} = t_k + h_k. \quad (2.24)$$

Proceeding as in the proofs of the preceding theorems, one obtains

$$\|x^{k+1} - x^k\| = \|M(x^k)^{-1}F(x^k)\|$$

$$= \left\|M(x^k)^{-1}\left[F(x^k) - \left(F(x^{k-1}) + M(x^{k-1})(x^k - x^{k-1})\right)\right]\right\|$$

$$\leq \left\|M(x^k)^{-1}\left[F(x^k) - F(x^{k-1}) - F'(x^{k-1})(x^k - x^{k-1})\right]\right\|$$

$$+ \left\|M(x^k)^{-1}\left[F'(x^{k-1}) - M(x^{k-1})\right](x^k - x^{k-1})\right\| .$$

The perturbation lemma yields:

$$\|M(x^k)^{-1}M(x^0)\| \leq 1 \ / \ (1 - \delta_2\|x^k - x^0\|) .$$

Combining these intermediate results and using $\overline{\delta}_i := \delta_i/\overline{\omega}_0$, $i = 1,2$, then supplies

$$\overline{\omega}_0\|x^{k+1} - x^k\| \leq \frac{1}{1 - \overline{\delta}_2 t_k}\left[\tfrac{1}{2}h_{k-1}^2 + \left(\delta_0 + \overline{\delta}_1 t_{k-1}\right)h_{k-1}\right] .$$

Thus one ends up with the second majorant equation

$$h_k = \left[\left(\delta_0 + \overline{\delta}_1 t_{k-1}\right)h_{k-1} + \tfrac{1}{2}h_{k-1}^2\right] \ / \ \left(1 - \overline{\delta}_2 t_k\right) .$$

Reformulation in view of a possible application of the Ortega technique leads to

$$\left(1 - \overline{\delta}_2 t_k\right)h_k - \left(1 - \overline{\delta}_2 t_{k-1}\right)h_{k-1}$$

$$= \tfrac{1}{2}\left(t_k^2 - t_{k-1}^2\right) - (1 - \delta_0)(t_k - t_{k-1}) \tag{2.25}$$

$$+ \left(\overline{\delta}_1 + \overline{\delta}_2 - 1\right)\left(t_k t_{k-1} - t_{k-1}^2\right) .$$

Obviously, this technique is only applicable, if one requires that

$$\overline{\delta}_1 + \overline{\delta}_2 = 1 \ , \ \text{i.e.,} \ \delta_1 + \delta_2 = \overline{\omega}_0 ,$$

which will not be the case in general. However, by defining σ as in assumption (2.23) and redefining

$$\sigma\|\delta x^k\| \ \leq \ h_k , \qquad h_0 := \alpha\sigma ,$$

$$\sigma\|x^k - x^0\| \ \leq \ t_k , \qquad t_0 := 0 ,$$

$$\overline{\delta}_i \ := \ \delta_i/\sigma , \qquad i = 1,2$$

the disturbing term in (2.25) will vanish. Insertion of (2.24) then gives

$$\left(1 - \overline{\delta}_2 t_k\right)(t_{k+1} - t_k) + (1 - \delta_0)t_k - \tfrac{1}{2}t_k^2 = t_1 = \alpha\sigma ,$$

which can be rewritten in the form

$$t_{k+1} - t_k = \frac{h_0 - (1 - \delta_0)t_k + \frac{1}{2}t_k^2}{1 - \bar{\delta}_2 t_k}.$$

This iteration can be interpreted as a Newton-like iteration in $\mathbb{R}^1$ for the solution of

$$g(t) := h_0 - (1 - \delta_0)t + \tfrac{1}{2}t^2 = 0.$$

The associated two roots

$$t^* = (1 - \delta_0)\left(1 - \sqrt{1 - \frac{2\alpha\sigma}{(1 - \delta_0)^2}}\right),$$

$$t^{**} = (1 - \delta_0)\left(1 + \sqrt{1 - \frac{2\alpha\sigma}{(1 - \delta_0)^2}}\right)$$

are real if

$$\frac{2\alpha\sigma}{(1 - \delta_0)^2} \leq 1.$$

This is just assumption (2.23). The remaining part of the proof essentially follows the lines of the proof of Theorem 2.5 and is therefore omitted here. $\square$

The above theorem does not supply any direct advice towards algorithmic realization. In practical applications, however, additional structure on the approximations $M(x)$ will be given—often as a dependence on an additional parameter, which can be manipulated in such a way that convergence criteria can be met. A typical version of Newton-like methods is the deliberate dropping of 'weak couplings' in the derivative, which can be neglected on the basis of insight into the specific underlying problem. In finite dimensions, deliberate 'sparsing' can be used, which means dropping of 'small' entries in a large Jacobian matrix; this technique works efficiently, if the vague term 'small' can be made sufficiently precise from the application context. Needless to say that 'sparsing' nicely goes with sparse matrix techniques.

2.1.4 Broyden's 'good' rank-1 updates

In order to derive an *error oriented quasi-Newton method*, we start by rewriting the secant condition (1.17) strictly in *affine covariant* terms of quantities in the domain space of F. This leads to

$$E_k(J)\,\delta x_k = \overline{\delta x}_{k+1} = -J_k^{-1}F_{k+1}$$

in terms of the affine covariant update change matrix

$$E_k(J) := I - J_k^{-1}J.$$

Any Jacobian rank-1 update of the kind

$$\tilde{J}_{k+1} = J_k \left(I - \frac{\overline{\delta x}_{k+1} v^T}{v^T \delta x_k} \right) , \quad v \in \mathbb{R}^n, \ v \neq 0$$

with v some vector in the domain space of F will both satisfy the secant condition and exhibit the here desired affine covariance property. The update with $v = \delta x_k$ is known in the literature as 'good Broyden update' [40].

Auxiliary results. The following theorem will collect a bunch of useful results for a single iterative step of the thus defined quasi-Newton method.

Theorem 2.7 *In the notation just introduced, let*

$$J_{k+1} = J_k \left(I - \frac{\overline{\delta x}_{k+1} \delta x_k^T}{\|\delta x_k\|_2^2} \right) \tag{2.26}$$

denote an affine covariant Jacobian rank-1 update and assume the local contraction condition

$$\Theta_k = \frac{\|\overline{\delta x}_{k+1}\|_2}{\|\delta x_k\|_2} < \tfrac{1}{2}.$$

Then:

(I) *The update matrix J_{k+1} is a least change update in the sense that*

$$\|E_k(J_{k+1})\|_2 \ \leq \ \|E_k(J)\|_2, \ \forall \ J \in \mathcal{S}_k,$$
$$\|E_k(J_{k+1})\|_2 \ \leq \ \Theta_k.$$

(II) *The update matrix J_{k+1} is nonsingular whenever J_k is nonsingular, and its inverse can be represented in the form*

$$J_{k+1}^{-1} = \left(I + \frac{\overline{\delta x}_{k+1} \delta x_k^T}{(1 - \alpha_{k+1})\|\delta x_k\|_2^2} \right) J_k^{-1} \tag{2.27}$$

with

$$\alpha_{k+1} = \frac{\delta x_k^T \overline{\delta x}_{k+1}}{\|\delta x_k\|_2^2} < \tfrac{1}{2}.$$

(III) *The next quasi-Newton correction is*

$$\delta x_{k+1} = -J_{k+1}^{-1} F_{k+1} = \frac{\overline{\delta x}_{k+1}}{1 - \alpha_{k+1}}.$$

(IV) *Iterative contraction in terms of quasi-Newton corrections shows up as*

$$\frac{\|\delta x_{k+1}\|_2}{\|\delta x_k\|_2} = \frac{\Theta_k}{1 - \alpha_{k+1}} < 1.$$

Proof. For the rank-1 update we directly have

$$E_k(J_{k+1}) = \frac{\overline{\delta x}_{k+1} \delta x_k^T}{\|\delta x_k\|_2^2} \implies \|E_k(J_{k+1})\|_2 \leq \Theta_k \,.$$

As for the least change update property, we obtain

$$\|E_k(J_{k+1})\|_2 = \left\|\frac{\overline{\delta x}_{k+1} \delta x_k^T}{\|\delta x_k\|_2^2}\right\|_2 = \left\|E_k(J)\frac{\delta x_k \delta x_k^T}{\|\delta x_k\|_2^2}\right\|_2 \leq \|E_k(J)\|_2 \,,$$

which confirms statement I. By application of the Sherman-Morrison formula (see, for instance, the book of A.S. Householder [121]), we directly verify the statements II and III. In order to show IV, we apply the Cauchy-Schwarz inequality to see that

$$|\alpha_{k+1}| \leq \Theta_k,$$

which, for $\Theta_k < 1/2$, implies

$$\frac{\|\delta x_{k+1}\|}{\|\delta x_k\|} = \frac{\Theta_k}{1 - \alpha_{k+1}} \leq \frac{\Theta_k}{1 - \Theta_k} < 1\,.$$

$\square$

Algorithmic realization. The result (2.27) may be rewritten as

$$J_{k+1}^{-1} = \left(I + \frac{\delta x_{k+1} \delta x_k^T}{\|\delta x_k\|_2^2}\right) J_k^{-1}\,.$$

This recursion cannot be used directly for the computation of δx_{k+1}. However, the product representation

$$J_k^{-1} = \left(I + \frac{\delta x_k \delta x_{k-1}^T}{\|\delta x_{k-1}\|_2^2}\right) \cdot \ldots \cdot \left(I + \frac{\delta x_1 \delta x_0^T}{\|\delta x_0\|_2^2}\right) J_0^{-1}\,.$$

can be applied up to the correction δx_k. This consideration leads to a rather economic recursive 'good' Broyden algorithm, which has been used for quite a while in the public domain code NLEQ1 [161]. It essentially requires the $k_{\max} + 1$ quasi-Newton corrections $\delta x_0, \ldots, \delta x_k$ as extra array storage.

Discrete norms for differential equations. Inner products $\langle u, v \rangle$ other than the Euclidean inner product $u^T v$ may be used in view of the underlying problem—such as (discrete) Sobolev inner products for discretized differential equations. By all means, *scaling in the domain space* of F should be carefully considered. This means that any corrections δx arising in the above inner products should actually be implemented as $D^{-1} \delta x$ with appropriate diagonal scaling matrix D. If D is chosen in agreement with a *relative error* concept, then in this way *scaling invariance* of the algorithm can be assured.

Condition number monitor. Recursive implementations based on the above rank-1 factorization have often been outruled with the argument that some hidden ill-conditioning in the arising Jacobian updates might occur. In order to derive a some monitor, we may use

$$\text{cond}_2(J_{k+1}) \leq \text{cond}_2 \left(I + \frac{\delta x_{k+1} \delta x_k^T}{\|\delta x_k\|_2^2} \right) \text{cond}_2(J_k) \,.$$

In this context, the following technical lemma may be helpful.

Lemma 2.8 *Given a rank-1 matrix*

$$A = I - \frac{uv^T}{v^T v} \ \text{with} \ \Theta := \frac{\|u\|_2}{\|v\|_2} < 1 \,,$$

its condition number can be bounded as

$$\text{cond}_2(A) \leq \frac{1+\Theta}{1-\Theta} \,.$$

Proof. We just use the two bounds

$$\|A\| \leq 1 + \left\| \frac{uv^T}{v^T v} \right\| \leq 1 + \Theta, \quad \|A^{-1}\| \leq \left(1 - \left\| \frac{uv^T}{v^T v} \right\| \right)^{-1} \leq (1-\Theta)^{-1}$$

and insert into the definition $\text{cond}_2(A) = \|A\| \, \|A^{-1}\|$.   □

With this result and $\Theta_k < 1/2$ we are certainly able to assure that

$$\text{cond}_2(J_{k+1}) \leq \frac{1+\Theta_k}{1-\Theta_k} \, \text{cond}_2(J_k) < 3 \ \text{cond}_2(J_k) \,.$$

Convergence monitor. In accordance with the above theoretical results, we impose the condition $\Theta_k < 1/2$ throughout the whole iteration. Note that this is an extension of the local convergence domain compared with the simplified Newton method where $\Theta_0 \leq 1/4$ has to be required. With these preparations we are now ready to state the 'good Broyden algorithm' QNERR (for ERRor oriented Quasi-Newton method) in the usual informal manner.

Algorithm QNERR.

For given $\ \ x^0: \quad F_0 = F(x^0) \qquad$ evaluation and store

$$J_0 \delta x_0 = -F_0 \quad \text{linear system solve}$$

$$\sigma_0 = \|\delta x_0\|_2^2 \quad \text{store } \delta x_0, \sigma_0$$

For $k = 0, \ldots, k_{\max}$**:**

I. $x^{k+1} = x^k + \delta x_k$ new iterate

$F_{k+1} = F(x^{k+1})$ evaluation

$J_0 v = -F_{k+1}$ linear system solve

II.   If $k > 0$: for $i = 1, \ldots, k$

$$\overline{\alpha} := \frac{v^T \delta x_{i-1}}{\sigma_{i-1}},$$

$$v := v + \overline{\alpha} \delta x_i$$

III.   Compute

$$\alpha_{k+1} := \frac{v^T \delta x_k}{\sigma_k}, \quad \Theta_k = \left(\frac{v^T v}{\sigma_k} \right)^{1/2} \qquad \text{store}$$

If $\Theta_k > \frac{1}{2}$: **stop, no convergence**

IV. $\delta x_{k+1} = \dfrac{v}{1 - \alpha_{k+1}},$ store

$\sigma_{k+1} = \|\delta x_{k+1}\|_2^2$ store

If $\sqrt{\sigma_{k+1}} \leq \text{XTOL}$:

solution $x^* = x^{k+1} + \delta x_{k+1}$

Else: no convergence within $k_{\max}$ iterations.

Convergence analysis. The above Theorem 2.7 does not give conditions, under which the contraction condition $\Theta_k < 1/2$ is assured *throughout the whole iteration*. This will be the topic of the next theorem.

Theorem 2.9 *For $F : D \longrightarrow \mathbb{R}^n$ be a continuously differentiable mapping with D open and convex. Let $x^* \in D$ denote a unique solution point of F with $F'(x^*)$ nonsingular. Assume that the following affine covariant Lipschitz condition holds:*

$$\|F'(x^*)^{-1}(F'(x) - F'(x^*))v\| \leq \omega \|x - x^*\| \cdot \|v\|$$

for $x, x + v \in D$ and $0 \leq \omega < \infty$. Consider the quasi-Newton iteration as defined in Theorem 2.7. For some $\overline{\Theta}$ in the range $0 < \overline{\Theta} < 1$ assume that:

(I)   *the initial approximate Jacobian J_0 satisfies*

$$\delta_0 := \left\| F'(x^*)^{-1}(J_0 - F'(x^0)) \right\| < \overline{\Theta}/(1 + \overline{\Theta}), \qquad (2.28)$$

(II)   *the initial guess x^0 satisfies*

$$t_0 := \omega \|x^0 - x^*\| \leq \frac{1 - \overline{\Theta}}{2 - \overline{\Theta}} \left(\frac{\overline{\Theta}}{1 + \overline{\Theta}} - \delta_0 \right). \qquad (2.29)$$

Then the quasi-Newton iterates $\{x^k\}$ converge to x^* in terms of errors as

$$\|x^{k+1} - x^*\| < \overline{\Theta} \|x^k - x^*\|, \qquad (2.30)$$

or, in terms of corrections as

$$\|\delta x^{k+1}\| \leq \overline{\Theta} \|\delta x^k\|. \qquad (2.31)$$

The convergence is superlinear with

$$\lim_{k \to \infty} \frac{\|\delta x^{k+1}\|}{\|\delta x^k\|} = 0.$$

As for the Jacobian rank-1 updates, the 'bounded deterioration property' holds in the form

$$\|E_k\| := \|F'(x^*)^{-1} J_k - I\| \leq \frac{\overline{\Theta}}{1 + \overline{\Theta}} < \tfrac{1}{2} \qquad (2.32)$$

together with the asymptotic property

$$\lim_{k \to \infty} \frac{\|E_k \delta x_k\|}{\|\delta x_k\|} = 0. \qquad (2.33)$$

Proof. Let $\| \cdot \|$ be $\| \cdot \|_2$ throughout. For ease of writing we characterize the Jacobian update approximation by

$$\eta_k = \frac{\|E_k \delta x_k\|}{\|\delta x_k\|}, \, \overline{\eta}_k = \|E_k\| = \|E_k^T\| = \max_{v \neq 0} \frac{\|E_k^{(T)} v\|}{\|v\|}.$$

By definition, $\eta_k \leq \overline{\eta}_k$. For the convergence analysis we introduce

$$t_k = \omega \|e_k\|, \, e_k := x^k - x^*.$$

As usual [57], the proof is performed in two basic steps: first *linear* convergence, then *superlinear* convergence.

I. To begin with, exploit the *Lipschitz condition* in the form

$$\|F'(x^*)^{-1} F(x^{k+1})\| \leq \int_{s=0}^{1} \|F'(x^*)^{-1} (F'(x^k + s \delta x_k) - F'(x^*)) \delta x_k\| \, ds$$

$$+ \|E_k \delta x_k\|$$

$$\leq \left(\tfrac{1}{2}(t_{k+1} + t_k) + \eta_k \right) \|\delta x_k\|.$$

Under the assumption $\eta_{k+1} < 1$ we may estimate

$$\|F'(x^*)^{-1}F(x^{k+1})\| = \|(I + E_{k+1})\delta x_{k+1}\| \geq (1 - \eta_{k+1})\|\delta x_{k+1}\|$$

so that

$$\frac{\|\delta x_{k+1}\|}{\|\delta x_k\|} \leq \frac{\eta_k + \bar{t}_k}{1 - \eta_{k+1}}, \quad \text{where} \quad \bar{t}_k := \tfrac{1}{2}(t_k + t_{k+1}). \tag{2.34}$$

As for the iterative errors e_k, we may derive the relation

$$
\begin{aligned}
e_{k+1} &= e_k - J_k^{-1}F(x^k) \\[2mm]
&= (I + E_k)^{-1}\left(E_k e_k - F'(x^*)^{-1}\int\limits_{s=0}^{1}\big(F'(x^* + se_k) - F'(x^*)\big)e_k\,ds\right),
\end{aligned}
$$

from which we obtain the estimate (let $\bar{\eta}_k < 1$)

$$t_{k+1} \leq \frac{\bar{\eta}_k + \tfrac{1}{2}t_k}{1 - \bar{\eta}_k}t_k. \tag{2.35}$$

Upon comparing the right hand upper bounds in (2.35) and (2.34) we are led to define the majorant

$$\overline{\Theta} := \frac{\bar{\eta}_k + \bar{t}_k}{1 - \bar{\eta}_k}, \tag{2.36}$$

which implies that

$$t_{k+1} < \overline{\Theta}t_k. \tag{2.37}$$

II. Next, we study the *approximation properties* of the Jacobian updates. With E_k as defined, the above rank-1 update may be rewritten in the form

$$E_{k+1} = E_k + F'(x^*)^{-1}\frac{F_{k+1}\delta x_k^T}{\|\delta x_k\|_2^2}.$$

If we insert

$$F'(x^*)^{-1}F_{k+1} = (D_{k+1} - E_k)\delta x_k,$$

wherein

$$D_{k+1} := F'(x^*)^{-1}\int\limits_{s=0}^{1}\big(F'(x^k + s\delta x_k) - F'(x^*)\big)ds$$

and introduce the orthogonal projections

$$Q_k^{\perp} = I - Q_k = \frac{\delta x_k \delta x_k^T}{\|\delta x_k\|^2},$$

then we arrive at the decomposition

$$E_{k+1} = E_k Q_k + D_{k+1}Q_k^{\perp}$$

and its transpose ($v \neq 0$ arbitrary)

$$E_{k+1}^T v = Q_k E_k^T v + Q_k^\perp D_{k+1}^T v . \tag{2.38}$$

Note that

$$\frac{\|E_{k+1}\delta x_k\|}{\|\delta x_k\|} = \frac{\|D_{k+1}\delta x_k\|}{\|\delta x_k\|} \leq \bar{t}_k . \tag{2.39}$$

III. In order to prove *linear* convergence, equation (2.38) is used for the quite rough estimate

$$\bar{\eta}_{k+1} = \max_{v \neq 0} \frac{\|E_{k+1}^T v\|}{\|v\|} \leq \max_{v \neq 0} \frac{\|E_k^T v\|}{\|v\|} + \max_{v \neq 0} \frac{|\langle D_{k+1}\delta x_k, v\rangle|}{\|\delta x_k\|\|v\|} \leq \bar{\eta}_k + \bar{t}_k . \tag{2.40}$$

Assume now that we have *uniform* upper bounds

$$\bar{\Theta} \leq \overline{\Theta} < 1, \ \bar{\eta}_k \leq \bar{\eta} < 1 .$$

Then (2.37) can be replaced by

$$t_{k+1} < \overline{\Theta} t_k < t_k$$

and (2.36) leads to the natural definition

$$\bar{\Theta} \leq \frac{\bar{\eta} + \bar{t}_0}{1 - \bar{\eta}} =: \overline{\Theta} . \tag{2.41}$$

As for the definition of $\bar{\eta}$, we apply (2.40) to obtain

$$\bar{\eta}_{k+1} < \bar{\eta}_0 + \sum_{l=0}^{k} \bar{t}_l < \bar{\eta}_0 + \frac{\bar{t}_0}{1 - \overline{\Theta}} =: \bar{\eta} . \tag{2.42}$$

Insertion of $\bar{\eta}$ into (2.41) then eventually yields after some calculation:

$$\bar{t}_0 \leq (1 - \overline{\Theta}) \left(\frac{\overline{\Theta}}{1 + \overline{\Theta}} - \bar{\eta}_0 \right) , \tag{2.43}$$

which obviously requires

$$\bar{\eta}_0 < \frac{\overline{\Theta}}{1 + \overline{\Theta}} < \tfrac{1}{2} \text{ for } \overline{\Theta} < 1 .$$

Observe now that by mere triangle inequality, with δ_0 as defined in (2.28), we have $\bar{\eta}_0 \leq \bar{t}_0 + \delta_0$. Therefore, the assumption (2.43) can finally be replaced by the above two assumptions (2.28) and (2.29). Once such a $\overline{\Theta} < 1$ exists, we have (2.30) directly from $t_{k+1} < \overline{\Theta} t_k$ and (2.31) from inserting $\bar{\eta}$ into (2.34). The bounded deterioration property (2.32) follows by construction and insertion of (2.29) into (2.42).

IV. In order to show *superlinear* convergence, we use (2.38) in a more subtle manner. In terms of the *Euclidean* inner product $\langle \cdot, \cdot \rangle$, some short calculation supplies the equation

$$\|E_{k+1}^T v\|^2 = \|E_k^T v\|^2 - \frac{\langle E_k \delta x_k, v \rangle^2}{\|\delta x_k\|^2} + \frac{\langle D_{k+1} \delta x_k, v \rangle^2}{\|\delta x_k\|^2}.$$

Summing over the indices k, we arrive at

$$\sum_{k=0}^{l} \frac{\langle E_k \delta x_k, v \rangle^2}{\|v\|^2 \|\delta x_k\|^2} = \frac{\|E_0^T v\|^2}{\|v\|^2} - \frac{\|E_{l+1}^T v\|^2}{\|v\|^2} + \sum_{k=0}^{l} \frac{\langle D_{k+1} \delta x_k, v \rangle^2}{\|\delta x_k\|^2 \|v\|^2}.$$

Upon dropping the negative right hand term, letting $l \to \infty$, and using (2.39) with $\bar{t}_{k+1} < \overline{\Theta} \cdot \bar{t}_k$, we end up with the estimate

$$\sum_{k=0}^{\infty} \frac{\langle E_k \delta x_k, v \rangle^2}{\|v\|^2 \|\delta x_k\|^2} \le \overline{\eta}_0^2 + \frac{1}{2} \frac{1 + \overline{\Theta}}{1 - \overline{\Theta}} t_0^2.$$

Since the right hand side is bounded, we immediately conclude that

$$\lim_{k \to \infty} \frac{\langle E_k \delta x_k, v \rangle^2}{\|\delta x_k\|^2 \|v\|^2} = 0 \quad \forall v \in \mathbb{R}^n.$$

As a consequence, with

$$\xi_k := \frac{\delta x_k}{\|\delta x_k\|},$$

we must have

$$\lim_{k \to \infty} E_k \xi_k = 0$$

from which statement (2.33) follows. Finally, with (2.34), we have proved superlinear convergence. □

BIBLIOGRAPHICAL NOTE. Quasi-Newton methods are described, e.g., in the classical optimization book [57] by J.E. Dennis and R.B. Schnabel or, more recently, in the textbook [132] by C.T. Kelley. These methods essentially started with the pioneering paper [40] by C.G. Broyden. For quite a time, the convergence of the 'good' Broyden method was not at all clear. A breakthrough in its convergence analysis came by the paper [41] of C.G. Broyden, J.E. Dennis, and J.J. Moré, where local and superlinear convergence has been shown on the basis of condition (2.33), the meanwhile so-called *Dennis-Moré condition* (see [55]). To the most part, the present section is an affine covariant reformulation of well-known material spread over a huge literature—see, e.g., the original papers [56] by J.E. Dennis and R.B. Schnabel or [58] by J.E. Dennis and H.F. Walker.

The above quasi-Newton algorithm is realized within the earlier code NLEQ1 and its update NLEQ-ERR.

2.1.5 Inexact Newton-ERR methods

Inexact Newton methods consist of a combination of an outer iteration, the Newton iteration, and an inner iteration such that (dropping the inner iteration index i)

$$F'(x^k)(\delta x^k - \Delta x^k) = r^k\,, \quad x^{k+1} = x^k + \delta x^k\,, \quad k = 0, 1, \ldots\,.$$

Here the inner residual r^k gives rise to the difference between the exact Newton correction Δx^k and the inexact Newton correction δx^k. Among the possible inner iterative solvers we will concentrate on those that reduce the *Euclidean error norms* $\|\delta x^k - \Delta x^k\|$, which leads us to CGNE (compare Section 1.4.3) and to GBIT (compare Section 1.4.4). In both cases, the perturbation will be measured by the relative difference between the exact Newton correction Δx^k and the inexact Newton correction δx^k via

$$\delta_k = \frac{\|\delta x^k - \Delta x^k\|}{\|\delta x^k\|}\,, \quad k = 0, 1, \ldots\,. \tag{2.44}$$

As a guiding principle for convergence, we will focus on contraction in terms of the (not actually computed) *exact* Newton corrections

$$\Theta_k = \frac{\|\Delta x^{k+1}\|}{\|\Delta x^k\|}\,,$$

subject to the perturbation coming from the truncation of the inner iteration.

Convergence analysis—CGNE. First we work out details for the error *minimizing* case, exemplified by CGNE specifying the norm $\|\cdot\|$ to be the Euclidean norm $\|\cdot\|_2$. Upon recalling (1.28), the starting value $\delta x_0^k = 0$ for the CGNE iteration implies that

$$\|\Delta x^k\| = \|\delta x^k\|\sqrt{1 + \delta_k^2} \geq \|\delta x^k\|\,.$$

Moreover, from (1.29) and (1.30) we conclude that δ_k is monotonically decreasing in the course of the inner iteration so that eventually any threshold condition of the type $\delta_k \leq \bar{\delta}$ can be met. With this preparation, we are now ready to state our convergence result.

Theorem 2.10 *Let $F : D \longrightarrow \mathbb{R}^n$ be a continuously differentiable mapping with $D \subset \mathbb{R}^n$ open, convex, and sufficiently large. Suppose that $F'(x)$ is invertible for each $x \in D$. Assume that the following affine covariant Lipschitz condition holds:*

$$\|F'(z)^{-1}\big(F'(y) - F(x)\big)v\| \leq \omega\|y - x\| \cdot \|v\|$$

for collinear $x, y, z \in D$.

Let $x^0 \in D$ denote a given starting point for a Newton-CGNE iteration. At an iterate x^k, let δ_k as defined in (2.44) denote the relative error of the inexact Newton correction δx^k. Let the inner CGNE iteration be started with $\delta x_0^k = 0$, which gives rise to the following relations between the Kantorovich quantities

$$h_k := \omega \|\Delta x^k\| \quad and \quad h_k^\delta := \omega \|\delta x^k\| = \frac{h_k}{\sqrt{1 + \delta_k^2}}.$$

Let $x^* \in D$ be the unique solution point.

I. Linear convergence mode. *Assume that an initial guess x^0 has been chosen such that*

$$h_0 < 2\overline{\Theta} < 2$$

for some $\overline{\Theta} < 1$. Let $\delta_{k+1} \geq \delta_k$ be realized throughout the inexact Newton iteration and control the inner iteration such that

$$\vartheta(h_k, \delta_k) = \frac{\frac{1}{2} h_k^\delta + \delta_k(1 + h_k^\delta)}{\sqrt{1 + \delta_k^2}} \leq \overline{\Theta},$$

which assures that

$$\delta_k \leq \frac{\overline{\Theta}}{\sqrt{1 - \overline{\Theta}^2}}. \tag{2.45}$$

Then this implies the exact monotonicity

$$\frac{\|\Delta x^{k+1}\|}{\|\Delta x^k\|} \leq \overline{\Theta}$$

and the inexact monotonicity

$$\frac{\|\delta x^{k+1}\|}{\|\delta x^k\|} \leq \sqrt{\frac{1 + \delta_k^2}{1 + \delta_{k+1}^2}} \overline{\Theta} \leq \overline{\Theta}.$$

The iterates $\{x^k\}$ remain in $\overline{S}(x^0, \rho)$ with $\rho = \|\delta x^0\|/(1 - \overline{\Theta})$ and converge at least linearly to x^*.

II. Quadratic convergence mode. *For some $\rho > 0$, let the initial guess x^0 satisfy*

$$h_0 < \frac{2}{1 + \rho} \tag{2.46}$$

and control the inner iteration such that

$$\delta_k \leq \frac{\rho}{2} \frac{h_k^\delta}{1 + h_k^\delta}, \tag{2.47}$$

which requires that

$$\rho > \frac{3\delta_0}{1-\delta_0} \tag{2.48}$$

be chosen. Then the inexact Newton iterates remain in $\overline{S}(x^0, \overline{\rho})$ with

$$\overline{\rho} = \|\delta x^0\| / \left(1 - \frac{1+\rho}{2}h_0\right)$$

and converge quadratically to x^* with

$$\|\Delta x^{k+1}\| \le \frac{1+\rho}{2}\omega\|\Delta x^k\|^2$$

and

$$\|\delta x^{k+1}\| \le \frac{1+\rho}{2}\omega\|\delta x^k\|^2.$$

Proof. First we show that

$$\|\Delta x^{k+1}\| \le \int_{t=0}^{1} \|F'(x^{k+1})^{-1}\left(F'(x^k+t\delta x^k)-F'(x^k)\right)\delta x^k\|dt + \|F'(x^{k+1})^{-1}r^k\|.$$

For the first term we just apply the Lipschitz condition in standard form. For the second term we may use the same condition plus the triangle inequality to obtain

$$\|F'(x^{k+1})^{-1}r^k\| = \|F'(x^{k+1})^{-1}F'(x^k)(\delta x^k - \Delta x^k)\| \le (1+h_k^\delta)\|\delta x^k - \Delta x^k\|.$$

With definition (2.44), this gives

$$\frac{\|\Delta x^{k+1}\|}{\|\delta x^k\|} \le \tfrac{1}{2}h_k^\delta + \delta_k(1+h_k^\delta). \tag{2.49}$$

With $h_k^\delta = h_k/\sqrt{1+\delta_k^2}$ we then arrive at

$$\frac{\|\Delta x^{k+1}\|}{\|\Delta x^k\|} \le \vartheta(h_k,\delta_k)) = \frac{\tfrac{1}{2}h_k^\delta + \delta_k(1+h_k^\delta)}{\sqrt{1+\delta_k^2}}.$$

In order to prove *linear* convergence, we might require $\vartheta(h_k,\delta_k) = \overline{\Theta} < 1$, which implies that δ_k monotonically *increases* as h_k monotonically decreases—which would automatically lead to $\delta_{k+1} \ge \delta_k$ when $h_{k+1} \le h_k$. However, since strict equality cannot be realized within CGNE, we have to assume the two separate inequalities $\vartheta \le \overline{\Theta}$ and $\delta_{k+1} \ge \delta_k$, as done in the theorem. Note that a necessary condition for $\vartheta(h_k,\delta_k) \le \overline{\Theta}$ with some $\delta_k > 0$ is that it holds at least for $\delta_k = 0$, which yields $h_0 < 2\overline{\Theta}$, the assumption made in the theorem. As for the contraction in terms of the inexact Newton corrections, we then obtain

$$\frac{\|\delta x^{k+1}\|}{\|\delta x^k\|} = \sqrt{\frac{1 + \delta_k^2}{1 + \delta_{k+1}^2}} \frac{\|\Delta x^{k+1}\|}{\|\Delta x^k\|} \leq \sqrt{\frac{1 + \delta_k^2}{1 + \delta_{k+1}^2}} \overline{\Theta} \leq \overline{\Theta}.$$

Usual linear convergence results then imply that $\{x^k\}$ remains in $\overline{S}(x^0, \rho)$ with $\rho = \|\delta x^0\|/(1 - \overline{\Theta})$, if only $\overline{S}(x^0, \rho) \subset D$, which we assumed by D to be 'sufficiently large'. Asymptotically we thus assure that $\vartheta(0, \delta_k) \leq \overline{\Theta}$, which is equivalent to (2.45).

For the *quadratic* convergence case we require that the first term in $\vartheta(h_k, \delta_k)$ originating from the outer iteration exceeds the second term, which brings us to (2.47). Note that now $h_{k+1} \leq h_k$ implies $\delta_{k+1} \leq \delta_k$ and $h_k \to 0$ also $\delta_k \to 0$—a behavior that differs from the linear convergence case. Insertion of (2.47) into $\vartheta(h_k, \delta_k)$ then directly leads to

$$\frac{\|\Delta x^{k+1}\|}{\|\Delta x^k\|} \leq \frac{1 + \rho}{2} \frac{h_k}{1 + \delta_k^2} \leq \frac{1 + \rho}{2} h_k$$

and to

$$\frac{\|\delta x^{k+1}\|}{\|\delta x^k\|} \leq \frac{1 + \rho}{2} \frac{h_k^{\delta}}{\sqrt{1 + \delta_{k+1}^2}} \leq \frac{1 + \rho}{2} h_k^{\delta}.$$

Upon applying the usual quadratic convergence results, we have to require the sufficient condition

$$\frac{1 + \rho}{2} h_0^{\delta} \leq \frac{1 + \rho}{2} h_0 < 1$$

and then, assuming that D is 'sufficiently large', obtain convergence within the ball

$$\overline{S}(x^0, \overline{\rho}), \ \overline{\rho} = \frac{\|\delta x^0\|}{\left(1 - \frac{1 + \rho}{2} h_0\right)}$$

as stated above. Finally, upon inserting (2.46) into (2.47) and using $h_0^{\delta} \leq h_0$, the result (2.48) is readily confirmed. $\qquad\qquad\qquad\qquad\qquad$ □

Convergence analysis—GBIT. By a slight modification of Theorem 2.10, the Newton-GBIT iteration can also be shown to converge.

Theorem 2.11 *Let $\delta_k < \frac{1}{2}$ in (2.44) and replace the Kantorovich quantities h_k^{δ} in Theorem 2.10 by their upper bounds such that*

$$h_k^{\delta} = \frac{h_k}{1 - \delta_k}.$$

Then we obtain the results:

I. Linear convergence mode. *Let δ_k in each inner iteration be controlled such that*

$$\vartheta(h_k, \delta_k) = \frac{\frac{1}{2}h_k^\delta + \delta_k(1 + h_k^\delta)}{1 - \delta_k} \leq \overline{\Theta},$$

which assures that

$$\delta_k \leq \frac{\overline{\Theta}}{1 + \overline{\Theta}}. \tag{2.50}$$

Then this implies the inexact monotonicity test

$$\frac{\|\delta x^{k+1}\|}{\|\delta x^k\|} \leq \frac{1 - \delta_k}{1 - \delta_{k+1}}\overline{\Theta} \tag{2.51}$$

and the exact monotonicity test

$$\frac{\|\Delta x^{k+1}\|}{\|\Delta x^k\|} \leq \overline{\Theta}.$$

II. Quadratic convergence mode. *Let the inner iteration be controlled according to (2.47) and*

$$h_0 < \frac{2(1 - \delta_0)^2}{1 + \rho}. \tag{2.52}$$

Then (2.48) needs to be replaced by

$$\rho > \frac{\delta_0(3 - 2\delta_0)}{1 - 2\delta_0}. \tag{2.53}$$

The exact Newton corrections behave like

$$\|\Delta x^{k+1}\| \leq \frac{1}{2}\frac{1 + \rho}{(1 - \delta_k)^2}\omega\|\Delta x^k\|^2$$

and the inexact Newton corrections like

$$\|\delta x^{k+1}\| \leq \frac{1}{2}\frac{1 + \rho}{1 - \delta_{k+1}}\omega\|\delta x^k\|^2.$$

Proof. The main difference to the previous theorem is that now we can only apply the triangle inequality

$$\big|\,\|\Delta x^k\| - \|\delta x^k - \Delta x^k\|\,\big| \leq \|\delta x^k\| \leq \|\delta x^k - \Delta x^k\| + \|\Delta x^k\|.$$

Assuming $\delta_k < 1$ in definition (2.44), we obtain

$$\frac{\|\Delta x^k\|}{1 + \delta_k} \leq \|\delta x^k\| \leq \frac{\|\Delta x^k\|}{1 - \delta_k},$$

which motivates the majorant $\omega\|\delta x^k\| \leq h_k^\delta$ as stated in the theorem. Upon revisiting the proof of Theorem 2.10, the result (2.49) is seen to still hold, which is

$$\frac{\|\Delta x^{k+1}\|}{\|\delta x^k\|} \leq \tfrac{1}{2}h_k^\delta + \delta_k(1 + h_k^\delta). \tag{2.54}$$

From this, we obtain the modified estimate for the exact Newton corrections

$$\frac{\|\Delta x^{k+1}\|}{\|\Delta x^k\|} \leq \frac{\tfrac{1}{2}h_k^\delta + \delta_k(1 + h_k^\delta)}{1 - \delta_k} = \frac{\tfrac{1}{2}h_k + \delta_k(1 - \delta_k + h_k)}{(1 - \delta_k)^2} = \vartheta(h_k, \delta_k).$$

In a similar way, we obtain for the inexact Newton corrections

$$\frac{\|\delta x^{k+1}\|}{\|\delta x^k\|} \leq \frac{\tfrac{1}{2}h_k^\delta + \delta_k(1 + h_k^\delta)}{1 - \delta_{k+1}} = \frac{1 - \delta_k}{1 - \delta_{k+1}}\vartheta(h_k, \delta_k).$$

For the *linear* convergence mode, we adapt δ_k such that

$$\vartheta(h_k, \delta_k) \leq \overline{\Theta}.$$

Asymptotically we thus assure that $\vartheta(0, \delta_k) \leq \overline{\Theta}$, equivalent to (2.50).

For the *quadratic* convergence mode, we again require (2.47) (with h_k^δ in the present meaning, of course), i.e.

$$\delta_k \leq \tfrac{1}{2}\rho\frac{h_k^\delta}{1 + h_k^\delta}.$$

With this choice we arrive at

$$\frac{\|\Delta x^{k+1}\|}{\|\Delta x^k\|} \leq \tfrac{1}{2}\frac{1 + \rho}{1 - \delta_k}h_k^\delta = \tfrac{1}{2}\frac{1 + \rho}{(1 - \delta_k)^2}\omega\|\Delta x^k\|$$

for the *exact* Newton contraction, which requires (2.52) as a necessary condition. Upon combining (2.47) and (2.52), we obtain

$$\delta_0 \leq \tfrac{1}{2}\rho\frac{h_0^\delta}{1 + h_0^\delta} < \frac{\rho(1 - \delta_0)}{1 + \rho + 2(1 - \delta_0)}.$$

Given ρ, this condition would lead to some uneasy quadratic root. Given δ_0, we merely have the linear inequality

$$\rho > \frac{\delta_0(3 - 2\delta_0)}{1 - 2\delta_0},$$

which is (2.53); it necessarily requires $\delta_0 < 1/2$ in agreement with the assumption $\delta_k < 1/2$ of the theorem.

The corresponding bound for the *inexact* Newton corrections is

$$\frac{\|\delta x^{k+1}\|}{\|\delta x^k\|} \leq \tfrac{1}{2}\frac{1 + \rho}{1 - \delta_{k+1}}\omega\|\delta x^k\|,$$

which completes the proof.   □

Convergence monitor. Assume that the quantity $\overline{\Theta} < 1$ in the linear convergence mode or the quadratic convergence mode have been specified; in view of (2.12), we may require that $\overline{\Theta} \le 1/2$. The desirable convergence criterion would be

$$\Theta_k := \frac{\|\Delta x^{k+1}\|_2}{\|\Delta x^k\|_2} \le \overline{\Theta}.$$

Since this criterion cannot be directly implemented, Θ_k needs to be substituted by a computationally available $\widetilde{\Theta}_k \approx \Theta_k$.

For CGNE with $\delta x_0^k = 0$, this leads to the inexact monotonicity test

$$\widetilde{\Theta}_k = \sqrt{\frac{1 + \bar{\delta}_{k+1}^2}{1 + \bar{\delta}_k^2}} \cdot \frac{\|\delta x^{k+1}\|_2}{\|\delta x^k\|_2} \le \overline{\Theta}, \qquad (2.55)$$

where the quantities $\bar{\delta}_k, \bar{\delta}_{k+1}$ are the computationally available estimates for the otherwise unavailable quantities δ_k, δ_{k+1} as given in (1.32).

For GBIT, the result (2.51) suggests the following inexact monotonicity test

$$\widetilde{\Theta}_k = \frac{1 - \bar{\delta}_{k+1}}{1 - \bar{\delta}_k} \cdot \frac{\|\delta x^{k+1}\|}{\|\delta x^k\|} \le \overline{\Theta}. \qquad (2.56)$$

As an alternative, we may also consider the weaker *necessary* condition

$$\widetilde{\Theta}_k = \frac{1 - \bar{\delta}_{k+1}}{1 + \bar{\delta}_k} \cdot \frac{\|\delta x^{k+1}\|}{\|\delta x^k\|} \le \Theta_k \le \overline{\Theta} \qquad (2.57)$$

or the stronger *sufficient* condition

$$\Theta_k \le \widetilde{\Theta}_k = \frac{1 + \bar{\delta}_{k+1}}{1 - \bar{\delta}_k} \cdot \frac{\|\delta x^{k+1}\|}{\|\delta x^k\|} \le \overline{\Theta} \qquad (2.58)$$

for use within the convergence monitor.

Preconditioning. In order to speed up the inner iteration, preconditioning from the left or/and from the right may be used. This means solving

$$\left(C_L F'(x^k) C_R \right) C_R^{-1} \left(\delta x^k - \Delta x^k \right) = C_L r^k .$$

In such a case, we will define

$$\delta_k = \frac{\|C_R^{-1}(\Delta x^k - \delta x^k)\|}{\|C_R^{-1} \delta x^k\|} .$$

Of course, in this case the preconditioned error norm is reduced by the inner iteration, whereas C_L only affects its rate of convergence. Consequently, any adaptive strategy should then, in principle, be based upon the contraction factors

$$\Theta_k = \frac{\|C_R^{-1} \Delta x^{k+1}\|}{\|C_R^{-1} \Delta x^k\|}$$

and its corresponding scaled estimate $\widetilde{\Theta}_k \approx \Theta_k$ as in (2.55) for CGNE or any choice between (2.56), (2.57), and (2.58) for GBIT.

Termination criterion. In the same spirit as above, we mimic the termination criterion (2.14) for the exact Newton iteration by requiring for CGNE the substitute condition

$$\frac{\sqrt{1 + \bar{\delta}_k^2}}{1 - \widetilde{\Theta}_{k-1}^2} \|\delta x^k\|_2 \leq \text{XTOL}$$

and for GBIT the sufficient condition

$$\frac{1 + \bar{\delta}_k}{1 - \widetilde{\Theta}_{k-1}^2} \|\delta x^k\| \leq \text{XTOL},$$

each for the finally accepted iterate x^{k+1}, where XTOL is a user prescribed absolute *error tolerance* (to be replaced by some relative or some scaled error criterion).

Estimation of Kantorovich quantities. In order to deal successfully with the question of *how to match inner and outer iterations*, the above theory obviously requires the theoretical quantities $h_k^\delta = \omega \|\delta x^k\|$—which, however, are not directly available. In the spirit of the whole book we aim at replacing these quantities by *computational estimates* $[h_k^\delta]$. Recalling Section 2.1.1, we aim at estimating the a-priori estimates $[h_k] = 2\Theta_{k-1}^2 \leq h_k$ for $k \geq 1$.

For CGNE with initial correction $\delta x_0^k = 0$, we replace the relative errors δ_k by their estimates $\bar{\delta}_k$ from Section 1.4.3 and thus arrive at the a-priori estimates

$$[h_k^\delta] = [h_k]/\sqrt{1 + \bar{\delta}_k^2}, \quad [h_k] = 2\widetilde{\Theta}_{k-1}^2 \leq h_k, \quad k = 1, 2, \dots, \qquad (2.59)$$

where $\widetilde{\Theta}_{k-1}$ from (2.55) is inserted.

For GBIT, we get the a-priori estimates

$$[h_k^\delta] = \frac{[h_k]}{(1 - \bar{\delta}_k)}, \quad [h_k] = 2\widetilde{\Theta}_{k-1}^2 \leq h_k, \quad k = 1, 2, \dots, \qquad (2.60)$$

where $\widetilde{\Theta}_{k-1}$ from (2.57) is inserted.

In both CGNE and GBIT, we may alternatively use the a-posteriori estimates

$$[h_{k-1}]_1 = 2\widetilde{\Theta}_{k-1}$$

and insert them either into (2.59) or into (2.60), respectively, to obtain $[h_{k-1}^\delta]_1$. From this, we may construct the a-priori estimates (for $k \geq 1$)

$$[h_k^\delta] = [h_{k-1}^\delta]_1 \frac{\|\delta x^k\|}{\|\delta x^{k-1}\|}.$$

Note that in CGNE this formula inherits the saturation property.

For $k = 0$, we cannot but choose any 'sufficiently small' δ_0—as stated in the quadratic convergence mode to follow next.

Standard convergence mode. In this mode the inner iteration is terminated whenever

$$\delta_k \leq \bar{\delta} \qquad\qquad (2.61)$$

for some default value $\bar{\delta} < 1$ to be chosen. In this case, *asymptotic linear convergence* is obtained.

For CGNE, Theorem 2.10 requires

$$\bar{\delta}/\sqrt{1 + \bar{\delta}^2} < \overline{\Theta},$$

which for $\overline{\Theta} = \frac{1}{2}$ leads to the restriction $\bar{\delta} < \sqrt{3}/3 \approx 0.577$. For GBIT, Theorem 2.11 requires

$$\bar{\delta}/(1 - \bar{\delta}) < \overline{\Theta},$$

which leads to $\bar{\delta} < 1/3$. In any case, we recommend to choose $\bar{\delta} \leq 1/4$ to assure at least two binary digits.

Quadratic convergence mode. In CGNE, we set $\delta_0 = \frac{1}{4}$ in (2.48) and obtain $\rho > 1$—thus assuring at least the first binary digit. In GBIT, we also set $\delta_0 = \frac{1}{4}$ and apply the inequality (2.53) thus arriving at $\rho > \frac{5}{4}$.

As for the adaptive termination of the inner iteration, we want to satisfy condition (2.47) for $k \geq 1$. Following our paradigm, we will replace the computationally unavailable quantity h_k^δ therein by its computational estimate $[h_k^\delta]$, which yields, for both CGNE and GBIT, the substitute condition

$$\bar{\delta}_k \leq \frac{1}{2}\rho \cdot \frac{[h_k^\delta]}{1 + [h_k^\delta]} . \qquad\qquad (2.62)$$

Whenever $\delta_k \leq \bar{\delta}_k$, the above monotone *increasing* right side as a function of $[h_k^\delta]$ and the relation $[h_k^\delta] \leq h_k^\delta$ imply that the theoretical condition (2.47) is actually *assured* with (2.62). Based on the a-priori estimates (2.59) or (2.60), respectively, we obtain a simple nonlinear scalar equation for an upper bound of δ_k.

Note that $\delta_k \to 0$ is enforced when $k \to \infty$, which means: *the closer the iterates come to the solution point, the more work needs to be done in the inner iteration to assure quadratic convergence of the outer iteration.*

Linear convergence mode. Once the approximated contraction factor $\widetilde{\Theta}_k$ is sufficiently below some prescribed threshold value $\overline{\Theta} \leq 1/2$, we may switch to the linear convergence mode described in either of the above two convergence theorems. As for the termination of the inner iteration, we recall the theoretical condition

$$\vartheta(h_k, \delta_k) \leq \overline{\Theta}.$$

Since the quantity ϑ is unavailable, we will replace it by the computationally available estimate

$$[\vartheta(h_k, \delta_k)] = \vartheta([h_k], \delta_k) \leq \vartheta(h_k, \delta_k).$$

As this mode occurs only for $k > 0$, we can just insert the a-priori estimates (2.59) or (2.60), respectively. Since the above right hand side is a monotone *increasing* function of h_k and $[h_k] \leq h_k$, this estimate may be 'too small' and therefore lead to some δ_k, which is 'too large'. Fortunately, the difference between computational estimate and theoretical quantity can be ignored asymptotically. In any case, we require the monotonicity (2.55) for CGNE or (2.56), (2.57), or (2.58) for GBIT and run the inner iteration at each step k until either the actual value of δ_k obtained in the course of the inner iteration satisfies the condition above or divergence occurs with $\widetilde{\Theta}_k > 2\overline{\Theta}$.

In CGNE, we observe that in this mode *the closer the iterates come to the solution point, the less work is necessary within the inner iteration to assure linear convergence of the outer iteration.* In GBIT, this process continues only until the upper bound (2.50) for δ_k has been reached.

The here described error oriented local inexact Newton algorithms are self–contained and similar in spirit, but not identical with the local parts of the global inexact Newton codes GIANT-CGNE and GIANT-GBIT, which are worked out in detail in Section 3.3.4 below.

BIBLIOGRAPHICAL NOTE. A first affine covariant convergence analysis of a local inexact Newton method has been given by T.J. Ypma [203]. The first affine covariant inexact Newton code has been GIANT, developed by P. Deuflhard and U. Nowak [67, 160] in 1990. That code had also used a former version of GBIT for the inner iteration.

2.2 Residual Based Algorithms

In most algorithmic realizations of Newton's method iterative values of the *residual* norms are used for a check of convergence. An associated convergence analysis will start from *affine contravariant* Lipschitz conditions of the type (1.8) and lead to results in terms of residual norms only, which are tacitly assumed to be scaled. As explained in Section 1.2.2 above, such an analysis will not touch upon the question of local uniqueness of the solution.

2.2.1 Ordinary Newton method

Recall the notation of the ordinary Newton method

$$F'(x^k)\Delta x^k = -F(x^k), \ x^{k+1} = x^k + \Delta x^k, \quad k = 0, 1, \ldots. \tag{2.63}$$

Convergence analysis. Analyzing the iterative residuals leads to an affine contravariant version of the well-known Newton-Mysovskikh theorem.

Theorem 2.12 *Let $F : D \to \mathbb{R}^n$ be a differentiable mapping with $D \subset \mathbb{R}^n$ open and convex. Let $F'(x)$ be invertible for all $x \in D$. Assume that the following affine contravariant Lipschitz condition holds:*

$$\left\| (F'(y) - F'(x))(y - x) \right\| \le \omega \| F'(x)(y - x) \|^2 \text{ for } x, y \in D .$$

Define the open level set $\mathcal{L}_\omega = \{ x \in D \mid \| F(x) \| < \frac{2}{\omega} \}$ and let $\overline{\mathcal{L}}_\omega \subset D$ be bounded. For a given initial guess x^0 of an unknown solution x^* let

$$h_0 := \omega \| F(x^0) \| < 2 , \text{ i.e. } x^0 \in \mathcal{L}_\omega . \tag{2.64}$$

Then the ordinary Newton iterates $\{x^k\}$ defined by (2.63) remain in $\mathcal{L}_\omega$ and converge to some solution point $x^* \in \mathcal{L}_\omega$ with $F(x^*) = 0$. The iterative residuals $\{F(x^k)\}$ converge to zero at an estimated rate

$$\| F(x^{k+1}) \| \le \tfrac{1}{2} \omega \| F(x^k) \|^2 . \tag{2.65}$$

Proof. To show that $x^{k+1} \in D$ we apply the integral form of the mean value theorem and the above Lipschitz condition and obtain

$$
\begin{aligned}
\| F(x^k + \lambda \Delta x^k) \| &= \left\| F(x^k) + \int_{t=0}^{\lambda} F'(x^k + t\Delta x^k) \Delta x^k \, dt \right\| \\[2mm]
&= \left\| \int_{t=0}^{\lambda} \left(F'(x^k + t\Delta x^k) - F'(x^k) \right) \Delta x^k \right. \\
&\qquad \left. + (1 - \lambda) F(x^k) \, dt \right\| \\[2mm]
&\le \int_{t=0}^{\lambda} \left\| \left(F'(x^k + t\Delta x^k) - F'(x^k) \right) \Delta x^k \right\| \, dt \\
&\qquad + (1 - \lambda) \| F(x^k) \| \\[2mm]
&\le \omega \int_{t=0}^{\lambda} \| F'(x^k) \Delta x^k \|^2 t \, dt + (1 - \lambda) \| F(x^k) \| \\[2mm]
&= \left(1 - \lambda + \tfrac{1}{2} \omega \lambda^2 \| F(x^k) \| \right) \| F(x^k) \|
\end{aligned}
$$

for each $\lambda \in [0, 1]$ such that $x^k + t\Delta x^k \in \mathcal{L}_\omega$ for $t \in [0, \lambda]$. Now assume that $x^{k+1} \notin \mathcal{L}_\omega$. Then there exists a minimal $\bar{\lambda} \in \,]0, 1]$ with $x^k + \bar{\lambda} \Delta x^k \in \partial \mathcal{L}_\omega$ and $\| F(x^k + \bar{\lambda} \Delta x^k) \| < (1 - \bar{\lambda} + \bar{\lambda}^2) \| F(x^k) \| < 2/\omega$, which is a contradiction. For $\lambda = 1$ we get relation (2.65). In terms of the residual oriented so-called Kantorovich quantities

$$h_k := \omega \| F(x^k) \| \tag{2.66}$$

we may obtain the quadratic recursion

$$h_{k+1} \le \tfrac{1}{2} h_k^2 = (\tfrac{1}{2} h_k) h_k . \tag{2.67}$$

With assumption (2.64), $h_0 < 2$, we obtain $h_1 < h_0 < 2$ for $k = 0$ and, by repeated induction over k, then

$$h_{k+1} < h_k < 2, \ k = 0, 1, \ldots \quad \Rightarrow \quad \lim_{k \to \infty} h_k = 0 \, .$$

This can be also written in terms of the residuals as

$$\|F(x^{k+1})\| < \|F(x^k)\| < \frac{2}{\omega} \quad \Rightarrow \quad \lim_{k \to \infty} \|F(x^k)\| = 0 \, .$$

In terms of the iterates we have

$$\{x^k\} \subset \mathcal{L}_\omega \subset D \, .$$

Since $\mathcal{L}_w$ is bounded, there exists an accumulation point x^* of $\{x^k\}$ with $F(x^*) = 0$, i.e. x^* is a solution point, but not necessarily unique in $\mathcal{L}_\omega$. □

This theorem also holds for underdetermined nonlinear systems—compare Exercise 4.10.

Convergence monitor. We now want to exploit Theorem 2.12 for actual computation. For this purpose, we introduce the contraction factors

$$\Theta_k := \frac{\|F(x^{k+1})\|}{\|F(x^k)\|}$$

and write (2.67) in the equivalent form

$$\Theta_k = \frac{h_{k+1}}{h_k} \le \tfrac{1}{2} h_k \, . \tag{2.68}$$

For $k = 0$, assumption (2.64) assures *residual monotonicity*

$$\Theta_0 < 1 \, . \tag{2.69}$$

Whenever $\Theta_0 \ge 1$, the assumption (2.64) is certainly violated, which means that the initial guess x^0 is not 'sufficiently close' to the solution point x^* in the sense of the above theorem. Suppose now that the test $\Theta_0 < 1$ has been passed. For the construction of a *quadratic convergence monitor* we introduce *computationally available estimates* $[h_k]$ for the unknown theoretical quantities h_k from (2.66). In view of (2.68) we may define the computational *a-posteriori* estimate

$$[h_k]_1 = 2\Theta_k \le h_k$$

and, since $h_{k+1} = \Theta_k h_k$, also the *a-priori* estimate

$$[h_{k+1}] = \Theta_k [h_k]_1 = 2\Theta_k^2 \le h_{k+1} \, .$$

Upon roughly identifying $[h_{k+1}]_1 \approx [h_{k+1}]$, we arrive at the approximate recursion ($k = 0, 1, \ldots$):

$$\Theta_{k+1} \approx \Theta_k^2 \leq \Theta_0 < 1.$$

Violation of this recursion at least in the mild sense

$$\Theta_{k+1} > \Theta_0$$

or the stricter sense

$$\Theta_{k+1} \geq 2\Theta_k^2$$

may be used to terminate the ordinary Newton iteration as 'not convergent'.

Termination criterion. This affine contravariant theory agrees with a termination criterion of the form

$$\|F(\hat{x})\| \leq \text{FTOL},\tag{2.70}$$

where FTOL is a user prescribed *residual error tolerance*.

Computational complexity. A short calculation shows that, for a given starting point x^0, the number q of iterations such that $\hat{x} = x^{q+1}$ meets the above termination requirement satisfies roughly

$$q \approx \text{ld}\, \frac{\log(\text{FTOL}/\|F(x^0)\|)}{\log \Theta_0}.\tag{2.71}$$

The proof is left as Exercise 2.1. In other words, with 'sufficiently good' initial guesses x^0 of the solution x^* at hand, the computational complexity of the nonlinear problem is comparable to the one of the linearized problem. Such problems are sometimes called *mildly nonlinear*.

2.2.2 Simplified Newton method

Recall the notation of the simplified Newton iteration

$$F'(x^0)\overline{\Delta x}^k = -F(x^k),\ \ x^{k+1} = x^k + \overline{\Delta x}^k,\ \ k = 0, 1, \ldots.\tag{2.72}$$

Convergence analysis. Here we study convergence in terms of iterative residuals obtaining an affine contravariant variant of the Newton-Kantorovich theorem—without any uniqueness results, of course.

Theorem 2.13 *Let $F : D \to \mathbb{R}^n$ be $C^1(D)$ for $D \subset \mathbb{R}^n$ convex. Moreover, let $x^0 \in D$ denote a given starting point for the simplified Newton iteration (2.72). Assume that the following affine contravariant Lipschitz condition holds:*

$$\left\|(F'(x) - F'(x^0))v\right\| \le \omega\|F'(x^0)(x - x^0)\| \cdot \|F'(x^0)v\| \qquad (2.73)$$

for x, $x^0 \in D$, $v \in \mathbb{R}^n$ and $0 \le \omega < \infty$. Define the level set

$$\mathcal{L}_\omega := \left\{x \in \mathbb{R}^n \,\middle|\, \|F(x)\| \le \frac{1}{2\omega}\right\}$$

and let $\overline{\mathcal{L}}_\omega \subseteq D$ be bounded. Assume that $x^0 \in \mathcal{L}_\omega$, which is

$$h_0 := \omega\|F(x^0)\| \le \tfrac{1}{2}. \qquad (2.74)$$

Then the iterates remain in $\mathcal{L}_\omega$ and converge to a solution point x^*. The iterative residual norms converge to zero at an estimated rate

$$\frac{\|F(x^{k+1})\|}{\|F(x^k)\|} \le \tfrac{1}{2}(t_k + t_{k+1}) < 1 - \sqrt{1 - 2h_o},$$

wherein the $\{t_k\}$ are defined by $t_0 = 0$ and

$$t_{k+1} = h_0 + \tfrac{1}{2}t_k^2, \quad k = 0, 1, \dots.$$

Proof. We apply the Lipschitz condition (2.73) to obtain

$$\begin{aligned}
\|F(x^{k+1})\| &= \left\| \int_{t=0}^{1} \left(F'(x^k + t\overline{\Delta x}^k) - F'(x^0)\right)\overline{\Delta x}^k \, dt \right\| \\
&\le \omega\|F'(x^0)\overline{\Delta x}^k\| \cdot \int_{t=0}^{1} \|F'(x^0)(x^k - x^0 + t\overline{\Delta x}^k)\| \, dt
\end{aligned}$$

and, by triangle inequality:

$$\|F(x^{k+1})\| \le \omega\|F(x^k)\|\left(\|F'(x^0)(x^k - x^0)\| + \tfrac{1}{2}\|F(x^k)\|\right). \qquad (2.75)$$

We therefore introduce the *majorants*

$$\omega\|F'(x^0)(x^k - x^0)\| \le t_k$$
$$\omega\|F'(x^0)(x^{k+1} - x^k)\| = \omega\|F(x^k)\| \le h_k$$

with initial values $t_0 = 0$, $h_0 \le \tfrac{1}{2}$. Because of

$$\|F'(x^0)(x^{k+1} - x^0)\| \le \|F'(x^0)(x^k - x^0)\| + \|F'(x^0)(x^{k+1} - x^k)\|$$

and the above relation (2.75), we obtain the same two majorant equations as in Section 2.1.2

$$t_{k+1} = t_k + h_k, \quad h_k = h_{k-1}\left(t_{k-1} + \tfrac{1}{2}h_{k-1}\right)$$

and from these a single equation of the form

$$t_{k+1} - t_k = (t_k - t_{k-1})\left(t_{k-1} + \tfrac{1}{2}(t_k - t_{k-1})\right) = \tfrac{1}{2}(t_k^2 - t_{k-1}^2).$$

Rearrangement of this equation permits the application of the *Ortega trick*

$$t_{k+1} - \tfrac{1}{2}t_k^2 = t_1 - \tfrac{1}{2}t_0^2 = h_0,$$

which once again may be interpreted as the simplified Newton iteration

$$t_{k+1} - t_k = -\frac{g(t_k)}{g'(t_0)} = g(t_k)$$

for the scalar equation

$$g(t) = h_0 - t + \tfrac{1}{2}t^2 = 0.$$

As can be seen from the above Figure 2.1, here also we obtain $g(t_{k+1}) < g(t_k)$, which is equivalent to $h_{k+1} < h_k$ and therefore

$$\|F(x^{k+1})\| < \|F(x^k)\| \le \frac{1}{2\omega}.$$

This assures that all simplified Newton iterates remain in $\mathcal{L}_\omega \subset D$. As for the convergence to some (not necessarily unique) solution point $x^* \in \mathcal{L}_\omega \subset D$, arguments similar to the ones used for Theorem 2.12 can be applied. As for the convergence rate, we go back to (2.75) and derive

$$\frac{\|F(x^{k+1})\|}{\|F(x^k)\|} \le t_k + \tfrac{1}{2}h_k = \tfrac{1}{2}(t_k + t_{k+1}) < t^* = 1 - \sqrt{1 - 2h_0},$$

which completes the proof.   □

Convergence monitor. In order to exploit this theorem for actual implementation, we define the *residual contraction factors* $(k = 0, 1, \ldots)$

$$\Theta_k := \frac{\|F(x^{k+1})\|}{\|F(x^k)\|} \le \tfrac{1}{2}(t_k + t_{k+1}).$$

For $k = 0$, the local convergence domain is characterized by

$$\Theta_0 \le \tfrac{1}{2}h_0 \le \tfrac{1}{4}, \tag{2.76}$$

which is clearly more restrictive than the comparable condition $\Theta_0 < 1$ for the ordinary Newton method—compare (2.69).

2.2.3 Broyden's 'bad' rank-1 updates

In this section, we deal with a quasi-Newton update already discussed by C.G. Broyden in his seminal paper [40] and classified there, on the basis of his numerical experiments, as being 'bad'. This method can actually be derived in terms of affine contravariance. As stated before, only image space quantities like the residuals $F_k := F(x^k)$ are of interest in this frame. With $\delta F_{k+1} = F_{k+1} - F_k$, we rewrite the secant condition (1.17) here as

$$E_k(J)\delta F_{k+1} = F_{k+1} \tag{2.77}$$

in terms of the affine contravariant update change matrix

$$E_k(J) := I - J_k J^{-1}.$$

Any Jacobian rank-1 update satisfying

$$J_{k+1}^{-1} = J_k^{-1}\left(I - \frac{F_{k+1}v^T}{v^T \delta F_{k+1}}\right), \quad v \in R^n, \ v \neq 0$$

with v some vector in the image space of F will both satisfy the secant condition and reflect affine contravariance. As an example, the so-called 'bad' Broyden method is characterized by setting $v = \delta F_{k+1}$.

Convergence analysis. We start with an analysis of one quasi–Newton step of this kind.

Theorem 2.14 *Let*

$$J_{k+1}^{-1} = J_k^{-1}\left(I - \frac{F_{k+1}\delta F_{k+1}^T}{\|\delta F_{k+1}\|^2}\right) \tag{2.78}$$

denote the affine contravariant 'bad' Broyden rank-1 update and assume residual contraction

$$\Theta_k := \frac{\|F_{k+1}\|}{\|F_k\|} < 1.$$

Then:

1. *The update matrix J_{k+1} is a least change update in the sense that*

$$\|E_k(J_{k+1})\| \leq \|E_k(J)\| \qquad \forall J \in S_k$$
$$\|E_k(J_{k+1})\| \leq \frac{\Theta_k}{1-\Theta_k}.$$

2. *The update matrix J_{k+1} is nonsingular whenever J_k is nonsingular and can be represented by*

$$J_{k+1} = \left(I - \frac{F_{k+1}\delta F_{k+1}^T}{\delta F_{k+1}^T F_k}\right)J_k.$$

3. With $\overline{\delta x}_{k+1} = -J_k^{-1} F_{k+1}$, the next quasi-Newton correction is

$$\delta x_{k+1} = -J_{k+1}^{-1} F_{k+1} = \left(1 - \frac{\delta F_{k+1}^T F_{k+1}}{\|\delta F_{k+1}\|^2}\right) \overline{\delta x}_{k+1}.$$

Proof. For the above rank-1 update we have

$$E_k(J_{k+1}) = \frac{F_{k+1} \delta F_{k+1}^T}{\|\delta F_{k+1}\|^2}$$

and therefore

$$\|E_k(J_{k+1})\| = \frac{\|E_k(J_{k+1}) \delta F_{k+1}\|}{\|\delta F_{k+1}\|} = \frac{\|F_{k+1}\|}{\|\delta F_{k+1}\|} = \frac{\|E_k(J) \delta F_{k+1}\|}{\|\delta F_{k+1}\|} \leq \|E_k(J)\|$$

for all J satisfying the secant condition (2.77). Further, for $\Theta_k < 1$, we obtain

$$\|E_k(J_{k+1})\| = \frac{\|F_{k+1}\|}{\|\delta F_{k+1}\|} \leq \frac{\Theta_k}{1 - \Theta_k},$$

which confirms the above statement 1. Statements 2 and 3 are direct consequences of the Sherman-Morrison formula.

□

The above Theorem 2.14 only deals with the situation within one iterative step. The iteration as a whole is studied next.

Theorem 2.15 For $F \in C^1(D)$, $F : D \subset R^n \to R^n$, D convex, let x^* denote a unique solution point of F with $F'(x^*)$ nonsingular. Assume that for some $\omega < \infty$ the affine contravariant Lipschitz condition

$$\|(F'(x) - F'(x^*))(y - x)\| \leq \omega \|F'(x^*)(x - x^*)\| \, \|F'(x^*)(y - x)\| \quad (2.79)$$

holds for $x, y \in D$. Consider the quasi-Newton iteration as defined in Theorem 2.14. For some $\overline{\Theta}$ in the range $0 < \overline{\Theta} < 1$ assume that:

1. in terms of the affine contravariant deterioration matrix

$$E_k := I - F'(x^*) J_k^{-1}$$

the initial approximate Jacobian satisfies

$$\overline{\eta}_0 := \|E_0\| < \overline{\Theta},$$

2. the initial guess x^0 satisfies

$$t_0 := \omega \|F'(x^*)(x^0 - x^*)\| \leq \frac{\overline{\Theta} - \overline{\eta}_0}{1 + \overline{\eta}_0 + \frac{4}{3}(1 - \overline{\Theta})^{-1}}.$$

Then the quasi-Newton iterates x^k converge to x^* in terms of errors as

$$\|F'(x^*)(x^{k+1} - x^*)\| \leq \overline{\Theta} \, \|F'(x^*)(x^k - x^*)\|$$

or, in terms of residuals as

$$\|F_{k+1}\| \leq \overline{\Theta} \, \|F_k\| \,.$$

The convergence is superlinear with

$$\lim_{k \to \infty} \frac{\|F_{k+1}\|}{\|F_k\|} = 0 \,. \tag{2.80}$$

As for the Jacobian rank-1 updates, the 'bounded deterioration property' holds in the form

$$\|E_k\| \leq \overline{\eta}_0 + \frac{t_0}{(1 - t_0)(1 - \overline{\Theta})} \leq \overline{\Theta}$$

together with the asymptotic property

$$\lim_{k \to \infty} \frac{\|E_k \delta F_{k+1}\|}{\|\delta F_{k+1}\|} = 0 \,.$$

Proof. For ease of writing we characterize the Jacobian update approximation by

$$\eta_k := \frac{\|E_k \delta F_{k+1}\|}{\|\delta F_{k+1}\|} \,, \quad \overline{\eta}_k := \|E_k\| \geq \eta_k \,.$$

For the convergence analysis we introduce

$$f_k := F'(x^*)(x^k - x^*) \text{ and } t_k := \omega \, \|f_k\| \,.$$

I. To begin with, we analyze the behavior of the iterative residuals:

$$
\begin{aligned}
F_{k+1} &= F_k + \int_{s=0}^{1} F'(x^k + s\delta x_k)\delta x_k \, ds \\
&= \int_{s=0}^{1} (F'(x^k + s\delta x_k) - F'(x^*))\delta x_k \, ds + (F'(x^*) - J_k)\delta x_k \,.
\end{aligned}
$$

Applying the Lipschitz condition (2.79) yields

$$\|F_{k+1}\| \leq \int_{s=0}^{1} \|(F'(x^k + s\delta x_k) - F'(x^*))\delta x_k\| \, ds + \|(F'(x^*)J_k^{-1} - I)F_k\|$$

$$\leq \int_{s=0}^{1} \omega \|F'(x^*)(x^k + s\delta x_k - x^*)\| \, \|F'(x^*)\delta x_k\| \, ds + \|E_k F_k\|$$

$$\leq \int_{s=0}^{1} \omega \big(\|F'(x^*)(1-s)(x^k - x^*)\| $$
$$+ \|F'(x^*)s(x^{k+1} - x^*)\| \big) \, \|F'(x^*)\delta x_k\| \, ds + \overline{\eta}_k \|F_k\|$$

$$= \tfrac{1}{2}(t_k + t_{k+1})\|F'(x^*)\delta x_k\| + \overline{\eta}_k \|F_k\| .$$

Defining $\bar{t}_k := \tfrac{1}{2}(t_k + t_{k+1})$, we get

$$\|F_{k+1}\| \leq \bar{t}_k\|(E_k - I)F_k\| + \overline{\eta}_k\|F_k\|$$
$$\leq (\bar{t}_k(1 + \overline{\eta}_k) + \overline{\eta}_k)\|F_k\| . \qquad (2.81)$$

As for the iterative errors f_k, we may derive the relation

$$f_{k+1} = f_k - F'(x^*)J_k^{-1}F_k = F'(x^*)(x^k - x^*) - F_k + E_k F_k$$
$$= \int_{s=0}^{1} \left(F'(x^*) - F'(x^* + s(x^k - x^*)) \right) (x^k - x^*) \, ds + E_k F_k ,$$

from which we obtain the estimate

$$\|f_{k+1}\| \leq \int_{s=0}^{1} s\omega \|F'(x^*)(x^k - x^*)\| \, \|F'(x^*)(x^k - x^*)\| \, ds + \overline{\eta}_k\|F_k\|$$
$$\leq \frac{\omega}{2}\|f_k\|^2 + \overline{\eta}_k(\|f_k - F_k\| + \|f_k\|) .$$

By multiplication with ω and proceeding as above, this can be further reduced to yield

$$t_{k+1} \leq \tfrac{1}{2}t_k^2 + \overline{\eta}_k \left(\tfrac{1}{2}t_k^2 + t_k \right) = \left(\overline{\eta}_k + \frac{1 + \overline{\eta}_k}{2}t_k \right) t_k . \qquad (2.82)$$

II. Next, we study the approximation properties of the Jacobian updates. Introducing the orthogonal projection

$$Q_k := \frac{\delta F_{k+1}\delta F_{k+1}^T}{\|\delta F_{k+1}\|^2}$$

onto the secant direction δF_{k+1}, the deterioration matrix may be written as

$$E_{k+1} = E_k Q_k^{\perp} + E_{k+1}Q_k , \qquad (2.83)$$

yielding, as in the 'good' Broyden proof,

$$\overline{\eta}_{k+1} = \|E_{k+1}\| \le \|E_k Q_k^{\perp}\| + \|E_{k+1}Q_k\| \le \|E_k\| + \frac{\|E_{k+1}\delta F_{k+1}\|}{\|\delta F_{k+1}\|}.$$

Using the secant condition (2.77), we get for the numerator of the second right hand term:

$$
\begin{aligned}
E_{k+1}\delta F_{k+1} &= \delta F_{k+1} - F'(x^*)J_{k+1}^{-1}\delta F_{k+1} = \delta F_{k+1} - F'(x^*)\delta x_k \\
&= \int_{s=0}^{1} (F'(x^k + s\delta x_k) - F'(x^*))\delta x_k .
\end{aligned}
$$

This can be estimated as above as follows

$$
\begin{aligned}
\|E_{k+1}\delta F_{k+1}\| &\le \bar{t}_k \|F'(x^*)\delta x_k\| \\
&= \bar{t}_k \|E_{k+1}\delta F_{k+1} - \delta F_{k+1}\| \\
&\le \bar{t}_k (\|E_{k+1}\delta F_{k+1}\| + \|\delta F_{k+1}\|)
\end{aligned}
$$

in order to get

$$\|E_{k+1}\delta F_{k+1}\| \le \frac{\bar{t}_k}{1 - \bar{t}_k} \|\delta F_{k+1}\|. \qquad (2.84)$$

Inserting this estimate into (2.83) yields the quite rough estimate

$$\overline{\eta}_{k+1} \le \overline{\eta}_k + \frac{\bar{t}_k}{1 - \bar{t}_k}.$$

III. For the purpose of repeated induction assume that we have

$$\overline{\eta}_k \le \overline{\eta}_0 + \frac{\sum_{i=0}^{k-1}\overline{\Theta}^i t_0}{1 - t_0} \le \overline{\eta}$$

with

$$\overline{\eta} := \overline{\eta}_0 + \frac{t_0}{(1 - t_0)(1 - \overline{\Theta})}$$

and

$$t_k \le \overline{\Theta}^k t_0.$$

Then by (2.82) and by the subsequent technical Lemma 2.16 below

$$t_{k+1} \le (\overline{\eta} + (1 + \overline{\eta})t_0)t_k \le \overline{\Theta}t_k \le \overline{\Theta}^{k+1}t_0$$

and thus

$$\overline{\eta}_{k+1} \le \overline{\eta}_k + \frac{t_k}{1 - t_0} \le \overline{\eta}_0 + \frac{\sum_{i=0}^{k-1}\overline{\Theta}^i t_0}{1 - t_0} + \frac{\overline{\Theta}^{k+1}t_0}{1 - t_0} \le \overline{\eta}_0 + \frac{\sum_{i=0}^{k}\overline{\Theta}^i t_0}{1 - t_0} \le \overline{\eta}.$$

By induction we have the 'bounded deterioration property'

$$\overline{\eta}_k \leq \overline{\eta}$$

and the error contraction

$$t_{k+1} \leq t_k$$

for any k. Obviously, by (2.81) and the subsequent technical Lemma 2.16 we also have contraction of the residuals:

$$\|F_{k+1}\| \leq \overline{\Theta}\|F_k\|$$

IV. In order to show *superlinear* convergence, we use the orthogonal splitting provided by (2.83) in a more subtle manner. Since

$$Q_k E_k^T v = \delta F_{k+1} \frac{\langle \delta F_{k+1}, E_k^T v\rangle}{\|\delta F_{k+1}\|^2} = \delta F_{k+1} \frac{\langle E_k \delta F_{k+1}, v\rangle}{\|\delta F_{k+1}\|^2},$$

some short calculation supplies the equation

$$\begin{aligned}
\|E_{k+1}^T v\|^2 &= \|Q_k^{\perp} E_k^T v\|^2 + \|Q_k E_{k+1}^T v\|^2 \\
&= \|E_k^T v\|^2 - \|Q_k E_k^T v\|^2 + \|Q_k E_{k+1}^T v\|^2 \\
&= \|E_k^T v\|^2 - \frac{\langle E_k \delta F_{k+1}, v\rangle^2}{\|\delta F_{k+1}\|^2} + \frac{\langle E_{k+1}\delta F_{k+1}, v\rangle^2}{\|\delta F_{k+1}\|^2}.
\end{aligned}$$

Summing over the indices k, we arrive at

$$\sum_{k=0}^{l} \frac{\langle E_k\delta F_{k+1}, v\rangle^2}{\|\delta F_{k+1}\|^2\|v\|^2} = \frac{\|E_0^T v\|^2}{\|v\|^2} - \frac{\|E_{l+1}^T v\|}{\|v\|^2} + \sum_{k=0}^{l} \frac{\langle E_{k+1}\delta F_{k+1}, v\rangle^2}{\|\delta F_{k+1}\|^2\|v\|^2}.$$

Upon dropping the negative right hand term, letting $l \to \infty$, and using (2.84), we end up with the estimate

$$\sum_{k=0}^{l} \frac{\langle E_k\delta F_{k+1}, v\rangle^2}{\|\delta F_{k+1}\|^2\|v\|^2} \leq \overline{\eta}_0^2 + \sum_{k=0}^{l} \left(\frac{t_k}{1-t_k}\right)^2 \leq \overline{\eta}_0^2 + \frac{t_0^2}{(1-t_0)^2(1-\overline{\Theta}^2)}.$$

Since the right hand side is bounded, we immediately conclude that

$$\lim_{k\to\infty} \frac{\langle E_k\delta F_{k+1}, v\rangle^2}{\|\delta F_{k+1}\|^2\|v\|^2} = 0$$

for all $v \in R^n$. As a consequence, we must have

$$\lim_{k\to\infty} \eta_k = 0.$$

In order to prove the superlinear convergence statement (2.80), we may collect some estimates from above and proceed as

$$\begin{aligned}
\|F'(x^*)J_k^{-1}F_{k+1}\| &= \|E_{k+1}\delta F_{k+1} - E_k\delta F_{k+1}\| \\
&\leq \bar{t}_k(1+\bar{\eta}_k)\|F_k\| + \eta_k\|\delta F_{k+1}\| \\
&\leq (\bar{t}_k(1+\bar{\eta}_k) + \eta_k(1+\overline{\Theta}))\|F_k\|\,.
\end{aligned}$$

Finally, with

$$\|F_{k+1}\| - \|F'(x^*)J_k^{-1}F_{k+1}\| \leq \|E_kF_{k+1}\| \leq \bar{\eta}_k\|F_{k+1}\|$$

$$\Rightarrow \qquad \|F_{k+1}\| \leq \frac{\|F'(x^*)J_k^{-1}F_{k+1}\|}{1-\bar{\eta}_k}\,,$$

we get

$$\|F_{k+1}\| \leq \frac{\bar{t}_k(1+\bar{\eta}_k) + \eta_k(1+\overline{\Theta})}{1-\bar{\eta}\eta_k}\|F_k\|\,.$$

Since $\bar{t}_k \to 0$ and $\eta_k \to 0$, superlinear convergence is easily verified. $\qquad\square$

For ease of the above derivation, the following technical lemma has been postponed.

Lemma 2.16 *Assume* $0 < \Theta < 1$, $0 \leq \eta_0 < \Theta$ *and*

$$t \leq \frac{\Theta - \eta_0}{1 + \eta_0 + \frac{4}{3}(1-\Theta)^{-1}}\,.$$

Then, with $\eta = \eta_0 + \dfrac{t}{(1-t)(1-\Theta)}$, *we have*

$$\eta + (1+\eta)t \leq \Theta\,.$$

Proof. Under the given assumptions, a short calculation shows that $t < \frac{1}{7}$. Therefore we can proceed as

$$\begin{aligned}
\Theta &\geq \eta_0 + \left(1 + \eta_0 + \tfrac{4}{3}(1-\Theta)^{-1}\right)t \\
&= \eta_0 + \frac{\frac{7}{6}t}{1-\Theta} + \left(1 + \eta_0 + \tfrac{1}{6}(1-\Theta)^{-1}\right)t \\
&\geq \eta_0 + \frac{t}{(1-t)(1-\Theta)} + \left(1 + \eta_0 + \frac{t}{(1-t)(1-\Theta)}\right)t \\
&= \eta + (1+\eta)t\,.
\end{aligned}$$

$$\square$$

Algorithmic realization. From representation (2.78) we again have a product form for the Jacobian update inverses. As a *condition number monitor* for the possible occurrence of ill-conditioning of the recursive Jacobian rank-1 updates, Lemma 2.8 may once more be applied, here to:

$$\text{cond}_2(J_{k+1}) \leq \text{cond}_2\left(I - \frac{F_{k+1}\delta F_{k+1}^T}{\|\delta F_{k+1}\|^2}\right)\text{cond}_2(J_k).$$

In the present context, we obtain for $\Theta_k < 1/2$:

$$\text{cond}_2(J_{k+1}) \leq \frac{1}{1 - 2\Theta_k}\text{cond}_2(J_k).$$

As a consequence, a restriction such as

$$\Theta_k \leq \Theta_{\max} < \tfrac{1}{2}$$

with, say $\Theta_{\max} = 1/4$, will be necessary. With these preparations, we are now ready to present the 'bad Broyden' algorithm QNRES (the acronym stands for **RES**idual based **Q**uasi-**N**ewton algorithm).

Algorithm QNRES.

$F_0 := F(x^0)$	evaluation and store
$\sigma_0 := \|F_0\|^2$	store
$J_0\delta x_0 = -F_0$	linear system solve
$\kappa := 1$	

For $k := 0, 1, \ldots, k_{\max}$:

$$x^{k+1} := x^k + \delta x^k$$

$$F_{k+1} := F(x^{k+1})$$

$$\delta F_{k+1} := F_{k+1} - F_k$$

$$\sigma_{k+1} := \|F_{k+1}\|^2$$

If $\sigma_{k+1} \leq \text{FTOL}^2$:

   solution found: $x^* = x^{k+1}$

$$\Theta_k := \sqrt{\sigma_{k+1}/\sigma_k}$$

If $\Theta_k \geq \Theta_{\max}$:

   stop: no convergence

$$w := \delta F_{k+1}$$

$$\gamma_k := \|w\|^2$$

$$\kappa := \kappa/(1 - 2\Theta_k)$$

If $\kappa \geq \kappa_{\max}$:

   stop: ill-conditioned update

$v := (1 - \langle w, F_{k+1}\rangle/\gamma_k)F_{k+1}$

For $j = k - 1, \ldots, 0$:

   $\beta := \langle \delta F_{j+1}, v\rangle/\gamma_j$

   $v = v - \beta F_{j+1}$

$J_0\delta x_{k+1} = -v$

stop: no convergence within $k_{\max}$ iterations

The above algorithm merely requires to store the residuals $F_0, \ldots, F_{k+1}$, and the differences $\delta F_1, \ldots, \delta F_{k+1}$, which means an extra array storage of up to $2(k_{\max} + 2)$ vectors of length n. Note that there is a probably machine-dependent tradeoff between computation and storage: the vectors δF_{j+1} can be either stored or recomputed. Moreover, careful considerations about *residual scaling* in the inner product $\langle \cdot, \cdot\rangle$ are recommended.

2.2.4 Inexact Newton-RES method

Recall inexact Newton methods with inner and outer iteration formally written as (dropping the inner iteration index i)

$$F'(x^k)\delta x^k = -F(x^k) + r^k\,, \quad x^{k+1} = x^k + \delta x^k\,, \quad k = 0, 1, \ldots. \qquad (2.85)$$

In what follows, we will work out details for GMRES as inner iteration (see Section 1.4.1). For ease of presentation, we fix the initial values

$$\delta x_0^k = 0 \quad \text{and} \quad r_0^k = F(x^k)\,,$$

which, during the inner iteration $(i = 0, 1, \ldots)$, implies in the generic case that

$$\eta_i = \frac{\|r_i^k\|}{\|F(x^k)\|} \leq 1 \quad \text{and} \quad \eta_{i+1} < \eta_i\,, \quad \text{if } \eta_i \neq 0\,.$$

In what follows, we will denote the final value obtained from the inner iteration in each outer iteration step k by η_k, again dropping the inner iteration index i.

Convergence analysis. For the inexact Newton-GMRES iteration, we may state the following convergence theorem.

Theorem 2.17 *Let $F : D \to \mathbb{R}^n$, $F \in C^1(D)$, $D \subset \mathbb{R}^n$ convex. Let $x^0 \in D$ denote a given starting point for an inexact Newton iteration (2.85). Assume the affine contravariant Lipschitz condition*

$$\left\|\left(F'(y) - F'(x)\right)(y - x)\right\| \leq \omega \|F'(x)(y - x)\|^2$$
for $0 \leq \omega < \infty$, and $x, y \in D$.

Let the level set $\mathcal{L}_0 := \left\{x \in \mathbb{R}^n \mid \|F(x)\| \leq \|F(x^0)\|\right\} \subseteq D$ be compact. For each well-defined iterate $x^k \in D$ define $h_k := \omega \|F(x^k)\|$. Then the outer residual norms can be bounded as

$$\|F(x^{k+1})\| \leq \left(\eta_k + \tfrac{1}{2}(1 - \eta_k^2)h_k\right)\|F(x^k)\|. \tag{2.86}$$

The convergence rate can be estimated as follows:

I. Linear convergence mode. Assume that the initial guess x^o gives rise to
$$h_0 < 2.$$
Then some $\overline{\Theta}$ in the range $h_0/2 < \overline{\Theta} < 1$ can be chosen. Let the inner GMRES iteration be controlled such that

$$\eta_k \leq \overline{\Theta} - \tfrac{1}{2}h_k. \tag{2.87}$$

Then the Newton-GMRES iterates $\{x^k\}$ converge at least linearly to some solution point $x^* \in \mathcal{L}_0$ at an estimated rate

$$\|F(x^{k+1})\| \leq \overline{\Theta}\|F(x^k)\|.$$

II. Quadratic convergence mode. If, for some $\rho > 0$, the initial guess x^0 guarantees that
$$h_0 < 2/(1 + \rho)$$
and the inner iteration is controlled such that

$$\frac{\eta_k}{1 - \eta_k^2} \leq \tfrac{1}{2}\rho h_k, \tag{2.88}$$

then the convergence is quadratic at an estimated rate

$$\|F(x^{k+1})\| \leq \tfrac{1}{2}\omega(1 + \rho)(1 - \eta_k^2)\|F(x^k)\|^2. \tag{2.89}$$

Proof. Proceeding as in earlier proofs, we obtain

$$
\begin{aligned}
\|F(x^{k+1})\| &= \left\| \int\limits_0^1 \left(F'(x^k + t\delta x^k) - F'(x^k)\right)\delta x^k\, dt + r^k \right\| \\
&\leq \int\limits_0^1 \left\|\left(F'(x^k + t\delta x^k) - F'(x^k)\right)\delta x^k\right\| dt + \|r^k\| \\
&\leq \tfrac{1}{2}\omega\|F(x^k) - r^k\|^2 + \|r^k\|.
\end{aligned}
$$

By use of (1.20), this is seen to be just (2.86). Under the assumption (2.87) with $\overline{\Theta} < 1$ and $\eta_k < 1$ from GMRES, we obtain

$$\|F(x^{k+1})\| \le \overline{\Theta}\|F(x^k)\|$$

and by repeated induction

$$\{x^k\} \subset \mathcal{L}_0 \subset D\,,$$

from which the convergence to $x^* \in \mathcal{L}_0$ is concluded. Quadratic convergence as in (2.89) is shown by mere insertion of (2.88) into (2.86). $\qquad\square$

Convergence monitor. Throughout the inexact Newton iteration we will check for *residual monotonicity*

$$\Theta_k := \frac{\|F(x^{k+1})\|}{\|F(x^k)\|} \le \overline{\Theta} < 1\,,\ k = 0,1,\dots\,,$$

introducing certain default parameters $\overline{\Theta}$ in accordance with the above Theorem 2.17. We will regard an iteration as *divergent*, whenever $\Theta_k \ge \overline{\Theta}$ holds.

Termination criterion. As in the exact Newton iteration, the finally accepted iterate $\hat{x}$ is required to satisfy

$$\|F(\hat{x})\| \le \mathrm{FTOL}$$

with FTOL a user prescribed *residual error tolerance*.

Standard convergence mode. If $\eta_k \le \bar{\eta} < 1$ is prescribed by the user, then (2.86) implies that $\Theta_k \to \bar{\eta}$ and *asymptotic linear convergence* occurs—as already shown in the early pioneering paper [51].

Quadratic convergence mode. Assume that for $k = 0$ some value η_0 is prescribed; from numerical experiments, we know that this value should be sufficiently small—compare, e.g., Table 8.3 in Section 8.2 below. For $k \ge 0$, (2.89) suggests the *a-posteriori estimate*

$$[h_k]_2 := \frac{2\Theta_k}{(1+\rho)(1-\eta_k^2)} \le h_k$$

and, since $h_{k+1} = \Theta_k h_k$, also the *a-priori estimate:*

$$[h_{k+1}] := \Theta_k[h_k]_2 \le h_{k+1}\,.$$

For $k > 0$, shifting the index $k+1$ now back to k, we therefore require that

$$\frac{\eta_k}{1-\eta_k^2} \le \tfrac{1}{2}\rho[h_k] \le \tfrac{1}{2}\rho h_k\,, \tag{2.90}$$

which can be assured in the course of the iterative computation of δx^k and r^k. For the parameter ρ some value $\rho \approx 1$ seems to be appropriate. Note that asymptotically this choice leads to $\eta_k \to \rho[h_k] \to 0$.

Linear convergence mode. Once the local contraction factor Θ_k is sufficiently below some prescribed value $\overline{\Theta}$, we may switch to the linear convergence mode described in the above Theorem 2.17. Careful examination of the proof shows that

$$\|F(x^{k+1}) - r^k\| \le \frac{\omega}{2}\|F(x^k) - r^k\|^2 = \tfrac{1}{2}(1 - \eta_k^2)h_k\|F(x^k)\| \,.$$

From this we may derive the *a-posteriori estimate*

$$[h_k]_1 := \frac{2\|F(x^{k+1}) - r^k\|}{(1 - \eta_k^2)\|F(x^k)\|} \le h_k$$

and, since $h_{k+1} = \Theta_k h_k$, also the *a-priori estimate*

$$[h_{k+1}] := \Theta_k [h_k]_1 \le h_{k+1} \,.$$

As a preparation of the next Newton step, we define

$$\overline{\eta}_{k+1} = \overline{\Theta} - \tfrac{1}{2}[h_{k+1}]$$

in terms of the above a-priori estimate. If this value is smaller than the value obtained from (2.90), then we continue the iteration in the quadratic convergence mode. Else, we realize the linear convergence mode in Newton step $k + 1$ with some

$$\eta_{k+1} \le \overline{\eta}_{k+1} \,.$$

Asymptotically, this strategy leads to $\eta_{k+1} \to \overline{\Theta}$.

Preconditioning. In order to speed up the inner iteration, preconditioning from the left or/and from the right may be used. This means solving

$$\left(C_L F'(x^k) C_R\right)\left(C_R^{-1}\delta x^k\right) = C_L\left(-F(x^k) + r^k\right)$$

instead of (2.85). In such a case, the norm of the *preconditioned residuals* $\bar{r}^k = C_L r^k$ is minimized in GMRES, whereas C_R only affects the rate of convergence via the Krylov subspace

$$\mathcal{K}_i(\bar{r}_0, \overline{A}) \text{ with } \overline{A} = C_L F'(x^k) C_R \,.$$

Consequently, the above strategy should be based on the contraction factors

$$\Theta_k = \frac{\|C_L F(x^{k+1})\|_2}{\|C_L F(x^k)\|_2}$$

for the outer iteration. Note, however, that C_L should *not* depend on the iterate x^k in this theoretical setting.

If strict residual minimization is aimed at, then only *right* preconditioning should be implemented (i.e., $C_L = I$).

The here described local Newton-GMRES algorithm is part of the global Newton code GIANT-GMRES, which will be described in Section 3.2.3 below.

Remark 2.2   If GMRES were replaced by some other *residual norm reducing* (but *not minimizing*) iterative linear solver, then a similar accuracy matching strategy can be worked out (left as Exercise 2.9).

BIBLIOGRAPHICAL NOTE. The concept of local inexact Newton methods—sometimes also called *truncated* Newton methods—seems to have first been published in 1982 by R.S. Dembo, S.C. Eisenstat, and T. Steihaug [51]; they presented an asymptotic analysis in terms of the residuals. In 1981, R.E. Bank and D.J. Rose [19] worked out details of an inexact Newton algorithm on the basis of residual control including certain algorithmic heuristics. In 1996, S.C. Eisenstat and H.F. Walker [91] suggested a further strategy to choose the η_k, which they call 'forcing terms'; their strategy is also based on convergence analysis results, but different from the one presented here.

2.3 Convex Optimization

In this section we consider the problem of minimizing a strictly convex functional $f : D \subset \mathbb{R}^n \longrightarrow \mathbb{R}^1$. Then $F(x) = f'(x)^T$ is a gradient mapping and $F'(x) = f''(x)$ is symmetric positive definite. We want to solve $F(x) = 0$, a system of n nonlinear equations, by local Newton methods. The convergence analysis will start from *affine conjugate* Lipschitz conditions of the type (1.9) and lead to results in terms of iterative functional values and energy norms of corrections or errors.

2.3.1 Ordinary Newton method

Recall the ordinary Newton method in the notation ($k = 0, 1, \ldots$)

$$F'(x^k)\Delta x^k = -F(x^k), \; x^{k+1} = x^k + \Delta x^k .$$

Convergence analysis. We analyze its convergence behavior in terms of iterative values of the functional to be minimized and energy norms of the Newton corrections. Thus we arrive at an affine conjugate variant of the Newton-Mysovskikh theorem.

Theorem 2.18 *Let $f : D \to \mathbb{R}^1$ be a strictly convex C^2-functional to be minimized over some open and convex domain $D \subset \mathbb{R}^n$. Let $F(x) = f'(x)^T$ and $F'(x) = f''(x)$, which is symmetric and assumed to be strictly positive definite. Assume that the following affine conjugate Lipschitz condition holds:*

$$\left\| F'(z)^{-1/2}\big(F'(y) - F'(x)\big)(y-x) \right\| \leq \omega \| F'(x)^{1/2}(y-x) \|^2 \qquad (2.91)$$

for collinear $x,\, y,\, z \in D$ *with* $0 \leq \omega < \infty$. *For the initial guess* x^0 *assume that*

$$h_0 = \omega \| F'(x^0)^{1/2} \Delta x^0 \| < 2 \qquad (2.92)$$

and that the level set $\mathcal{L}_0 := \{ x \in D \,|\, f(x) \leq f(x^0) \}$ *is compact. Then the ordinary Newton iterates remain in* $\mathcal{L}_0$ *and converge to the minimum point* x^* *at a rate estimated by*

$$\| F'(x^{k+1})^{1/2} \Delta x^{k+1} \| \leq \tfrac{1}{2}\omega \| F'(x^k)^{1/2} \Delta x^k \|^2 \qquad (2.93)$$

or, with $\epsilon_k := \| F'(x^k)^{1/2} \Delta x^k \|^2$ *and* $h_k := \omega \| F'(x^k)^{1/2} \Delta x^k \|$, *by*

$$
\begin{aligned}
-\tfrac{1}{6} h_k \epsilon_k &\leq& f(x^k) - f(x^{k+1}) - \tfrac{1}{2}\epsilon_k &\leq& \tfrac{1}{6} h_k \epsilon_k \\
\tfrac{1}{6}\epsilon_k &\leq& f(x^k) - f(x^{k+1}) &\leq& \tfrac{5}{6}\epsilon_k \,.
\end{aligned}
\qquad (2.94)
$$

The distance to the minimum can be bounded as

$$f(x^0) - f(x^*) \leq \frac{\tfrac{5}{6}\epsilon_0}{1 - h_0/2} \,.$$

Proof. With the Lipschitz condition (2.91) for $z = x^{k+1}$, $y = x^k + t\Delta x^k$, $x = x^k$, the result (2.93), which is equivalent to $h_{k+1} \leq h_k^2/2$, is proven just as before in Theorem 2.2. The fact that $x^{k+1} \in \mathcal{L}_0$ can be seen by applying the same technique as in the proof of Theorem 2.12 above. To derive (2.94), we verify that

$$f(x^{k+1}) - f(x^k) + \tfrac{1}{2}\| F'(x^k)^{1/2}\Delta x^k \|^2 = \int\limits_{s=0}^{1} s \int\limits_{t=0}^{1} \langle \Delta x^k, w \rangle \, dt \, ds \,, \qquad (2.95)$$

where $w = \big(F'(x^k + st\Delta x^k) - F'(x^k)\big)\Delta x^k$

with $\langle \cdot, \cdot \rangle$ the Euclidean inner product. The integrand term is estimated as

$$
\begin{aligned}
\langle \Delta x^k, w \rangle &\leq& | \langle F'(x^k)^{1/2}\Delta x^k, F'(x^k)^{-1/2}w \rangle | \\
&\leq& \| F'(x^k)^{1/2}\Delta x^k \| \cdot \omega st \| F'(x^k)^{1/2}\Delta x^k \|^2
\end{aligned}
$$

by the Cauchy-Schwarz inequality and (2.91) with $x = z = x^k$, $y = x^k + st\Delta x^k$. With $h_k < 2$ this is the left side of (2.94). Consequently, the iterates converge to x^*. Note that x^* is anyway unique in D under the assumptions made.

In order to obtain the right hand side of (2.94), we go up to (2.95), but this time apply Cauchy-Schwarz in the other direction, which yields:

$$0 \leq f(x^k) - f(x^{k+1}) \leq \big(\tfrac{1}{2} + \tfrac{1}{6}h_k\big) \| F'(x^k)^{1/2}\Delta x^k \|^2 < \tfrac{5}{6}\epsilon_k \,.$$

Summing over all $k = 0, 1, \ldots$ we get

$$0 \leq \omega^2 \big(f(x^0) - f(x^*)\big) \leq \sum_{k=0}^{\infty} \big(\tfrac{1}{2}h_k^2 + \tfrac{1}{6}h_k^3\big) < \tfrac{5}{6} \sum_{k=0}^{\infty} h_k^2 .$$

By using

$$\tfrac{1}{2}h_{k+1} \leq \big(\tfrac{1}{2}h_k\big)^2 \leq \tfrac{1}{2}h_k < 1$$

the right hand upper bound can be further treated to obtain

$$(\tfrac{1}{2}h_0)^2 + (\tfrac{1}{2}h_1)^2 + \cdots \quad \leq \quad (\tfrac{1}{2}h_0)^2 + (\tfrac{1}{2}h_0)^4 + (\tfrac{1}{2}h_1)^4 + \cdots$$

$$< \quad \tfrac{1}{4}h_0^2 \sum_{k=0}^{\infty} (\tfrac{1}{2}h_0)^k = \frac{\tfrac{1}{4}h_0^2}{1 - \tfrac{1}{2}h_0} ,$$

so that

$$\omega^2 \big(f(x^0) - f(x^*)\big) < \frac{\tfrac{5}{6}h_0^2}{1 - \tfrac{1}{2}h_0} .$$

This is the last statement of the theorem. $\square$

Convergence monitor. We now study the consequences of the above convergence theorem for actual implementation. Let ϵ_k, Θ_k be defined as

$$\epsilon_k = \|F'(x^k)^{1/2}\Delta x^k\|_2^2 = |\langle F(x^k), \Delta x^k\rangle| , \quad \Theta_k = \left(\frac{\epsilon_{k+1}}{\epsilon_k}\right)^{1/2} .$$

Then the basic convergence result is

$$\Theta_k = \frac{h_{k+1}}{h_k} \leq \tfrac{1}{2}h_k < 1$$

and

$$f(x^{k+1}) - f(x^k) < -\tfrac{1}{6}\epsilon_k .$$

For $k = 0$, we must have

$$\Theta_0 < 1$$

to assure that x^0 is within the local convergence domain. For $k > 0$, in a similar way as in the two cases before, we derive the approximate recursion $(k = 0, 1, \ldots)$

$$\Theta_{k+1} \approx \Theta_k^2 < \Theta_0 < 1 .$$

From this, we may terminate the iteration as 'divergent' whenever

$$f(x^{k+1}) - f(x^k) \geq -\tfrac{1}{6}\epsilon_k$$

or, since this criterion is prone to suffer from rounding errors, either

$$\Theta_k \geq \Theta_0 \quad (k > 0),$$

or

$$\Theta_{k+1} \geq \frac{\Theta_k^2}{\Theta_0} .$$

Termination criterion. We may terminate the iteration whenever either

$$\epsilon_k \leq \mathrm{ETOL}^2$$

or, recalling that asymptotically

$$f(x^{k+1}) - f(x^k) \doteq -\tfrac{1}{2}\epsilon_k\,,$$

whenever

$$f(x^k) - f(x^{k+1}) \leq \tfrac{1}{2}\mathrm{ETOL}^2$$

with ETOL a user prescribed *energy error tolerance*.

2.3.2 Simplified Newton method

Recall the notation of the simplified Newton iteration

$$F'(x^0)\overline{\Delta x}^k = -F(x^k)\,, \quad x^{k+1} = x^k + \overline{\Delta x}^k\,, \quad k = 0, 1, \ldots\,.$$

Convergence analysis. We now want to study its functional minimization properties, when the Jacobian matrix is kept throughout the Newton iteration.

Theorem 2.19 *Let* $f : D \to \mathbb{R}^1$ *be a strictly convex* C^2-*functional to be minimized over some convex domain* $D \subset \mathbb{R}^n$. *Let* $F(x) = f'(x)^T$ *and* $F'(x) = f''(x)$, *which is then symmetric positive definite. Let* $x^0 \in D$ *be some given starting point for a simplified Newton iteration. Assume that the following affine conjugate Lipschitz condition holds:*

$$\|F'(x^0)^{-1/2}(F'(z) - F'(x^0))v\| \leq \omega\|F'(x^0)^{1/2}(z - x^0)\| \cdot \|F'(x^0)^{1/2}v\|$$

for $z \in D$. *Let*

$$h_0 := \omega\|F'(x^0)^{1/2}\,\overline{\Delta x}^0\| \leq \tfrac{1}{2}$$

and define $t^* = 1 - \sqrt{1 - 2h_0}$. *Then, with* $\epsilon_k := \|F'(x^0)^{1/2}\overline{\Delta x}^k\|^2$, *the simplified Newton iteration converges to some* x^* *with*

$$\omega\|x^* - x^0\| \leq t^*\,.$$

The convergence rate can be estimated in terms of the functional by

$$-\tfrac{1}{6}\epsilon_k(t_{k+1} + 2t_k) \leq f(x^k) - f(x^{k+1}) - \tfrac{1}{2}\epsilon_k \leq \tfrac{1}{6}\epsilon_k(t_{k+1} + 2t_k) \quad (2.96)$$

or in terms of energy norms of the simplified Newton corrections by

$$\Theta_k = \left(\frac{\epsilon_{k+1}}{\epsilon_k}\right)^{1/2} \leq \tfrac{1}{2}(t_{k+1} + t_k)\,,$$

wherein $\{t_k\}$ *is defined from* $t_0 = 0$ *and*

$$t_{k+1} = h_0 + \tfrac{1}{2}t_k^2 < t^*\,, \quad k = 0, 1, \ldots\,.$$

Proof. The proof is similar to the previous proofs of Theorem 2.5 and Theorem 2.13 and will therefore only be sketched here. With the definition for ϵ_k and the majorants

$$\omega\|F'(x^0)^{1/2}(x^k - x^0)\| \le t_k \, , \quad \omega\|F'(x^0)^{1/2}\,\overline{\Delta x}^k\| \le h_k$$

we obtain for the functional decrease

$$f(x^{k+1}) - f(x^k) + \tfrac{1}{2}\epsilon_k =$$

$$= \int\limits_{s=0}^{1} s \int\limits_{t=0}^{1} \left\langle \overline{\Delta x}^k \, , \, \left(F'(x^k + ts\overline{\Delta x}^k) - F'(x^0)\right)\overline{\Delta x}^k \right\rangle dt ds$$

$$\le \omega\epsilon_k \int\limits_{s=0}^{1} s \int\limits_{t=0}^{1} \left((1 - ts)\|F'(x^0)^{1/2}(x^k - x^0)\| + \right.$$

$$\left. + ts\|F'(x^0)^{1/2}(x^{k+1} - x^0)\|\right) dt ds$$

$$\le \tfrac{1}{6}\epsilon_k(t_{k+1} + 2t_k) \, .$$

This is the basis for (2.96). The energy norm contraction factor arises as

$$\Theta_k = \left(\frac{\epsilon_{k+1}}{\epsilon_k}\right)^{1/2} \le \tfrac{1}{2}(t_k + t_{k+1}) =: \frac{h_{k+1}}{h_k} \, .$$

With $t_0 = 0$, $t_{k+1} = t_k + h_k$ and the usual 'Ortega trick' the results above are essentially established. □

Convergence monitor. For actual computation, we also have

$$\Theta_0 \le \tfrac{1}{2}h_0 \le \tfrac{1}{4} \, .$$

Note that for the *simplified* Newton iteration, the asymptotic property $f(x^*) - f(x^k) \approx \tfrac{1}{2}\epsilon_k$ does *not* hold—compare (2.96). Mutatis mutandis, essentially just replacing norms by energy norms in the contraction factors Θ_k, the techniques already worked out in Section 2.1.2 carry over.

Termination criterion. This also can be directly copied from Section 2.1.2 with the proper replacement of norms by energy norms.

2.3.3 Inexact Newton-PCG method

We next study *inexact* Newton methods (dropping, as usual, the inner iteration index i)

$$F'(x^k)(\delta x^k - \Delta x_k) = r^k \, , \quad x^{k+1} = x^k + \delta x^k \, , \quad k = 0, 1, \dots . \quad (2.97)$$

In the context of (strictly) convex optimization the Jacobian matrices can be assumed to be symmetric positive definite, so that the outstanding candidate for an inner iteration will be the *preconditioned conjugate gradient* (PCG). Throughout this section we set $\delta x_0^k = 0$.

Convergence analysis. For the purpose of our analysis below, we recall the following *orthogonality condition*, which is equivalent to condition (1.21) independent of the selected preconditioner:

$$\langle \delta x^k, \, F'(x^k)(\delta x^k - \Delta x_k) \rangle = \langle \delta x^k, r^k \rangle = 0 \,. \tag{2.98}$$

As before, Δx^k denotes the associated exact Newton correction. After these preparations, we are now ready to derive a Newton-Mysovskikh type theorem, which meets our above *affine conjugacy* requirements.

Theorem 2.20 *Let* $f : D \to \mathbb{R}$ *be a strictly convex* C^2-*functional to be minimized over some open and convex domain* $D \subset \mathbb{R}^n$. *Let* $F'(x) := f''(x)$ *be symmetric positive definite and let* $\| \cdot \|$ *denote the Euclidean vector norm. In the above introduced notation assume the existence of some* $\omega < \infty$ *such that the following affine conjugate Lipschitz condition holds for collinear* x, y, $z \in D$:

$$\left\| F'(z)^{-1/2} \big(F'(y) - F'(x) \big) v \right\| \leq \omega \left\| F'(x)^{1/2} (y - x) \right\| \cdot \left\| F'(x)^{1/2} v \right\|.$$

Consider an inexact Newton-PCG iteration (2.97) *satisfying* (2.98) *and started with* $\delta x_0^k = 0$. *At any well-defined iterate* x^k, *define the exact Newton terms*

$$\epsilon_k := \| F'(x^k)^{1/2} \Delta x^k \|^2 \quad \text{and} \quad h_k := \omega \, \| F'(x^k)^{1/2} \Delta x^k \|$$

and, subject to inner iteration errors characterized by

$$\delta_k := \frac{\| F'(x^k)^{1/2} (\delta x^k - \Delta x^k) \|}{\| F'(x^k)^{1/2} \delta x^k \|} \,,$$

the associated inexact Newton terms

$$c_k^\delta := \| F'(x^k)^{1/2} \delta x^k \|^2 = \frac{\epsilon_k}{1 + \delta_k^2} \quad \text{and} \quad h_k^\delta := \omega \, \| F'(x^k)^{1/2} \delta x^k \| = \frac{h_k}{\sqrt{1 + \delta_k^2}} \,.$$

For a given initial guess $x^0 \in D$ *assume that the level set* $\mathcal{L}_0 := \{ x \in D \mid f(x) \leq f(x^0) \}$ *is closed and bounded. Then the following results hold:*

I. Linear convergence mode. *Assume that* x^0 *satisfies*

$$h_0 < 2\overline{\Theta} < 2 \tag{2.99}$$

for some $\overline{\Theta} < 1$. *Let* $\delta_{k+1} \geq \delta_k$ *throughout the inexact Newton iteration. Moreover, let the inner iteration be controlled such that*

$$\vartheta(h_k^\delta, \delta_k) := \frac{h_k^\delta + \delta_k \left(h_k^\delta + \sqrt{4 + (h_k^\delta)^2} \right)}{2\sqrt{1 + \delta_k^2}} \leq \overline{\Theta} \,, \tag{2.100}$$

which assures that

$$\delta_k \leq \overline{\Theta}/\sqrt{1-\overline{\Theta}^2}.$$ (2.101)

Then the iterates x^k remain in $\mathcal{L}_0$ and converge at least linearly to the minimum point $x^* \in \mathcal{L}_0$ such that

$$\|F'(x^{k+1})^{1/2}\Delta x^{k+1}\| \leq \overline{\Theta}\,\|F'(x^k)^{1/2}\Delta x^k\|$$ (2.102)

and

$$\|F'(x^{k+1})^{1/2}\delta x^{k+1}\| \leq \overline{\Theta}\,\|F'(x^k)^{1/2}\delta x^k\|.$$

II. Quadratic convergence mode. *Let for some $\rho > 0$ the initial iterate x^0 satisfy*

$$h_0^\delta < \frac{2}{1+\rho}$$ (2.103)

and the inner iteration be controlled such that

$$\delta_k \leq \frac{\rho h_k^\delta}{h_k^\delta + \sqrt{4+(h_k^\delta)^2}},$$ (2.104)

which requires that

$$\delta_0 < \frac{\rho}{1+\sqrt{1+(1+\rho)^2}}.$$ (2.105)

Then the inexact Newton iterates x^k remain in $\mathcal{L}_0$ and converge quadratically to the minimum point $x^* \in \mathcal{L}_0$ such that

$$\|F'(x^{k+1})^{1/2}\Delta x^{k+1}\| \leq (1+\rho)\frac{\omega}{2}\|F'(x^k)^{1/2}\Delta x\|^2$$ (2.106)

and

$$\|F'(x^{k+1})^{1/2}\delta x^{k+1}\| \leq (1+\rho)\frac{\omega}{2}\|F'(x^k)^{1/2}\delta x\|^2.$$ (2.107)

III. Functional descent. *The convergence in terms of the functional can be estimated by*

$$-\tfrac{1}{6}h_k^\delta \epsilon_k^\delta \leq f(x^k) - f(x^{k+1}) - \tfrac{1}{2}\epsilon_k^\delta \leq \tfrac{1}{6}h_k^\delta \epsilon_k^\delta.$$ (2.108)

Proof. For the purpose of repeated induction, let $\mathcal{L}_k$ denote the level set defined in analogy to $\mathcal{L}_0$. First, in order to show that $x^{k+1} \in \mathcal{L}_k$, we start from the identity

$$f(x^k + \lambda\delta x^k) - f(x^k) + (\lambda - \tfrac{1}{2}\lambda^2)\,\epsilon_k^\delta$$

$$= \int_{s=0}^{\lambda} s \int_{t=0}^{\lambda} \langle \delta x^k, (F'(x^k + st\delta x^k) - F'(x^k))\delta x^k\rangle \, dt\, ds + \langle \delta x^k, r^k\rangle.$$

The second right hand term vanishes due to (2.98). The energy product in the first term can be bounded as

$$\langle \delta x^k, \ldots \rangle \leq \|F'(x^k)^{1/2}\delta x^k\| \; \omega st \|F'(x^k)^{1/2}\delta x^k\|^2 = sth_k^\delta \epsilon_k^\delta .$$

For the purpose of repeated induction, let $h_k < 2$ and $\epsilon_k \neq 0$, which then implies that

$$f(x^k + \lambda \delta x^k) \leq f(x^k) + \left(\tfrac{1}{3}\lambda^3 + \tfrac{1}{2}\lambda^2 - \lambda\right) \epsilon_k^\delta < f(x^k) \text{ for } \lambda \in \;]0,1] .$$

Therefore, the assumption $x^k + \delta x^k \notin \mathcal{L}_k$ would lead to a contradiction for some $\lambda \in \;]0,1]$.

For $\lambda = 1$, we get the left hand side of (2.108). Applying the Cauchy-Schwarz inequality in the other direction also yields the right hand side.

In order to monitor the behavior of the Kantorovich type quantities h_k, we estimate the local energy norms as

$$\|F'(x^{k+1})^{1/2}\Delta x^{k+1}\|$$
$$\leq \left\| F'(x^{k+1})^{-1/2} \left(\int\limits_{t=0}^{1} (F'(x^k + t\delta x^k) - F'(x^k))\delta x^k dt + r^k \right) \right\|$$
$$\leq \tfrac{1}{2}\omega \|F'(x^k)^{1/2}\delta x^k\|^2 + \|F'(x^{k+1})^{-1/2}r^k\| .$$

With $z = \delta x^k - \Delta x^k$, the second right hand term can be estimated implicitly by

$$\|F'(x^{k+1})^{-1/2}r^k\|^2 \leq \|F'(x^k)^{1/2}z\|^2 + h_k^\delta \|F'(x^k)^{1/2}z\| \; \|F'(x^{k+1})^{-1/2}r^k\| ,$$

which leads to the explicit bound

$$\|F'(x^{k+1})^{-1/2}r^k\| \leq \tfrac{1}{2} \left(h_k^\delta + \sqrt{4 + \left(h_k^\delta\right)^2} \right) \|F'(x^k)^{1/2}z\| .$$

Summarizing, we obtain the contraction factor bound

$$\Theta_k := \frac{\|F'(x^{k+1})^{1/2}\Delta x^{k+1}\|}{\|F'(x^k)^{1/2}\Delta x^k\|} \leq \vartheta(h_k^\delta, \delta_k) . \tag{2.109}$$

Herein *linear* convergence shows up via (2.100) and (2.102). The result (2.101) is obtained with $h_k = 0$. Obviously, $h_k < 2\overline{\Theta}$ is necessary to obtain $\Theta_k \leq \overline{\Theta}$ for some $\overline{\Theta} < 1$. As for the contraction of the inexact corrections, we apply $\delta_{k+1} \geq \delta_k$ and (1.26) to show that

$$\frac{\|F'(x^{k+1})^{1/2}\delta x^{k+1}\|}{\|F'(x^k)^{1/2}\delta x^k\|} = \sqrt{\frac{1 + \delta_k^2}{1 + \delta_{k+1}^2}} \Theta_k \leq \Theta_k \leq \overline{\Theta} .$$

Hence, we may complete the induction and conclude that the iterates x^k converge to x^*.

As for *quadratic* convergence, we impose condition (2.104) within (2.109) to obtain

$$\frac{\|F(x^{k+1})^{1/2}\Delta x^{k+1}\|}{\|F'(x^k)^{1/2}\Delta x^k\|} \leq \frac{1}{2\sqrt{1+\delta_k^2}}\left(h_k^\delta + \delta_k(h_k^\delta + \sqrt{4+(h_k^\delta)^2})\right)$$

$$\leq \tfrac{1}{2}(1+\rho)h_k^\delta ,$$

which, for $h_k^\delta \leq h_k \leq h_0$ assures the convergence relations (2.106) under the assumption (2.99). Upon inserting (2.103) into (2.104) we immediately verify (2.105). For the inexact corrections, we have equivalently

$$\frac{\|F(x^{k+1})^{1/2}\delta x^{k+1}\|}{\|F'(x^k)^{1/2}\delta x^k\|} \leq \frac{1}{2\sqrt{1+\delta_{k+1}^2}}\left(h_k^\delta + \delta_k(h_k^\delta + \sqrt{4+(h_k^\delta)^2})\right)$$

$$\leq \tfrac{1}{2}(1+\rho)h_k^\delta < 1 ,$$

which then assures the convergence relations (2.107). This finally completes the proof. □

Convergence monitor. Assume now that we have a reasonable (and cheap) estimate of the relative energy norm errors δ_k available from the inner PCG iteration. A new iterate x^{k+1} might be accepted whenever either

$$f(x^{k+1}) - f(x^k) \leq -\tfrac{1}{6}\epsilon_k = -\tfrac{1}{6}(1+\delta_k^2)\epsilon_k^\delta .$$

or, as a slight generalization of the situation of Theorem 2.20, the *inexact monotonicity criterion*

$$\Theta_k := \left(\frac{\epsilon_{k+1}}{\epsilon_k}\right)^{1/2} = \left(\frac{(1+\delta_{k+1}^2)\epsilon_{k+1}^\delta}{(1+\delta_k^2)\epsilon_k^\delta}\right)^{1/2} \leq \overline{\Theta}_k < 1$$

holds. We will regard the outer iteration as *divergent*, if none of the above criteria is met.

Termination criteria. We will terminate the iteration whenever

$$\epsilon_k = (1+\delta_k^2)\epsilon_k^\delta \leq \text{ETOL}^2 \quad \text{or} \quad f(x^k) - f(x^{k+1}) \leq \tfrac{1}{2}\text{ETOL}^2 . \qquad (2.110)$$

Standard convergence mode. If we just impose the inner iteration termination criterion $\delta_k \leq \bar{\delta}$ for some fixed default value $\bar{\delta}$, we obtain *asymptotic linear convergence*. If we set $\overline{\Theta} = \tfrac{1}{2}$, then (2.101) induces $\bar{\delta} < \sqrt{3}/3$. As in the other two cases, we recommend $\bar{\delta} = 1/4$ to assure at least two binary digits.

Quadratic convergence mode. Assume that $[h_0] < 2/(1 + \rho)$ for $\rho = 1$. Let δ_0 be given, say $\delta_0 = 1/4$ in agreement with (2.105). As for the adaptive termination of the inner iteration within the inexact local Newton method, we want to satisfy condition (2.104). Following our general paradigm, we will replace the unavailable upper bound therein by the computationally available condition in terms of computational estimates $[h_k]$ such that

$$\delta_k \leq \frac{\rho \, [h_k^\delta]}{[h_k^\delta] + \sqrt{4 + [h_k^\delta]^2}} . \tag{2.111}$$

Since the above right hand side is a monotone *increasing* function of $[h_k]$, the relation $[h_k] \leq h_k$ implies that the theoretical condition (2.104) is actually *assured* whenever (2.111) holds. Following our basic paradigm (compare Section 1.2), we apply (2.108) and define the computational *a-posteriori* estimates

$$[h_k^\delta]_2 = \frac{6}{\epsilon_k^\delta} |f(x^{k+1}) - f(x^k) + \tfrac{1}{2} \epsilon_k^\delta| , \quad [h_k]_2 = \sqrt{1 + \delta_k^2} [h_k^\delta]_2 .$$

From this, shifting the index $k + 1$ back to k, we may define the *a priori* estimate

$$[h_k] = \Theta_{k-1} [h_{k-1}]_2, \tag{2.112}$$

which we insert into (2.111) to obtain a simple implicit scalar equation for δ_k.

Note that $\delta_k \to 0$ is forced when $k \to \infty$. In words: *the closer the iterates come to the solution point, the more work needs to be done in the inner iteration to assure quadratic convergence of the outer iteration.*

Linear convergence mode. Once the local contraction factor Θ_k is sufficiently below some prescribed value $\overline{\Theta}$, we may switch to the linear convergence mode described by the above Theorem 2.20. As for the termination of the inner iteration, we would like to assure condition (2.100), briefly recalled as

$$\vartheta(h_k^\delta, \delta_k) \leq \overline{\Theta} .$$

Since the above quantity ϑ is unavailable, we will replace it by the computationally available estimate

$$[\vartheta(h_k^\delta, \delta_k)] := \vartheta([h_k^\delta], \delta_k) \leq \vartheta(h_k^\delta, \delta_k) .$$

For $k > 0$, we may again insert the a-priori estimate (2.112) above. In any case, we will run the inner iteration until the actual δ_k satisfies either condition (2.100) for the linear convergence mode or condition (2.111) for the quadratic convergence mode. Whenever $\Theta_k \geq 1$ occurs, then we switch to some global variant of this local inexact Newton method—see Section 3.4.3.

Note that asymptotically

$$\delta_k \to \overline{\Theta}/\sqrt{1 - \overline{\Theta}^2} \quad \text{as} \quad k \to \infty. \tag{2.113}$$

In other words: *the closer the iterates come to the solution point, the less work is necessary within the inner iteration to assure linear convergence of the outer iteration.*

The here described local inexact Newton algorithm for convex optimization is part of the global inexact Newton code GIANT-PCG worked out in detail in Section 3.4.3 below.

BIBLIOGRAPHICAL NOTE. The presentation in this chapter is a finite dimensional restriction of the affine conjugate convergence theory and the corresponding algorithmic concepts given by P. Deuflhard and M. Weiser [84] for nonlinear elliptic PDEs. Our here developed inexact Newton-PCG algorithm may be regarded as a competitor to nonlinear CG methods—both to the variant [93] due to R. Fletcher and C.M. Reeves and to the one due to E. Polak and R. Ribière [169, Section 2.3]. For the application of nonlinear CG to discrete partial differential equations see, e.g., the lecture notes [102] by R. Glowinski; from this perspective, our Newton-PCG method may be viewed as a nonlinear CG variant with Jacobian savings in a firm theoretical frame.

Exercises

Exercise 2.1   Derive the computational complexity bounds (2.71) in terms of number of iterations from Theorem 2.12.

Exercise 2.2   Let $M(x)$ denote a perturbed Jacobian matrix of the form $M(x^k) = F'(x^k) + \delta M(x^k)$. Derive a convergence theorem for a Newton-like method based on Theorem 2.10.

Exercise 2.3   As an illustration of the not affine covariant classical Newton-Mysovskikh theorem take $X = Y = \mathbb{R}^2$ and define

$$F(x) := \begin{pmatrix} x_1 - x_2 \\ (x_1 - 8)x_2 \end{pmatrix}.$$

Verify that here $h_F = \alpha_F \beta_F \gamma_F < 2$. The simple affine transformation

$$F \to G := \begin{pmatrix} 1 & 1 \\ 0 & \frac{1}{2} \end{pmatrix} F$$

induces the associated quantities α_G, β_G, γ_G, h_G. Once more, give best possible bounds and verify that now $h_G > 2$! Finally, prove that the affine invariant characterization from Theorem 2.2 yields $h_0 = \alpha\omega \ll 2$. Interpretation?

Hint: One obtains $h_F = 0.762$, $h_G = 2.159$, $h_0 = 0.127$.

Exercise 2.4   *Theorem of H.B. Keller.* Let $F : D \to \mathbb{R}^n$ be a continuously differentiable mapping with $D \subset \mathbb{R}^n$ convex. Suppose that $F'(x)$ is invertible for each $x \in D$ and satisfies the affine invariant Hölder continuity

$$\left\| F'(z)^{-1}\big(F'(y) - F'(x)\big)\right\| \leq \omega \|y - x\|^\gamma,$$

where $0 < \gamma \leq 1$.

a) Prove a variant of the affine covariant Newton-Mysovskikh theorem (Theorem 2.2).

b) Prove a variant of the affine covariant Newton-Kantorovich theorem (Theorem 2.1).

Exercise 2.5   *Theorem of L.B. Rall (improved by W.C. Rheinboldt).* Let $F : D \subseteq \mathbb{R}^n \to \mathbb{R}^n$, D open convex. Assume that there exists a unique solution $x^* \in D$ and that $F'(x^*)$ is invertible. Let

$$\left\| F'(x^*)^{-1}\big(F'(y) - F'(x)\big)\right\| \leq \omega_* \|y - x\| \text{ for } x, \, y \in D$$

denote a special affine covariant Lipschitz condition. Let

$$S(x^*, \rho) := \{x \in X | \ \|x - x^*\| < \rho\} \subset D .$$

By introduction of the majorants

$$\omega_* \left\| x^k - x^* \right\| \leq t_k$$

prove that for any starting point $x^0 \in S(x^*, \rho)$ with $\rho := \dfrac{2}{3\omega_*}$, the ordinary Newton iteration remains in S and converges to x^*. Give a convergence rate estimate.

Exercise 2.6   For convex optimization there are three popular symmetric Jacobian rank-2 updates

- *Broyden-Fletcher-Goldfarb-Shanno* (BFGS):

$$J_{k+1} = J_k - \frac{F_k F_k^T}{\delta x_k^T J_k \delta x_k} + \frac{(F_{k+1} - F_k)(F_{k+1} - F_k)^T}{(F_{k+1} - F_k)^T \delta x_k},$$

- *Davidon-Fletcher-Powell* (DFP):

$$J_{k+1} = J_k + \frac{F_{k+1}(F_{k+1} - F_k)^T + (F_{k+1} - F_k)F_{k+1}^T}{(F_{k+1} - F_k)^T \delta x_k} -$$

$$- \frac{F_{k+1}^T \delta x_k}{((F_{k+1} - F_k)^T \delta x_k)^2}(F_{k+1} - F_k)(F_{k+1} - F_k)^T,$$

• *Powell's symmetric Broyden* (PSB):

$$J_{k+1} = J_k + \frac{F_{k+1}\delta x_k^T + \delta x_k F_{k+1}^T}{\delta x_k^T \delta x_k} - \frac{F_{k+1}^T \delta x_k}{(\delta x_k^T \delta x_k)^2}\delta x_k \delta x_k^T.$$

a) Show that all updates satisfy the classical secant condition.

b) Which of these updates are defined in an affine conjugate way? For not affine conjugate updates: design an appropriate scaling so that at least scaling invariance is achieved.

c) Which of these updates can be interpreted as a least change secant update? Derive the associated error concept.

Exercise 2.7 *Rank-2 update formulas for convex optimization.* We consider several update formulas for convex optimization. Common basis for all these updates is the classical secant condition

$$J\delta x_k = F(x^k + \delta x_k) - F(x_k) = F_{k+1} - F_k = \delta F_{k+1}.$$

a) Show that u and v in the general symmetric positive definite update formula

$$J = (I - uv^T)\, J_k (I - vu^T)$$

cannot be specified such that both the secant condition is satisfied and the update is of full rank 2.

b) Verify that this can be achieved by the comparable representation, the DFP update:

$$J_{k+1} = \left(I - \frac{\delta F_{k+1}\delta x_k^T}{(\delta F_{k+1}^T \delta x_k)}\right) J_k \left(I - \frac{\delta x_k \delta F_{k+1}^T}{(\delta F_{k+1}^T \delta x_k)}\right) + \frac{\delta F_{k+1}\delta F_{k+1}^T}{(\delta F_{k+1}^T \delta x_k)}.$$

c) Verify that this can be also achieved by the inverse representation, the BFGS update:

$$J_{k+1}^{-1} = \left(I - \frac{\delta x_k \delta F_{k+1}^T}{(\delta F_{k+1}^T \delta x_k)}\right) J_k^{-1} \left(I - \frac{\delta F_{k+1}\delta x_k^T}{(\delta F_{k+1}^T \delta x_k)}\right) + \frac{\delta x \delta x_k^T}{\delta F_{k+1}^T \delta x_k}.$$

Exercise 2.8 Recall the notation for quasi-Newton methods as given in Section 2.1.4. With the majorant definitions

$$\frac{\|\overline{\Delta x}_{k+1}\|}{\|\Delta x_k\|} \leq \Theta_k < \tfrac{1}{2}, \ \|\Delta x_k\| \leq e_k,$$

$$\left\| J_k^{-1}\left[F'(x^k) - J_k\right] \right\| \leq \delta_k,$$

$$\left\| J_k^{-1}\left[F'(u) - F'(v)\right] \right\| \leq \omega_k \|u - v\|,$$

verify the following set of recursions:

$$
\begin{aligned}
\delta_{k+1} &= \left[\delta_k + \Theta_k + \omega_k e_k\right] / \left(1 - \Theta_k\right), \\
e_{k+1} &= \frac{\Theta_k}{1 - \Theta_k} e_k, \\
\omega_{k+1} &= \frac{\omega_k}{1 - \Theta_k}, \\
\Theta_{k+1} &= \delta_{k+1} + \tfrac{1}{2}\omega_{k+1} e_{k+1}.
\end{aligned}
$$

Under the additional assumption of 'bounded deterioration' in the form

$$
\delta_k \le \delta
$$

derive a Kantorovich-type local convergence theorem. Why is such a theorem unsatisfactory?

Exercise 2.9   Consider a residual based inexact Newton method, where the inner iteration is done by some *residual norm reducing*, but not *minimizing*, iterative solver—like the 'bad' Broyden algorithm BB for *linear* systems as described in [74]. Then the contraction results (2.86), which hold for the residual minimizer GMRES, must be replaced.

a) Show the alternative contraction result

$$
\Theta_k \le \eta_k + \tfrac{1}{2}(1 + \eta_k)^2 h_k.
$$

b) For the Kantorovich quantities h_k, find cheap and reliable a-posteriori and a-priori computational estimates $[h_k] \le h_k$.

c) Design accuracy matching strategies (standard, linear, and quadratic convergence mode) similar to those worked out for GMRES in Section 2.2.4.

Exercise 2.10   Consider two Newton sequences $\{x^k\}$, $\{y^k\}$ starting at different initial guesses x^0, y^0 and continuing as

$$
x^{k+1} = x^k + \Delta x^k, \qquad y^{k+1} = y^k + \Delta y^k,
$$

where $\Delta x^k, \Delta y^k$ are the corresponding ordinary Newton corrections. Upon using the affine covariant Lipschitz condition

$$
\|F'(u)^{-1}\left(F'(v) - F'(w)\right) u\| \le \omega \|v - w\| \|u\|
$$

verify the nonlinear perturbation result

$$
\|x^{k+1} - y^{k+1}\| \le \omega \left(\tfrac{1}{2}\|x^k - y^k\| + \|\Delta x^k\|\right) \|x^k - y^k\|.
$$

Is the result invariant under $x \leftrightarrow y$?

3 Systems of Equations: Global Newton Methods

As in the preceding chapter, the discussion here is also restricted to systems of n nonlinear equations, say

$$F(x) = 0,$$

where $F \in C^1(D)$, $D \subseteq \mathbb{R}^n$, $F : D \longrightarrow \mathbb{R}^n$ with Jacobian (n, n)-matrix $F'(x)$. In contrast to the preceding chapter, however, available initial guesses x^0 of the solution point x^* are no longer assumed to be 'sufficiently close' to x^*.

In order to specify the colloquial term 'sufficiently close', we recur to any of the local convergence conditions of the preceding chapter. Let ω denote an affine covariant Lipschitz constant. Then Theorem 2.3 presents an appropriate local convergence condition of the form

$$\|x^* - x^0\| < 2/\omega.$$

In the error oriented framework, Theorem 2.2 yields a characterization in terms of the Kantorovich quantity

$$h_0 := \|\Delta x^0\|\omega < 2,$$

which restricts the ordinary Newton correction Δx^0. Under any of these conditions local Newton methods are guaranteed to converge. Such problems are sometimes called *mildly* nonlinear. Their computational complexity is *a priori bounded* in terms of the computational complexity of solving linear problems of the same structure—see, for example, the bound (2.71).

In contrast to that, under a condition of the type $h_0 \gg 1$, which is equivalent to

$$\|\Delta x^0\| \gg 2/\omega \tag{3.1}$$

local Newton methods will not exhibit guaranteed convergence. In this situation, the computational complexity cannot be bounded a priori. Such problems are often called *highly* nonlinear. Nevertheless, local Newton methods may actually converge for some of these problems even in the situation of condition (3.1). A *guaranteed* convergence, however, will only occur, if additional *global structure* on F can be exploited: as an example, we treat *convex* nonlinear mappings in Section 3.1.1 below.

For general mapping F, a *globalization* of local Newton methods must be constructed. In Section 3.1 we survey globalization concepts such as

- steepest descent methods,
- trust region methods,
- the Levenberg-Marquardt method, and
- the Newton method with damping strategy.

In Section 3.1.4, a rather general geometric approach is taken: the idea is to derive a globalization concept without pre-occupation to any of the known iterative methods, just starting from the requirement of affine covariance as a 'first principle'. Surprisingly, this general approach leads to the derivation of Newton's method with damping strategy.

Monotonicity tests. Monotonicity tests serve the purpose to accept or reject a new iterate. We study different such tests, according to different affine invariance requirements:

- the most popular *residual* monotonicity test, which is based on affine contravariance (Section 3.2),
- the error oriented so-called *natural* monotonicity test, which is based on affine covariance (Section 3.3), and
- the convex functional test as the natural requirement in convex optimization, which reflects affine conjugacy (Section 3.4).

For each of these three affine invariance classes, *adaptive trust region strategies* are designed in view of an efficient choice of damping factors in Newton's method. They are all based on the *paradigm* already mentioned at the end of Section 1.2. On a theoretical basis, details of algorithmic realization in combination with either *direct* or *iterative* linear solvers are worked out. As it turns out, an efficient determination of the steplength factor in global Newton methods is intimately linked with the accuracy matching for affine invariant combinations of inner and outer iteration within various *inexact* Newton methods.

3.1 Globalization Concepts

Efficient iterative methods should be able to cope with 'bad' guesses x^0. In this section we survey methods that permit rather general initial guesses x^0, not only those sufficiently close to the solution point x^*. Of course, such methods should merge into local Newton techniques as soon as the iterates x^k come 'close to' the solution point x^*—to exploit the local *quadratic* or *superlinear* convergence property.

Parameter continuation methods. The simplest way of globalization of local Newton methods is to embed the given problem $F(x) = 0$ into a one-parameter family of problems, a so-called *homotopy*

$$F(x, \tau) = 0 , \quad \tau \in [0, 1] ,$$

such that the starting point x^0 is the solution for $\tau = 0$ and the desired solution point x^* is the solution point for $\tau = 1$. If we choose sufficiently many intermediate problems in the *discrete homotopy*

$$F(x, \tau_\nu) = 0 , \quad 0 = \tau_0 < \cdots < \tau_\nu < \cdots < \tau_N = 1 ,$$

then the solution point of one problem can serve as initial guess in a local Newton method for the next problem. In this way, global convergence can be assured under the assumption that existence and uniqueness of the solution along the *homotopy path* is guaranteed. In this context, questions like the adaptive choice of the stepsizes $\Delta\tau_\nu$ or the computation of *bifurcation diagrams* are of interest. An efficient choice of the embedding will exploit specific features of the given problem to be solved—with consequences for the local uniqueness of the solution along the homotopy path and for the computational speed of the discrete continuation process. The discussion of these and many related topics is postponed to Chapter 5.

Pseudo-transient continuation methods. Another continuation method uses the embedding of the algebraic equation into an *initial value problem* of the type

$$x' = F(x) , \quad x(0) = x^0 .$$

Discretization of this problem with respect to a timestep τ by the *explicit Euler method* leads to the fixed point iteration

$$x^{k+1} - x^k = \Delta x^k = \tau F(x^k)$$

or, by the *linearly implicit Euler method* to the iteration scheme

$$\left(I - \tau F'(x^k) \right) \Delta x^k = \tau F(x^k) .$$

Note that for $\tau \to \infty$ the latter scheme merges into the ordinary Newton method. The scheme reflects *affine similarity* as described in Section 1.2 and will be treated in detail in Section 6.4 in the context of so-called pseudo-transient continuation methods, which are a special realization of stiff integrators for ordinary differential equations.

3.1.1 Componentwise convex mappings

The Newton-Raphson method for scalar equations (see Figure 1.1) may be geometrically interpreted as taking the intersection of the local tangent with the axis—and repeating this process until sufficient accuracy is achieved. In this interpretation, the simplified Newton method just means to keep the initial tangent throughout the whole iterative process. From this it can be directly seen that both the ordinary and the simplified Newton method converge *globally* for *convex* (or concave) scalar functions. The convergence is *monotone*, i.e., the iterates x^k approach the solution point x^* from one side only. On the basis of this insight we are now interested in a generalization of such a monotonicity property to systems. General convex minimization problems, which lead to gradient mappings F, will be treated in the subsequent Section 3.4. Here we concentrate on some componentwise convexity as discussed in the textbook of J.M. Ortega and W.C. Rheinboldt [163].

Such componentwise convex mappings F may be characterized by one of the following equivalent properties (let $x, y \in D \subseteq \mathbb{R}^n$, D convex, $\lambda \in [0, 1]$):

$$F\big(\lambda x + (1 - \lambda)y\big) \;\leq\; \lambda F(x) + (1 - \lambda)F(y)\,, \qquad (3.2)$$

$$F(y) - F(x) \;\geq\; F'(x)(y - x)\,,$$

$$\big(F'(y) - F'(x)\big)(y - x) \;\geq\; 0\,. \qquad (3.3)$$

Herein, the inequalities are understood componentwise. Since the objects of interest will be the iterates, we miss *affine covariance* in the above formulation. In fact, Ortega and Rheinboldt show monotone convergence of the ordinary Newton method under the *additional* assumption

$$F'(z)^{-1} \geq 0\,, \; z \in D\,, \qquad (3.4)$$

which is essentially a global *M-matrix property* (cf. R.S. Varga [192]) for the Jacobian. Upon combining the above three equivalent convexity characterizations with (3.4), we obtain the three equivalent affine covariant formulations

$$F'(z)^{-1}F\big(\lambda x + (1 - \lambda)y\big) \leq F'(z)^{-1}\big(\lambda F(x) + (1 - \lambda)F(y)\big) \qquad (3.5)$$

$$F'(x)^{-1}\big(F(y) - F(x)\big) \;\geq\; y - x \qquad (3.6)$$

$$F'(z)^{-1}\big(F'(y) - F'(x)\big)(y - x) \;\geq\; 0\,. \qquad (3.7)$$

Note that these conditions cover any mapping F such that (3.2) up to (3.3) together with (3.4) hold for AF and AF' with some $A \in \mathrm{GL}(n)$.

Lemma 3.1 *Let* $F : D \longrightarrow \mathbb{R}^n$ *be a continuously differentiable mapping with* $D \subseteq \mathbb{R}^n$ *open and convex. Let this mapping satisfy one of the convexity characterizations* (3.5)-(3.7). *Then the ordinary Newton iteration starting at some* $x^0 \in D$ *converges monotonically and globally such that componentwise*

$$x^* \leq x^{k+1} \leq x^k\,, \quad k = 1, 2, \ldots\,. \qquad (3.8)$$

Proof. For the *ordinary* Newton iteration, one obtains:

$$x^{k+1} - x^k = -F'(x^k)^{-1}\big(F(x^k) - F(x^{k-1}) - F'(x^{k-1})(x^k - x^{k-1})\big)$$

$$= -F'(x^k)^{-1} \int_{\delta=0}^{1} \big[F'\big(x^{k-1} + \delta(x^k - x^{k-1})\big) - F'(x^{k-1})\big] (x^k - x^{k-1})d\delta \,.$$

Insertion of (3.7) for

$$z = x^k, x = x^{k-1}, y = x^{k-1} + \delta(x^k - x^{k-1}), x^k - x^{k-1} = (y - x)/\delta$$

leads to

$$x^{k+1} - x^k \leq 0 \text{ for } k \geq 1 \,.$$

In a similar way, one derives

$$x^{k+1} - x^* = (x^{k+1} - x^k) + (x^k - x^*) = F'(x^k)^{-1}\big(F(x^*) - F(x^k)\big) + x^k - x^* \,,$$

which, by application of (3.6), supplies

$$x^{k+1} - x^* \geq 0 \,, \ k \geq 0 \,.$$

The rest of the proof can be found in [163], p. 453.   □

Remark 3.1   An immediate generalization of this lemma is obtained by allowing *different* inequalities for different components in (3.5) to (3.7)—which directly leads to the corresponding inequalities in (3.8).

Note that the above results do *not* apply to the *simplified* Newton iteration, unless $n = 1$: following the lines of the above proof, the application of (3.7) here would lead to

$$x^{k+1} - x^k \leq -F'(x^0)^{-1}\big(F'(x^{k-1}) - F'(x^0)\big)(x^k - x^{k-1}) \,.$$

In order to apply (3.7) once more, a relation of the kind

$$x^{k-1} - x^0 = \Theta \cdot (x^k - x^{k-1})$$

for some $\Theta > 0$ would be required—which will only hold in $\mathbb{R}^1$.

In actual computation, the global monotone convergence property does not require any control in terms of some monotonicity test. Only reasonable componentwise termination criteria need to be implemented. It is worth mentioning that this special type of convergence of the ordinary Newton method does *not* mean global *quadratic* convergence: rather this type of convergence may be arbitrarily slow, as can be verified in simple scalar problems—see Exercise 1.3. Not even an a-priori estimation for the number of iterations needed to achieve a prescribed accuracy may be possible.

BIBLIOGRAPHICAL NOTE. The above componentwise monotonicity results are discussed in detail in the classical monograph [163] by J.M. Ortega and W.C. Rheinboldt, there in not affine invariant form. In 1987, F.A. Potra and W.C. Rheinboldt proved affine invariant conditions, under which the simplified Newton method and other Newton-like methods converge, see [172, 170].

3.1.2 Steepest descent methods

A *desirable requirement* for any iterative methods would be that the iterates x^k successively *approach* the solution point x^*—which may be written as

$$\|x^{k+1} - x^*\| < \|x^k - x^*\|, \quad \text{if } x^k \neq x^* .$$

Local Newton techniques implicitly realize such a criterion under affine covariant theoretical assumptions, as has been shown in detail in the preceding chapter. *Global* methods, however, require a *substitute approach criterion*, which may be based on the *residual level function*

$$T(x) := \tfrac{1}{2} \| F(x) \|_2^2 \equiv \tfrac{1}{2} F(x)^T F(x) . \tag{3.9}$$

Such a function has the property

$$
\begin{aligned}
T(x) &= 0 \Longleftrightarrow x = x^* , \\
T(x) &> 0 \Longleftrightarrow x \neq x^* .
\end{aligned}
\tag{3.10}
$$

In terms of this level function, the approach criterion may be formulated as a *monotonicity criterion*

$$T(x^{k+1}) < T(x^k), \quad \text{if } T(x^k) \neq 0 .$$

Associated with the level function are the so-called *level sets*

$$G(z) := \{ x \in D \subseteq \mathbb{R}^n | \, T(x) \leq T(z) \} .$$

Let $\overset{\circ}{G}$ denote the interior of G. Then property (3.10) implies

$$x^* \in G(x), \, x \in D$$

and the monotonicity criterion may be written in geometric terms as

$$x^{k+1} \in \overset{\circ}{G}(x^k), \quad \text{if } \overset{\circ}{G}(x^k) \neq \emptyset .$$

An intuitive approach based on this geometrical insight, which dates back even to A. Cauchy [44] in 1847, is to choose the *steepest descent* direction as the direction of the iterative correction—see Figure 3.1.

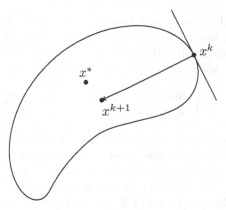

Fig. 3.1. Geometric interpretation: level set and steepest descent direction.

This idea leads to the following iterative method:

$$\Delta x^k \quad := \quad -\operatorname{grad} T(x^k) = -F'(x^k)^T F(x^k) \tag{3.11}$$

$$x^{k+1} \quad := \quad x^k + s_k \Delta x^k \tag{3.12}$$

$$s_k \quad > \quad 0 : \text{steplength parameter.}$$

Figure 3.1 also nicely shows the so-called *downhill property*.

Lemma 3.2 *Let $F : D \longrightarrow \mathbb{R}^n$ be a continuously differentiable mapping with $D \subseteq \mathbb{R}^n$. Dropping the iterative index k in the notation of (3.11), let $\Delta x \neq 0$. Then there exists a $\mu > 0$ such that*

$$T(x + s\Delta x) < T(x), \ 0 < s < \mu. \tag{3.13}$$

Proof. Define $\varphi(s) := T(x + s\Delta x)$. As $F \in C^1(D)$, one has $\varphi \in C^1(D_1)$, $D_1 \subseteq \mathbb{R}^1$. Then

$$\varphi'(0) = \left(\operatorname{grad} T(x + s \cdot \Delta x)^T \Delta x\right)\big|_{s=0} = -\|\Delta x\|_2^2 < 0.$$

With $\varphi \in C^1$, the result (3.13) is established. □

Steplength strategy. This result is the theoretical basis for a strategy to select the steplength in method (3.12). It necessarily consists of two parts: a reduction strategy and a prediction strategy. The *reduction strategy* applies whenever

$$T(x^k + s_k^0 \Delta x^k) > T(x^k)$$

for some given parameter s_k^0. In this case, the above monotonicity test is repeated with some

$$s_k^{i+1} \quad := \quad \kappa \cdot s_k^i, \quad i = 0, 1, \ldots, \quad \kappa < 1 \quad (\text{typically } \kappa = 1/2).$$

Lemma 3.2 assures that a *finite* number i^* of reductions will ultimately lead to a feasible steplength factor $s_k^* > 0$. The *prediction strategy* applies, when s_{k+1}^0 must be selected—usually based on an ad-hoc rule that takes the steplength history into account such as

$$s_{k+1}^0 := \begin{cases} \min\left(s_{\max}, s_k^*/\kappa\right), & \text{if } s_{k-1}^* \leq s_k^* \\ s_k^* & \text{else} \end{cases} \tag{3.14}$$

The possible increase from s_k to s_{k+1} helps to avoid inefficiency coming from 'too small' local corrections. One may also aim at implementing an optimal choice of s_k out of a sequence of sample values—a strategy, which is often called *optimal line search*. However, since s_k may range from 0 to ∞, a reasonable set of values to be sampled may be hard to define, if a sufficiently large class of problems is to be considered.

Convergence properties. An elementary convergence analysis shows that the iterative scheme (3.12) with steplength strategies like (3.14) converges *linearly* even for rather bad initial guesses x^0—however, possibly arbitrarily slow. Moreover, so-called 'pseudo-convergence' characterized by

$$\|F'(x)^T F(x)\| \quad \text{'small'}$$

may occur far from the solution point due to local ill-conditioning of the Jacobian matrix.

General level functions. In a large class of problems, the described difficulties are a consequence of the fact that the whole scheme is not affine covariant so that the choice of $T(x)$ as a level function appears to be rather arbitrary. In principle, *any* level function

$$T(x|A) := \tfrac{1}{2} \left\| AF(x) \right\|_2^2 \tag{3.15}$$

with arbitrary nonsingular (n, n)-matrix A could be used in the place of $T(x)$ above. To make things worse, even though the direction of steepest descent Δx is 'downhill' with respect to $T(x)$, there nearly always exists a matrix A such that Δx is 'uphill' with respect to $T(x|A)$, as will be shown in the following lemma.

Lemma 3.3 *Let $\Delta x = -\operatorname{grad} T(x)$ denote the direction of steepest descent with respect to the level function $T(x)$ as defined in (3.9). Then, unless*

$$F'(x)\Delta x = \chi \cdot F(x), \quad \text{for some } \chi < 0, \tag{3.16}$$

there exists a class of nonsingular matrices A such that

$$T(x + s\Delta x|A) > T(x|A), \ 0 < s < \nu,$$

for some $\nu > 0$.

Proof. Let $F = F(x), J = F'(x), \overline{J} = JJ^T, \overline{A} = A^T A$. Then

$$\Delta x^T \operatorname{grad} T(x|A) = -F^T \overline{J} \,\overline{A} F .$$

Now, choose

$$\overline{A} := \overline{J} + \mu y y^T$$

with some $\mu > 0$ to be specified later and $y \in \mathbb{R}^n$ such that

$$F^T(\overline{J} + I)y = 0 , \quad \text{but} \quad F^T y \neq 0 .$$

Here the assumption (3.16) enters for any $\chi \in \mathbb{R}^1$. By definition, however, the choice $\chi \geq 0$ is impossible, since (3.16) implies that

$$\chi = -\frac{\|J^T F\|^2}{\|F\|^2} < 0 .$$

Hence, for the above choice of $\overline{A}$, we obtain

$$\Delta x^T \operatorname{grad} T(x|A) = -\|\overline{J} F\|_2^2 + \mu (F^T y)^2 .$$

Then the specification

$$\mu > \|\overline{A} F\|_2^2 / (F^T y)^2$$

leads to

$$\Delta x^T \operatorname{grad} T(x|A) > 0 ,$$

which, in turn, implies the statement of the lemma. $\square$

Summarizing, even though the underlying geometrical idea of steepest descent methods is intriguing, the technical details of implementation cannot be handled in a theoretically satisfactory manner, let alone in an affine covariant setting.

3.1.3 Trust region concepts

As already shown in Section 1.1, the ordinary Newton method can be algebraically derived by linearization of the nonlinear equation around the solution point x^*. This kind of derivation supports the interpretation that the Newton correction is useful only in a close neighborhood of x^*. Far away from x^*, such a linearization might still be trusted in some 'trust region' around the current iterate x^k. In what follows we will present several models defining such a region. For a general survey of trust region methods in optimization see, e.g., the book [45] by A.R. Conn, N.I.M. Gould, and P.L. Toint.

Levenberg-Marquardt model. The above type of elementary considera-tion led K.A. Levenberg [143] and later D.W. Marquardt [147]) to suggest a modification of Newton's method for 'bad' initial guesses that merges into the ordinary Newton method close to the solution point. Following the pre-sentation by J.J. Moré in [152] we define a correction vector Δx (dropping the iteration index k) by the constrained quadratic minimization problem:

$$\|F(x) + F'(x)\Delta x\|_2 = \min$$

subject to the constraint

$$\|\Delta x\|_2 \leq \delta$$

in terms of some prescribed parameter $\delta > 0$, which may be understood to quantify the *trust region* in this approach.

The trust region constraint may be treated by the introduction of a *Lagrange multiplier* $p \geq 0$ subject to

$$p\left(\|\Delta x\|_2^2 - \delta^2\right) = 0,$$

which yields the equivalent unconstrained quadratic optimization problem

$$\|F(x^0) + F'(x^0)\Delta x\|_2^2 + p\|\Delta x\|_2^2 = \min .$$

After a short calculation and re-introduction of the iteration index k, we then end up with the *Levenberg-Marquardt method:*

$$\left(F'(x^k)^T F'(x^k) + pI\right)\Delta x^k = -F'(x^k)^T F(x^k), \ x^{k+1} := x^k + \Delta x^k . \quad (3.17)$$

The correction vector $\Delta x^k(p)$ has two interesting limiting cases:

$$p \to 0^+ \ : \ \Delta x^k(0) = -F'(x^k)^{-1}F(x^k), \ \text{if } F'(x^k) \text{ nonsingular}$$

$$p \to \infty \ : \ \Delta x^k(p) \to -\frac{1}{p}\,\text{grad}\,T(x^k) .$$

In other words: Close to the solution point, the method merges into the ordinary Newton method; far from the solution point, it turns into a steepest descent method with steplength parameter $1/p$.

Trust region strategies for the Levenberg-Marquardt method. All strategies to choose the parameter p or, equivalently, the parameter δ are based on the following simple lemma.

Lemma 3.4 *Under the usual assumptions of this section let $\Delta x(p) \neq 0$ de-note the Levenberg-Marquardt correction defined in (3.17). Then there exists a $p_{\min} \geq 0$ such that*

$$T\left(x + \Delta x(p)\right) < T(x), \ p > p_{\min} .$$

Proof. Substitute $q := 1/p$, $0 \leq q \leq \infty$, and define

$$\varphi(q) := T\big(x + \Delta x(1/q)\big), \quad \varphi(0) = T(x).$$

Then

$$\varphi'(0) = 0, \quad \varphi''(0) < 0.$$

Hence, there exists a $q_{\max} = 1/p_{\min}$ such that

$$\varphi(q) < \varphi(0), \quad 0 < q < q_{\max}.$$

$\square$

The method looks rather robust, since for any $p > 0$ the matrix $J^T J + pI$ is nonsingular, even when the Jacobian J itself is singular. Nevertheless, similar as the steepest descent method, the above iteration may also terminate at 'small' gradients, since for singular J the right-hand side of (3.17) also degenerates. This latter property is often overlooked both in the literature and by users of the method. Since the Levenberg-Marquardt method lacks affine invariance, special scaling methods are often recommended.

BIBLIOGRAPHICAL NOTE. Empirical trust region strategies for the Levenberg-Marquardt method have been worked out, e.g., by M.D. Hebden [118], by J.J. Moré [152], or by J.E. Dennis, D.M. Gay, and R. Welsch [54]. The associated codes are rather popular and included in several mathematical software libraries. However, as already stated above, these algorithms may terminate at a wrong solution with small gradient. When more than one solution exists locally, these algorithms might not indicate that. The latter feature is particularly undesirable in the application of the Levenberg-Marquardt method to nonlinear least squares problems—for details see Chapter 4 below.

Affine covariant trust region model. A straightforward affine covariant reformulation of the Levenberg-Marquardt model would be the following constrained quadratic optimization problem:

$$\|F'(x^0)^{-1}\big(F(x^0) + F'(x^0)\Delta x\big)\|_2 = \min$$

subject to the constraint

$$\|\Delta x\|_2 \leq \delta_0. \tag{3.18}$$

This problem can easily be solved geometrically—as shown in Figure 3.2, where the constraint (3.18) is represented by a sphere around the current iterate x^0 with radius δ_0. Whenever δ_0 exceeds the length of the ordinary Newton correction, which means that the constraint is *not active*, then Δx is just the ordinary Newton correction—and the quadratic functional vanishes. Whenever the constraint is *active*, then the direction of Δx still is the Newton direction, but now with reduced steplength. This leads to the Newton method with so-called *damping*

$$F'(x^k)\Delta x^k = -F(x^k)\,,\ x^{k+1} := x^k + \lambda_k \Delta x^k\,,$$

wherein the damping factor varies in the range $0 < \lambda_k \le 1$.

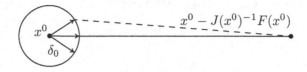

Fig. 3.2. Geometric interpretation: affine covariant trust region model.

Affine contravariant trust region model. An affine contravariant reformulation of the Levenberg-Marquardt model would lead to a constrained quadratic optimization problem of the form:

$$\|F(x^0) + F'(x^0)\Delta x\| = \min$$

subject to the constraint

$$\|F'(x^0)\Delta x\|_2 \le \delta_0\,.$$

Once again, the problem can be solved geometrically by Figure 3.2: only the terms of the domain space of F must be reinterpreted by the appropriate terms in the image space of F. As a consequence, the Newton method with damping is obtained again.

Damping strategies for Newton method. All strategies for choosing the above damping factors λ_k are based on the following insight.

Lemma 3.5 *Under the usual assumptions of this section let $F'(x)$ be nonsingular and $F \ne 0$. With Δx defined to be the Newton direction there exists some $\mu > 0$ such that*

$$T(x + \lambda\Delta x) < T(x)\,,\ 0 < \lambda < \mu\,.$$

Proof. As before, let $F = F(x), J = F'(x)$ and define $\varphi(\lambda) := T(x + \lambda\Delta x)$, which then yields

$$\varphi(0) = T(x)\,,\ \varphi'(0) = (J^T F)^T \Delta x = -F^T F = -2T(x) < 0\,.$$

$\square$

Among the most popular empirical damping strategies is the

Armijo strategy [7]. Let $\Lambda_k \subset \{1, \frac{1}{2}, \frac{1}{4}, \dots, \lambda_{\min}\}$ denote a sequence such that

$$T(x^k + \lambda_k \Delta x^k) \le \left(1 - \tfrac{1}{2}\lambda_k\right) T(x^k), \quad \lambda \in \Lambda_k \tag{3.19}$$

holds and define an optimal damping factor via

$$T(x^k + \lambda_k \Delta x^k) = \min_{\lambda \in \Lambda_k} T(x^k + \lambda \Delta x^k).$$

In order to avoid overflow in critical examples, the above evaluation of T will be sampled from the side of small values λ. In a neighborhood of x^*, this strategy will produce $\lambda = 1$. If $\lambda < \lambda_{\min}$ would be required, the iteration should be terminated with a warning. Unfortunately, the latter occurrence appears quite frequently in realistic problems of scientific computing, especially when the arising Jacobian matrices are ill-conditioned. This failure is a consequence of the fact that the choice $T(x)$ for the level function destroys the affine covariance of the local Newton methods—a consequence that will be analyzed in detail in Section 3.3 below.

3.1.4 Newton path

All globalization techniques described up to now were based on the requirement of local monotonicity with respect to the standard level function $T(x) = T(x|I)$ as defined in (3.9). In this section we will follow a more general approach, which covers general level functions $T(x|A)$ for *arbitrary* nonsingular matrix A as defined in (3.15). The associated *level sets* are written as

$$G(z|A) := \{x \in D \subseteq \mathbb{R}^n \mid T(x|A) \le T(z|A)\}. \tag{3.20}$$

With this notation, iterative monotonicity with respect to $T(x|A)$ can be written in the form

$$x^{k+1} \in \overset{\circ}{G}(x^k|A), \quad \text{if } \overset{\circ}{G}(x^k|A) \ne \emptyset.$$

We start from the observation that each choice of the matrix A could equally well serve within an iterative method. With the aim of getting rid of this somewhat arbitrary choice, we now focus on the intersection of all corresponding level sets:

$$\overline{G}(x) := \bigcap_{A \in \mathrm{GL}(n)} G(x|A). \tag{3.21}$$

By definition, the thus defined geometric object is affine covariant. Its nature will be revealed by the following theorem.

Theorem 3.6 *Let $F \in C^1(D)$, $D \subseteq \mathbb{R}^n$, $F'(x)$ nonsingular for all $x \in D$. For some $\widehat{A} \in \mathrm{GL}(n)$, let the path-connected component of $G(x^0|\widehat{A})$ in x^0 be compact and contained in D. Then the path-connected component of $\overline{G}(x^0)$*

as defined in (3.21) is a topological path $\bar{x} : [0, 2] \rightarrow \mathbb{R}^n$, the so-called Newton path, which satisfies

$$F(\bar{x}(\lambda)) = (1 - \lambda)F(x^0), \tag{3.22}$$

$$T(\bar{x}(\lambda)|A) = (1 - \lambda)^2 T(x^0|A), \tag{3.23}$$

$$\frac{d\bar{x}}{d\lambda} = -F'(\bar{x})^{-1}F(x^0), \tag{3.24}$$

$$\bar{x}(0) = x^0, \ \bar{x}(1) = x^*,$$

$$\left.\frac{d\bar{x}}{d\lambda}\right|_{\lambda=0} = -F'(x^0)^{-1}F(x^0) \equiv \Delta x^0, \tag{3.25}$$

where Δx^0 is the ordinary Newton correction.

Proof. Let $F_0 = F(x^0)$. In a *first stage of the proof*, level sets and their intersection are defined in the *image space* of F using the notation

$$H(x^0|A) := \{y \in \mathbb{R}^n \mid \|Ay\|_2^2 \leq \|AF_0\|_2^2\},$$

$$\overline{H}(x^0) := \bigcap_{A \in \mathrm{GL}(n)} H(x^0|A).$$

Let σ_i denote the singular values of A and q_i the eigenvectors of $A^T A$ such that

$$A^T A = \sum_{i=1}^n \sigma_i^2 q_i q_i^T.$$

Select those A with $q_1 := F_0/\|F_0\|_2$, which defines the matrix set:

$$\mathcal{A} := \{A \in \mathrm{GL}(n) | A^T A = \sum_{i=1}^n \sigma_i^2 q_i q_i^T, \ q_1 = F_0/\|F_0\|_2\}.$$

Then every $y \in \mathbb{R}^n$ can be represented by

$$y = \sum_{j=1}^n b_j q_j, \ b_j \in \mathbb{R}.$$

Hence

$$\|Ay\|_2^2 = y^T A^T A y = \sum_{i=1}^n \sigma_i^2 b_i^2,$$

$$\|AF_0\|_2^2 = \sigma_1^2 \|F_0\|_2^2,$$

which, for $A \in \mathcal{A}$, yields

$$H(x^0|A) = \left\{ y \mid \sum_{i=1}^{n} \sigma_i^2 b_i^2 \leq \sigma_1^2 \|F_0\|_2^2 \right\},$$

or, equivalently, the n-dimensional ellipsoids

$$\frac{1}{\|F_0\|_2^2} b_1^2 + \left(\frac{\sigma_2}{\sigma_1 \|F_0\|_2} \right)^2 b_2^2 + \cdots + \left(\frac{\sigma_n}{\sigma_1 \|F_0\|_2} \right)^2 b_n^2 \leq 1.$$

For $A \in \mathcal{A}$, all corresponding ellipsoids have a common b_1-axis of length $\|F_0\|_2$—see Figure 3.3. The other axes are arbitrary.

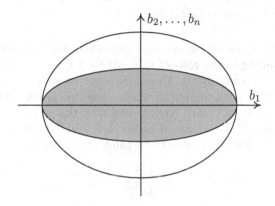

Fig. 3.3. Intersection of ellipsoids $H(x^0|A)$ for $A \in \mathcal{A}$.

Figure 3.3 directly shows that

$$
\begin{aligned}
\widehat{H}(x^0) \;\; &:= \;\; \bigcap_{A \in \mathcal{A}} H(x^0|A) = \{ y = b_1 q_1 \mid |b_1| \leq \|F_0\|_2 \} \\
&= \;\; \{ y \in \mathbb{R}^n \mid y = (1 - \lambda) F_0 , \; \lambda \in [0, 2] \} \\
&= \;\; \{ y \in \mathbb{R}^n \mid A y = (1 - \lambda) A F_0 , \; \lambda \in [0, 2] , \; A \in \mathrm{GL}(n) \}.
\end{aligned}
$$

As $\mathcal{A} \subset \mathrm{GL}(n) : \overline{H}(x^0) \subseteq \widehat{H}(x^0)$. On the other hand, for $y \in \widehat{H}(x^0)$, $A \in \mathrm{GL}(n)$, one has

$$\|A y\|_2^2 = (1 - \lambda)^2 \|A F_0\|_2^2 \leq \|A F_0\|_2^2 ,$$

which implies $\widehat{H}(x^0) \subseteq \overline{H}(x^0)$ and, in turn, confirms

$$\widehat{H}(x^0) = \overline{H}(x^0).$$

The *second stage of the proof* now involves 'lifting' of the path $\overline{H}(x^0)$ to $\overline{G}(x^0)$. This is done by means of the *homotopy*

$$\Phi(x,\lambda) := F(x) - (1-\lambda)F(x^0).$$

Note that

$$\Phi_x = F'(x), \quad \Phi_\lambda = F(x^0).$$

Hence, Φ_x is nonsingular for $x \in D$. As $D \supset G(x^0|\widehat{A}\,)$, local continuation starting at $\overline{x}(0) = x^0$ by means of the *implicit function theorem* finally establishes the existence of the path

$$\overline{x} \subset G(x^0|\widehat{A}\,) \subset D,$$

which is defined by (3.22) from $\Phi \equiv 0$. The differentiability of $\overline{x}$ follows, since $F \in C^1(D)$, which confirms (3.24) and (3.25). $\qquad\square$

The above theorem deserves some contemplation. The constructed Newton path $\overline{x}$ is outstanding in the respect that *all level functions $T(x|A)$ decrease along $\overline{x}$*—this is the result (3.22). Therefore, a rather natural approach would be just follow that path computationally—say, by numerical integration of the initial value problem (3.24). Arguments, why this is *not* a recommended method of choice, will be presented in Section 5 in a more general context. Rather, the local information about the tangent direction

$$\frac{\Delta x^0}{\|\Delta x^0\|}$$

should be used—which is just the Newton direction. In other words:

Even 'far away' from the solution point x^*, the Newton direction is an outstanding direction, only its length may be 'too large' for highly nonlinear problems.

Such an insight could not have been gained from the merely algebraic local linearization approach that had led to the ordinary Newton method.

The assumptions in the above theorem deliberately excluded the case that the Jacobian may be singular at some $\widehat{x}$ close to x^0. This case, however, may and will occur in practice. Application of the implicit function theorem in a more general situation shows that all Newton paths starting at points x^0 will end at one of the following three classes of points—see the schematic Figure 3.4:

- at the 'nearest' solution point x^*, or
- at some sufficiently close *critical point* $\widehat{x}$ with singular Jacobian, or
- at some point on the boundary ∂D of the domain of F.

The situation is also illustrated in the rather simple, but intuitive Example 3.2, see Figure 3.10 below.

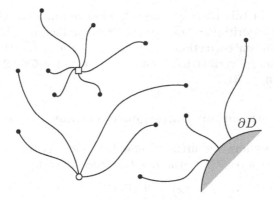

Fig. 3.4. Newton paths starting at initial points x^0 (•) will end at a solution point x^* (○), at a critical point $\hat{x}$ (□), or on the domain boundary ∂D.

BIBLIOGRAPHICAL NOTE. Standard derivations of the Newton path as a mathematical object had started from the so-called continuous analog of Newton's method, which is the ODE initial value problem (3.23)—see, e.g., the 1953 paper [48] by D. Davidenko. The geometric derivation of the Newton path from affine covariance as a 'first principle' dates back to the author's dissertation [59, 60] in 1972.

3.2 Residual Based Descent

In this section we study the *damped Newton iteration*

$$F'(x^k)\Delta x^k = -F(x^k), \qquad x^{k+1} = x^k + \lambda_k \Delta x^k, \qquad \lambda_k \in]0, 1]$$

under the requirement of *residual contraction*

$$\|F(x^{k+1})\| < \|F(x^k)\|,$$

which is certainly the most popular and the most widely used global convergence measure.

From Section 3.1.4 we perceive this iterative method as the tangent deviation from the *Newton path*, which connects the given initial guess x^0 to the unknown solution point x^*—under sufficient regularity assumptions on the Jacobian matrix, of course. The deviation is theoretically characterized by means of *affine contravariant* Lipschitz conditions as defined in the convergence theory for residual based *local* Newton methods in Section 2.2.

In what follows, we derive *theoretically optimal* iterative damping factors and prove global convergence within some range around these optimal factors

(Section 3.2.1). On this basis we then develop residual based trust region strategies for the algorithmic choice of the damping factors. This is first done for the *exact* Newton correction Δx as defined above (Section 3.2.2) and second for an *inexact* variant using the iterative solver GMRES for the inner iteration (Section 3.2.3).

3.2.1 Affine contravariant convergence analysis

From Lemma 3.5 above we already know that the Newton correction Δx^k points *downhill* with respect to the *residual level function*

$$T(x) := \tfrac{1}{2}\|F(x)\|_2^2$$

and therefore into the interior of the associated *residual level set*

$$G(x) := \{y \in D|T(y) \le T(x)\}\,.$$

At a given iterate x^k, we are certainly interested to determine some steplength (defined by the associated damping factor λ_k) along the Newton direction such that the residual reduction is in some sense *optimal*.

Theorem 3.7 *Let $F \in C^1(D)$ with $D \subset \mathbb{R}^n$ open convex and $F'(x)$ nonsingular for all $x \in D$. Assume the special affine contravariant Lipschitz condition*

$$\|(F'(y) - F'(x))(y - x)\| \le \omega\|F'(x)(y - x)\|^2 \text{ for } x, y \in D\,.$$

Then, with the convenient notation

$$h_k := \omega\|F(x^k)\|\,,$$

and $\lambda \in [0, \min{(1, 2/h_k)}]$ we have:

$$\|F(x^k + \lambda\Delta x^k)\|_2 \le t_k(\lambda)\|F(x^k)\|_2\,,$$

where

$$t_k(\lambda) := 1 - \lambda + \tfrac{1}{2}\lambda^2 h_k\,. \tag{3.26}$$

The optimal choice of damping factor in terms of this local estimate is

$$\overline{\lambda}_k := \min{(1,\ 1/h_k)}\,. \tag{3.27}$$

Proof. Dropping the superscript index k we may derive

$$\|F(x + \lambda\Delta x)\| = \|F(x + \lambda\Delta x) - F(x) - F'(x)\Delta x\|$$

$$= \left\| \int_{s=0}^{\lambda} \left(F'(x + s \cdot \Delta x) - F'(x)\right)\Delta x\, ds - (1 - \lambda)F'(x)\Delta x \right\|$$

$$\le (1 - \lambda)\|F(x)\| + O(\lambda^2) \text{ for } \lambda \in [0, 1]\,.$$

The arising $O(\lambda^2)$-term obviously characterizes the deviation from the Newton path and can be estimated as:

$$\left\| \int\limits_{s=0}^{\lambda} \left(F'(x + s \cdot \Delta x) - F'(x) \right) \Delta x \, ds \right\| \le \omega \cdot \tfrac{1}{2}\lambda^2 \|F'(x)\Delta x\|^2 = \tfrac{1}{2}\lambda^2 h_k \cdot \|F(x)\| .$$

Minimization of the above defined parabola t_k then directly yields $\overline{\lambda}_k$ with the a-priori restriction to the unit interval due to the underlying Newton path concept. $\square$

We are now ready to derive a global convergence theorem on the basis of this local descent result.

Theorem 3.8 *Notation and assumptions as in the preceding Theorem 3.7. In addition, let D_0 denote the path-connected component of $G(x^0)$ in x^0 and assume that $D_0 \subseteq D$ is compact. Let the Jacobian $F'(x)$ be nonsingular for all $x \in D_0$. Then the damped Newton iteration ($k = 0, 1, \ldots$) with damping factors in the range*

$$\lambda_k \in \left[\varepsilon, \; 2\overline{\lambda}_k - \varepsilon \right]$$

and sufficiently small $\varepsilon > 0$, which depends on D_0, converges to some solution point x^*.

Proof. The proof is by induction using the local results of the preceding theorem. In Figure 3.5, the estimation parabola t_k defined in (3.26) is depicted as a function of the damping factor λ together with the polygonal upper bound

$$t_k(\lambda) \le \begin{cases} 1 - \tfrac{1}{2}\lambda & , \quad 0 \le \lambda \le \dfrac{1}{h_k}, \\[2mm] 1 + \tfrac{1}{2}\lambda - \dfrac{1}{h_k} & , \quad \dfrac{1}{h_k} \le \lambda \le \dfrac{2}{h_k}. \end{cases}$$

Upon restricting λ to the range indicated in the present theorem, we immediately have

$$t_k(\lambda) \le 1 - \tfrac{1}{2}\varepsilon, \; 0 < \varepsilon \le \frac{1}{h_k}, \tag{3.28}$$

which induces *strict* reduction of the residual level function $T(x)$ in each iteration step k. In view of a proof of *global* convergence, a question left to discuss is whether there exists some global $\varepsilon > 0$. This follows from the fact that

$$\max_{x \in D_0} \|F(x)\| < \infty$$

under the *compactness* assumption on D_0. Hence, whenever $G(x^k) \subseteq D_0$, then (3.28) assures that

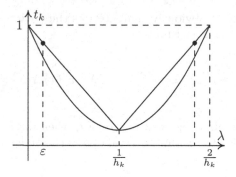

Fig. 3.5. Local reduction parabola t_k together with polygonal upper bounds.

$$G\big(x^{k+1}(\lambda)\big) \subset G(x^k) \subseteq D_0 .$$

With arguments similar as in the proof of Theorem 2.12, we finally conclude by induction that the defined damped Newton iteration converges towards some limit point x^* with $F(x^*) = 0$, which completes the proof.   □

3.2.2 Adaptive trust region strategies

The above derived theoretical damping strategy (3.27) cannot be implemented directly, since the arising Kantorovich quantities h_k are computationally unavailable due to the arising Lipschitz constant ω. The obtained theoretical results can nevertheless be exploited for the construction of computational strategies. Following the paradigm of Section 1.2.3, we may determine damping factors in the course of the iteration *as close to the convergence analysis as possible* replacing the unavailable Lipschitz constants ω by computational estimates $[\omega]$ and the unavailable Kantorovich quantities $h_k = \omega\|F(x^k)\|$ by computational estimates $[h_k] = [\omega]\|F(x^k)\|$. Such estimates can only be obtained by *pointwise sampling* of the domain dependent Lipschitz constants, which immediately implies that

$$[\omega] \le \omega , [h_k] \le h_k . \tag{3.29}$$

By definition, the estimates $[\cdot]$ will inherit the *affine contravariant* structure. As soon as we have iterative estimates $[h_k]$ at hand, associated estimates of the optimal damping factors may be naturally defined:

$$[\overline{\lambda}_k] := \min\left(1, 1/[h_k]\right) . \tag{3.30}$$

The relation (3.29) induces the equivalent relation

$$[\overline{\lambda}_k] \ge \overline{\lambda}_k .$$

This means that the estimated damping factors might be 'too large'—obviously *an unavoidable gap between analysis and algorithm*. As a consequence, repeated reductions might be necessary, which implies that any damping strategy to be derived will have to split into a *prediction strategy* and a *correction strategy*.

Bit counting lemma. As for the *required accuracy* of the computational estimates, the following lemma is important.

Lemma 3.9 *Notation as just introduced. Assume that the damped Newton method with damping factors as defined in (3.30) is realized. As for the accuracy of the computational estimates let*

$$0 \le h_k - [h_k] < \sigma \max(1, [h_k]) \ \textit{for some } \sigma < 1. \tag{3.31}$$

Then the residual monotonicity test will yield

$$\|F(x^{k+1})\| \le \left(1 - \tfrac{1}{2}(1 - \sigma)\lambda\right)\|F(x^k)\|.$$

Proof. We reformulate the relation (3.31) as

$$[h_k] \le h_k < (1 + \sigma)\max(1, [h_k]).$$

Then the above notation directly leads to the estimation

$$
\begin{aligned}
\frac{\|F(x^{k+1})\|}{\|F(x^k)\|} &\le \left[1 - \lambda + \tfrac{1}{2}\lambda^2 h_k\right]_{\lambda=[\overline{\lambda}_k]} \\
&< \left[1 - \lambda + \tfrac{1}{2}(1 + \sigma)\lambda^2 [h_k]\right]_{\lambda=[\overline{\lambda}_k]} \le 1 - \tfrac{1}{2}(1 - \sigma)\overline{\lambda}_k.
\end{aligned}
$$

□

For $\sigma < 1$, any computational estimates $[h_k]$ are just required to catch the *leading binary digit* of h_k, in order to assure residual monotonicity. For $\sigma \le \tfrac{1}{2}$, we arrive at the *restricted residual monotonicity test*

$$\|F(x^{k+1})\| \le \left(1 - \tfrac{1}{4}\lambda\right)\|F(x^k)\|. \tag{3.32}$$

This test nicely compares with the *Armijo strategy* (3.19), though derived by a different argument.

Computational estimates. After these preliminary considerations, we now proceed to identify *affine contravariant* computational estimates [·]—preferably those, which are cheap to evaluate in the course of the damped Newton iteration. In order to derive such estimates, we first recall from Section 3.1.4 that the damped Newton method may be interpreted as a deviation from the associated Newton path. Measuring the deviation in an affine contravariant setting leads us to the bound

$$\|F(x^{k+1}) - (1 - \lambda)F(x^k)\| \le \tfrac{1}{2}\lambda^2\omega\|F(x^k)\|^2\,,$$

which, in turn, leads to the following lower bound for the affine contravariant Kantorovich quantity:

$$[h_k] := \frac{2\|F(x^{k+1}) - (1 - \lambda)F(x^k)\|}{\lambda^2\|F(x^k)\|} \le h_k\,.$$

This estimate requires at least one trial value $x^{k+1} = x^k + \lambda_k^0 \Delta x^k$ so that it can only be exploited for the design of a *correction strategy* of the kind $(i = 0, 1, \ldots)$:

$$\lambda_k^{i+1} := \min\bigl(\tfrac{1}{2}\lambda_k^i, 1/[h_k^{i+1}]\bigr)\,. \tag{3.33}$$

In order to construct a theoretically backed initial estimate λ_k^0, we may apply the relation

$$h_{k+1} = \frac{\|F(x^{k+1})\|}{\|F(x^k)\|} h_k,$$

which directly inspires estimates of the kind

$$[h_{k+1}^0] = \frac{\|F(x^{k+1})\|}{\|F(x^k)\|} [h_k^{i_*}] < [h_k^{i_*}]\,,$$

wherein i_* indicates the final computable index within estimate (3.33) for the previous iterative step k. Thus we are led to the following *prediction strategy* for $k \ge 0$:

$$\lambda_{k+1}^0 := \min\bigl(1, 1/[h_{k+1}^0]\bigr)\,.$$

As can be seen, the only empirical choice left to be made is the starting value λ_0^0. It is recommended to set $\lambda_0^0 = 1$ for 'mildly nonlinear' problems and $\lambda_0^0 = \lambda_{\min} \ll 1$ for 'highly nonlinear' problems in a definition to be put in the hands of the users.

Intermediate quasi-Newton steps. Whenever $\lambda_k = 1$ and the residual monotonicity test yields

$$\Theta_k = \frac{\|F(x^{k+1})\|}{\|F(x^k)\|} \le \Theta_{\max} < 1$$

for some default value $\Theta_{\max}$, then the residual based quasi-Newton method of Section 2.2.3 may be applied—compare Theorem 2.14. This means that Jacobian evaluations are replaced by *residual rank-1 updates*. As for a possible switch back from quasi-Newton steps to Newton steps just look into the details of the informal quasi-Newton algorithm QNRES, also in Section 2.2.3.

The just described adaptive trust region strategy is realized in

Algorithm NLEQ-RES. Set a required *residual accuracy* ε sufficiently above the machine precision.

Guess an initial iterate x^0. Evaluate $F(x^0)$.

Set an initial damping factor either $\lambda_0 := 1$ or $\lambda_0 \ll 1$.

Norms are tacitly understood to be scaled smooth norms, such as $\|\bar{D}^{-1} \cdot \|_2$, where $\bar{D}$ is a diagonal scaling matrix, constant throughout the iteration.

For iteration index $k = 0, 1, \ldots$ do:

1. **Step k:**

    Convergence test: If $\|F(x^k)\| \le \varepsilon$: **stop.** Solution found $x^* := x^k$.

    Else: Evaluate Jacobian matrix $F'(x^k)$. Solve linear system

$$F'(x^k)\Delta x^k = -F(x^k)$$

   For $k > 0$: compute a prediction value for the damping factor

$$\lambda_k := \min(1, \mu_k), \quad \mu_k := \frac{\|F(x^{k-1})\|}{\|F(x^k)\|} \mu'_{k-1}.$$

   Regularity test: If $\lambda_k < \lambda_{\min}$: **stop.** Convergence failure.

2. **Else:** compute the trial iterate $x^{k+1} := x^k + \lambda_k \Delta x^k$ and evaluate $F(x^{k+1})$.

3. Compute the monitoring quantities

$$\Theta_k := \frac{\|F(x^{k+1})\|}{\|F(x^k)\|}, \quad \mu'_k := \frac{\frac{1}{2}\|F(x^k)\| \cdot \lambda_k^2}{\|F(x^{k+1}) - (1 - \lambda_k)F(x^k)\|}$$

   If $\Theta_k \ge 1$ (or, if **restricted**: $\Theta_k > 1 - \lambda_k/4$):

   then replace λ_k by $\lambda'_k := \min(\mu'_k, \frac{1}{2}\lambda_k)$. **Go to** Regularity test.

   Else: let $\lambda'_k := \min(1, \mu'_k)$.

   If $\lambda'_k = \lambda_k = 1$ **and** $\Theta_k < \Theta_{\max}$: switch to QNRES.

   Else: **If** $\lambda'_k \ge 4\lambda_k$: replace λ_k by λ'_k and **goto** 2.

   Else: accept x^{k+1} as new iterate. **Goto** 1 with $k \to k + 1$.

3.2.3 Inexact Newton-RES method

In this section we discuss the *inexact* global Newton method

$$x^{k+1} = x^k + \lambda_k \delta x^k, \; 0 < \lambda_k \le 1$$

realized by means of GMRES such that (dropping the inner iteration index i)

$$F'(x^k)\delta x^k = -F(x^k) + r^k.$$

Let $\delta x_0^k = 0$ and thus $r_0^k = F(x^k)$. The notation here follows Section 2.2.4 on local Newton-RES methods.

Convergence analysis. Before going into details of the analysis, we want to point out that the inexact Newton method with damping can be viewed as a tangent step in x^k for the *inexact Newton path*

$$F\left(\tilde{x}(\lambda)\right) - r^k = (1 - \lambda)\left(F(x^k) - r^k\right)$$

or, equivalently,

$$F\left(\tilde{x}(\lambda)\right) = (1 - \lambda)F(x^k) + \lambda r^k , \tag{3.34}$$

wherein $\tilde{x}(0) = x^k$, $\dot{\tilde{x}}(0) = \delta x^k$, but $\tilde{x}(1) \neq x^*$. Hence, when approaching x^*, we will have to assure that $r^k \to 0$. With this geometric interpretation in mind, we are now prepared to derive the following convergence statements.

Theorem 3.10 *Under the assumptions of Theorem 3.7 for the exact Newton iteration with damping, the inexact Newton-GMRES iteration can be shown to satisfy*

$$\Theta_k := \frac{\|F(x^{k+1})\|}{\|F(x^k)\|} \leq t_k(\lambda_k, \eta_k) \tag{3.35}$$

with

$$t_k(\lambda, \eta) = 1 - (1 - \eta)\lambda + \tfrac{1}{2}\lambda^2(1 - \eta^2)h_k , \quad \eta_k = \frac{\|r^k\|_2}{\|F(x^k)\|_2} < 1 .$$

The optimal choice of damping factor is

$$\overline{\lambda}_k := \min\left(1, \frac{1}{(1 + \eta_k)h_k}\right) . \tag{3.36}$$

Proof. Recall from Section 2.1.5 that for GMRES

$$\|F(x^k) - r^k\|_2^2 = \|F(x^k)\|_2^2 - \|r^k\|_2^2 = (1 - \eta_k^2)\|F(x^k)\|_2^2$$

for well-defined $\eta_k < 1$. Along the line $x^k + \lambda\delta x^k$ the descent behavior can be estimated using

$$F(x^k + \lambda\delta x^k) = (1 - \lambda)F(x^k) + \lambda r^k + \lambda \int_{s=0}^{1} \left(F'(x^k + s\lambda\delta x^k) - F'(x^k)\right)\delta x^k \, ds .$$

The last right hand term is directly comparable to the exact case: in the application of the affine contravariant Lipschitz condition we merely have to replace Δx^k by δx^k and, accordingly,

$$\|F'(x^k)\Delta x^k\|^2 = \|F(x^k)\|^2$$

by

$$\|F'(x^k)\delta x^k\|^2 = (1 - \eta_k^2)\|F(x^k)\|^2 .$$

With this modification, the result (3.35) is readily verified. The optimal damping factor follows from setting $t_k'(\lambda) = 0$. With $\lambda_k \leq 1$ as restriction, we have (3.36). $\square$

Adaptive trust region method. In order to exploit the above convergence analysis for the construction of an inexact Newton-GMRES algorithm, we will follow the usual *paradigm* and certainly aim at defining certain damping factors

$$[\lambda_k] := \min\left(1, \frac{1}{(1+\eta_k)[h_k]}\right)$$

in terms of affine contravariant computationally available estimates. First, if we once again apply the 'bit counting' Lemma 3.9, we arrive at the *inexact variant* of the *restricted residual monotonicity test*

$$\|F(x^{k+1})\|_2 \leq \left(1 - \frac{1-\eta_k}{4}\lambda_k\right)\|F(x^k)\|_2,$$

which here replaces (3.32). Next, upon returning to the above proof of Theorem 3.10, we readily observe that

$$\|F(x^{k+1}) - (1-\lambda_k)F(x^k) - \lambda_k r^k\|_2 \leq \frac{\lambda_k^2}{2}(1-\eta_k^2)h_k\|F(x^k)\|_2.$$

On this basis, we may simply define the *a-posteriori* estimates

$$[h_k](\lambda) := \frac{2\|F\left(x^{k+1}(\lambda)\right) - (1-\lambda)F(x^k) - \lambda r^k\|_2}{\lambda^2(1-\eta_k^2)\|F(x^k)\|_2} \leq h_k,$$

which give rise to the *correction strategy* $(i = 0, 1, \ldots, i_k^*)$

$$\lambda_k^{i+1} = \min\left(\tfrac{1}{2}\lambda_k^i, \frac{1}{(1+\eta_k)[h_k^i]}\right),$$

and the associated *a-priori* estimates

$$[h_{k+1}^0] := \Theta_k[h_k^{i_*}] \leq h_{k+1},$$

which induce the *prediction strategy* $(k = 0, 1, \ldots)$

$$\lambda_{k+1}^0 := \min\left(1, \frac{1}{(1+\eta_{k+1})[h_{k+1}^0]}\right).$$

As for the choice of η_k, we already have a strategy for $\lambda_k = \lambda_{k-1} = 1$—see Section 2.1.5. For $\lambda_k < 1$, some constant value $\eta_k \leq \eta$ with some sufficiently small threshold value η can be selected (and handed over to the local Newton method, see Section 2.2.4). Then only λ_0^0 remains to be set externally.

The just described residual based adaptive trust region strategy in combination with the strategy to match inner and outer iteration is realized in the code GIANT-GMRES.

BIBLIOGRAPHICAL NOTE. Residual based inexact Newton methods date back to R.S. Dembo, S.C. Eisenstat, and T. Steihaug [51]. Quite popular algorithmic heuristics have been worked out by R.E. Bank and D.J. Rose in [19] and are applied in a number of published algorithms. A different global convergence analysis has been given in [90, 91] by S.C. Eisenstat and H.F. Walker. Their strategies are implemented in the code NITSOL due to M. Pernice and H.F. Walker [166]. They differ from the ones presented here.

3.3 Error Oriented Descent

In this section we study the *damped Newton iteration*

$$F'(x^k)\Delta x^k = -F(x^k), \qquad x^{k+1} = x^k + \lambda_k \Delta x^k, \qquad \lambda_k \in]0,1]$$

in an error oriented framework, which aims at overcoming certain difficulties that are known to arise in the residual based framework, especially in situations where the Jacobian matrices are ill-conditioned—such as in discretized nonlinear partial differential equations. Once again, we treat the damped Newton method as a deviation from the Newton path, but this time we characterize the deviation by means of *affine covariant* Lipschitz conditions such as those used in the convergence theory for error oriented *local* Newton methods in Section 2.1.

The construction of an error oriented globalization of local Newton methods is slightly more complicated than in the residual based approach. For this reason, we first recur to the concept of *general level functions* $T(x|A)$ as already introduced in (3.15) for *arbitrary* nonsingular matrix A and study the descent behavior of the damped Newton method for the whole class of such functions in an affine covariant theoretical framework (Section 3.3.1). As it turns out, the obtained theoretically optimal damping factors actually reflect the observed difficulties of the residual based variants. Moreover, the analysis directly leads to the specific choice $A = F'(x^k)^{-1}$, which defines the so-called *natural* level function (Section 3.3.2). As a consequence for actual computation, the iterates are required to satisfy the so-called *natural monotonicity test*

$$\|\overline{\Delta x}^{k+1}\| < \|\Delta x^k\|,$$

wherein the *simplified* Newton correction $\overline{\Delta x}^{k+1}$ defined by

$$F'(x^k)\overline{\Delta x}^{k+1} = -F(x^{k+1})$$

is only computed to evaluate this test (and later also for an adaptive trust region method). As for a proof of global convergence, only a theorem covering a slightly different situation is available up to now—despite the convincing global convergence properties of the thus derived algorithm! From the associated theoretically optimal damping factors we develop computational trust region strategies—first for the *exact* damped Newton method as defined above (Section 3.3.3) and second for *inexact* variants using error reducing linear iterative solvers for the inner iteration (Section 3.3.4).

3.3.1 General level functions

In order to derive an affine covariant or error oriented variant of the damped Newton method we first recur to general level functions, which have been already defined in (3.15) as

$$T(x|A) := \tfrac{1}{2}\|AF(x)\|_2^2 .$$

Local descent. It is an easy task to verify that the Newton direction points 'downhill' with respect to *all* such level functions.

Lemma 3.11 *Let $F \in C^1(D)$ and let Δx denote the ordinary Newton correction (dropping the iteration index k). Then, for all $A \in \mathrm{GL}(n)$,*

$$\Delta x^T \operatorname{grad} T(x|A) = -2T(x|A) < 0 .$$

This is certainly a distinguishing feature to any other descent directions—compare Lemma 3.3. Hence, on the basis of *first* order information only, all monotonicity criteria look equally well-suited for the damped Newton method. The selection of a specific level function will therefore require *second* order information.

Theorem 3.12 *Let $F \in C^1(D)$ with $D \subset \mathbb{R}^n$ convex and $F'(x) = F'(x)$ nonsingular for all $x \in D$. For a given current iterate $x^k \in D$ let $G(x^k|A) \subset D$ for some $A \in \mathrm{GL}(n)$. For $x,y \in D$ assume that*

$$\|F'(x)^{-1}\big(F'(y) - F'(x)\big)(y - x)\| \le \omega\|y - x\|^2 .$$

Then, with the convenient notation

$$h_k := \|\Delta x^k\|\omega , \quad \overline{h}_k := h_k \operatorname{cond}\big(AF'(x^k)\big)$$

one obtains for $\lambda \in \big[0, \min\big(1, 2/\overline{h}_k\big)\big]$:

$$\|AF(x^k + \lambda\Delta x^k)\| \le t_k(\lambda|A)\|AF(x^k)\| , \tag{3.37}$$

where

$$t_k(\lambda|A) := 1 - \lambda + \tfrac{1}{2}\lambda^2\overline{h}_k . \tag{3.38}$$

The optimal choice of damping factor in terms of this local estimate is

$$\overline{\lambda}_k(A) := \min\big(1,\ 1/\overline{h}_k\big) .$$

Proof. Dropping the superscript index k we may derive

$$\|AF(x + \lambda\Delta x)\| = \|A\big(F(x + \lambda\Delta x) - F(x) - F'(x)\Delta x\big)\|$$

$$= \left\| A \left(\int\limits_{\delta=0}^{\lambda} \big(F'(x + \delta\Delta x) - F'(x)\big)\Delta x d\delta - (1-\lambda)F'(x)\Delta x \right) \right\|$$

$$\leq (1-\lambda)\|AF(x)\| + O(\lambda^2) \quad \text{for } \lambda \in [0,1].$$

The arising $O(\lambda^2)$-term obviously characterizes the deviation from the Newton path and can be estimated as:

$$\left\| AF'(x) \int\limits_{\delta=0}^{\lambda} F'(x)^{-1}\big(F'(x + \delta\Delta x) - F'(x)\big)\Delta x d\delta \right\|$$

$$\leq \|AF'(x)\|\omega\tfrac{1}{2}\lambda^2\|\Delta x\|^2 = \tfrac{1}{2}\lambda^2\|AF'(x)\|\|h_k\|\big(AF'(x)\big)^{-1}AF(x)\|$$

$$\leq \tfrac{1}{2}\lambda^2 h_k\|AF'(x)\|\|\big(AF'(x)\big)^{-1}\|\|AF(x)\| = \tfrac{1}{2}\lambda^2 \overline{h}_k\|AF(x)\|.$$

Minimization of the parabola t_k then directly yields $\overline{\lambda}_k(A)$ with the a-priori restriction to the unit interval due to the underlying Newton path concept. □

Global convergence. The above *local descent* result may now serve as a basis for the following global convergence theorem.

Theorem 3.13 *Notation and assumptions as in the preceding Theorem 3.12. In addition, let D_0 denote the path-connected component of $G(x^0|A)$ in x^0 and assume that $D_0 \subseteq D$ is compact. Let the Jacobian $F'(x)$ be nonsingular for all $x \in D_0$. Then the damped Newton iteration $(k = 0, 1, \ldots)$ with damping factors in the range*

$$\lambda_k \in \big[\varepsilon,\ 2\overline{\lambda}_k(A) - \varepsilon\big]$$

and sufficiently small $\varepsilon > 0$, which depends on D_0, converges to some solution point x^*.

Proof. The proof is by induction using the local results of the preceding theorem. Moreover, it is just a slight modification of the proof of Theorem 3.8 for the residual level function. In particular, Figure 3.5 shows the same type of estimation parabola t_k as defined here in (3.38): once again, the proper polygonal upper bound supplies the global upper bound

$$t_k(\lambda|A) \leq 1 - \tfrac{1}{2}\varepsilon,\ 0 < \varepsilon \leq \frac{1}{\overline{h}_k}, \tag{3.39}$$

which induces *strict* reduction of the general level function $T(x|A)$ in each iteration step k. We are now just left to discuss whether there exists some global $\varepsilon > 0$. This follows from the fact that

$$\max_{x \in D_0} \|F'(x)^{-1}F(x)\| \cdot \mathrm{cond}_2\big(AF'(x)\big) < \infty$$

under the *compactness* assumption on D_0. Hence, whenever $G(x^k|A) \subseteq D_0$, then (3.39) assures that

$$G\big(x^{k+1}(\lambda)|A\big) \subset G(x^k|A) \subseteq D_0.$$

We therefore conclude by induction that

$$\lim_{k \to \infty} x^k = x^*,$$

which completes the proof. $\qquad\qquad\qquad\qquad\qquad\qquad\qquad\qquad\square$

Algorithmic limitation of residual monotonicity. The above theorem offers an intriguing explanation, why the damped Newton method endowed with the traditional *residual monotonicity criterion*

$$T(x^{k+1}|I) \le T(x^k|I)$$

may fail in practical computation despite its proven global convergence property (compare Theorem 3.8): in fact, whenever the *Jacobian* is *ill-conditioned*, then the 'optimal' damping factors are bound to satisfy

$$\overline{\lambda}_k(I) = \Big(h_k \mathrm{cond}_2\big(F'(x^k)\big)\Big)^{-1} < \lambda_{\min} \ll 1. \qquad (3.40)$$

Therefore, in worst cases, also any computational damping strategies (no matter, how sophisticated they might be) will lead to a *practical termination of the iteration*, since then $x^{k+1} \approx x^k$, which means that the iteration 'stands still'. For illustration of this effect see Example 3.1 below, especially Fig. 3.9.

Another side of the same medal is the quite often reported observation that for 'well-chosen' initial guesses x^0 residual monotonicity may be violated over several initial iterative steps even though the ordinary Newton iteration converges when allowed to do so by skipping the residual monotonicity test. In fact, from the error oriented local convergence analysis of Section 2.1, one would expect to obtain the optimal value $\overline{\lambda}_k = 1$ roughly as soon as the iterates are contained in the 'neighborhood' of the solution x^*—say, as soon as for some iterate $h_k < 1$. A comparison with the above theorem, however, shows that a condition of the kind $\overline{h}_k = h_k \mathrm{cond}_2\big(F'(x^k)\big) < 1$ would be required in the residual framework. The effect is illustrated by Example 3.1 at the end of Section 3.3.2.

Summarizing, we have the following situation:

Combining any damping strategy with the residual monotonicity criterion may have the consequence that mildly nonlinear problems actually 'look like' highly nonlinear problems, especially in the situation (3.40); as a consequence, especially in the presence of ill-conditioned Jacobians, the Newton iteration with damping tends to terminate without the desired result—despite an underlying global convergence theorem like Theorem 3.13 for $A = I$.

3.3.2 Natural level function

The preceding section seemed to indicate that 'all level functions are equal'; here we want to point out that 'some animals are more equal than others' (compare George Orwell, *Animal Farm*).

As has been shown, the condition number $\mathrm{cond}_2\big(AF'(x^k)\big)$ plays a central role in the preceding analysis, at least in the worst case situation. Therefore, due to the well-known property

$$\mathrm{cond}_2\big(AF'(x^k)\big) \geq 1 = \mathrm{cond}_2(I),$$

the special choice

$$A_k := F'(x^k)^{-1}$$

seems to be *locally optimal* as a specification of the matrix in the general level function. The associated level function $T\big(x|F'(x^k)^{-1}\big)$ is called *natural level function* and the associated *natural monotonicity test* requires that

$$\|\overline{\Delta x}^{k+1}\|_2 \leq \|\Delta x^k\|_2 \tag{3.41}$$

in terms of the *ordinary* Newton correction Δx^k and the *simplified* Newton correction defined by

$$\overline{\Delta x}^{k+1} := -F'(x^k)^{-1}F(x^{k+1}).$$

This specification gives rise to several outstanding properties.

Extremal properties. For $A \in \mathrm{GL}(n)$ the reduction factors $t_k(\lambda|A)$ and the theoretical optimal damping factors $\overline{\lambda}_k(A)$ satisfy:

$$t_k(\lambda|A_k) = 1 - \lambda + \tfrac{1}{2}\lambda^2 h_k \leq t_k(\lambda|A)$$

$$\overline{\lambda}_k(A_k) = \min\left(1,\, 1/h_k\right) \geq \overline{\lambda}_k(A).$$

An associated graphical representation is given in Figure 3.6.

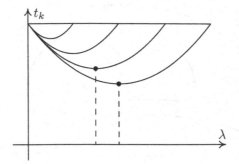

Fig. 3.6. Extremal properties of natural level function: reduction factors $t_k(\lambda|A)$ and optimal damping factors $\overline{\lambda}_k(A)$.

Steepest descent property. The steepest descent direction for $T(x|A)$ in x^k is

$$- \operatorname{grad} T(x^k|A) = - \left(AF'(x^k)\right)^T AF(x^k) \,.$$

With the specification $A = A_k$ this leads to

$$\varDelta x^k = - \operatorname{grad} T(x^k|A_k) \,,$$

which means that *the damped Newton method in x^k is a method of steepest descent for the natural level function $T(x|A_k)$*.

Merging property. The locally optimal damping factors nicely reflect the expected behavior in the contraction domain of the ordinary Newton method: in fact, we have

$$h_k \leq 1 \implies \overline{\lambda}_k(A_k) = 1 \,.$$

Hence, quadratic convergence is asymptotically achieved by the damping strategy based on $\overline{\lambda}_k$.

Asymptotic distance function. For $F \in C^2(D)$, we easily verify that

$$T\left(x|F'(x^*)^{-1}\right) = \tfrac{1}{2}\|x - x^*\|_2^2 + O(\|x - x^*\|^3) \,.$$

Hence, for $x^k \to x^*$, the natural monotonicity criterion asymptotically merges into a desirable distance criterion of the form

$$\|x^{k+1} - x^*\|_2 \stackrel{\cdot}{\leq} \|x^k - x^*\|_2 \,,$$

which is exact for *linear* problems. The situation is represented graphically in Figure 3.7. Far away from the solution point, this nice geometrical property survives in the form that the osculating ellipsoid to the level surface at the current iterate turns out to be an *osculating sphere*.

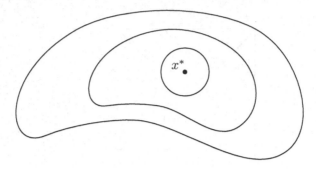

Fig. 3.7. Natural level sets: asymptotic distance spheres.

In the linear case, the Jacobian condition number represents the quotient of the largest over the smallest half-axis of the level ellipsoid. In the non-linear case, too, Jacobian ill-conditioning gives rise to cigar-shaped residual level sets, which, in general, are distorted ellipsoids. Therefore, geometrically speaking, the natural level function realizes some *nonlinear preconditioning*.

Local descent. Any damping strategy based on the natural monotonicity test is sufficiently characterized by Theorem 3.12: just insert $A = A_k$ into (3.37) and (3.38), which then yields

$$\|\overline{\Delta x}^{k+1}\| \le \left(1 - \lambda + \tfrac{1}{2}\lambda^2 h_k\right)\|\Delta x^k\|.$$

Global convergence. In the present situation, the above global convergence theorem for general level functions, Theorem 3.13, does *not* apply, since the choice A_k varies from step to step. In order to obtain an *affine covariant* global convergence theorem, the locally optimal choice $A = F'(x^k)^{-1}$ will now be *modelled* by the *fixed* choice $A = F'(x^*)^{-1}$—in view of the asymptotic distance function property.

Theorem 3.14 *Let $F : D \longrightarrow \mathbb{R}^n$ be a continuously differentiable mapping with $D \subseteq \mathbb{R}^n$ open convex. Assume that $x^0, x^* \in D$ with x^* unique solution of F in D and the Jacobian $F'(x^*)$ nonsingular. Furthermore, assume that*

(I) *$F'(x)$ is nonsingular for all $x \in D$,*

(II) *the path-connected component D_0 of $G\big(x^0|F'(x^*)^{-1}\big)$ in x^0 is compact and contained in D,*

(III) *the following affine covariant Lipschitz condition holds*

$$\big\| F'(x^*)^{-1}\big(F'(y) - F'(x)\big)(y - x) \big\| \le \omega_* \|y - x\|^2 \text{ for } y, x \in D,$$

(IV) *for any iterate $x^k \in D$ let $h_k^* := \omega_* \|\Delta x^k\| \cdot \|F'(x^k)^{-1}F'(x^*)\| < \infty$.*

As the locally optimal damping strategy we obtain

$$\lambda_k^* := \min\left(1,\, 1/h_k^*\right).$$

Then any damped Newton iteration with iterative damping factors in the range

$$\lambda_k \in \left[\varepsilon,\, 2\lambda_k^* - \varepsilon)\right] \quad for\; 0 < \varepsilon < 1/h_k^*$$

converges globally to x^*.

Proof. In the proof of Theorem 3.12 we had for $x = x^k$:

$$\|AF(x + \lambda\Delta x)\| \le (1 - \lambda)\|AF(x)\| + O(\lambda^2).$$

For $A = F'(x^*)^{-1}$ the $O(\lambda^2)$-term may now be treated differently as follows:

$$\left\| F'(x^*)^{-1} \int_{\delta=0}^{\lambda} \left(F'(x + \delta\Delta x) - F'(x)\right)\Delta x d\delta \right\| \le \int_{\delta=0}^{\lambda} \omega_* \cdot \delta\|\Delta x\|^2 d\delta$$

$$\le \omega_* \cdot \tfrac{1}{2}\lambda^2\|\Delta x\| \cdot \|F'(x)^{-1}F'(x^*)\| \cdot \|F'(x^*)^{-1}F(x)\|$$

$$= \tfrac{1}{2}\lambda^2 h_k^* \|F'(x^*)^{-1}F(x)\|.$$

On the basis of the thus modified local reduction property, global convergence in terms of the above specified level function can be shown along the same lines of argumentation as in Theorem 3.13. The above statements are just the proper copies of the statements of that theorem.   □

Corollary 3.15 *Under the assumptions of the preceding theorem with the replacement of* x^* *by an arbitrary* $z \in D_0$ *in the Jacobian inverse and the associated affine covariant Lipschitz condition*

$$\left\| F'(z)^{-1}\left(F'(y) - F'(x)\right)(y - x) \right\| \le \omega(z) \|y - x\|^2 for\; x, y, z \in D_0,$$

a local level function reduction of the form

$$T\left(x^k + \lambda\Delta x^k | F'(z)^{-1}\right) \le \left(1 - \lambda + \tfrac{1}{2}\lambda^2 h_k(z)\right)^2 T\left(x^k | F'(z)^{-1}\right)$$

in terms of

$$h_k(z) := \|\Delta x^k\|\, \omega(z)\, \|F'(x^k)^{-1}F'(z)\|$$

and a locally optimal damping factor

$$\overline{\lambda}_k(z) := \min\left(1, 1/h_k(z)\right)$$

can be shown to hold. Assuming further that the used matrix norm is sub-multiplicative, then we obtain for best possible estimates $\omega(z)$ *the extremal properties*

$$\|F'(x^k)^{-1}F'(z)\|\,\omega(z) \geq \omega(x^k)$$

and

$$h_k(x^k) \leq h_k(z)\,,\ \overline{\lambda}_k \geq \overline{\lambda}_k(z)\,,\ z \in D_0\,.$$

The corollary states that the locally optimal damping factors in terms of the locally defined natural level function are outstanding among all possible globally optimal damping factors in terms of any globally defined affine covariant level function. Our theoretical convergence analysis shows that we may substitute the global affine covariant Lipschitz constant ω by its more local counterpart $\omega_k = \omega(x^k)$ defined via

$$\big\| F'(x^k)^{-1}\big(F'(x) - F'(x^k)\big)(x - x^k) \big\| \leq \omega_k \|x - x^k\|^2 \text{ for } x, x^k \in D_0\,. \quad (3.42)$$

We have thus arrived at the following *theoretically optimal damping strategy* for the *exact* Newton method

$$x^{k+1} = x^k + \overline{\lambda}_k \Delta x^k\,, \quad \overline{\lambda}_k := \min(1, 1/h_k)\,, \quad h_k = \omega_k \|\Delta x^k\|\,. \quad (3.43)$$

We must state again that this Newton method with damping based on the natural monotonicity test does not have the comfort of an accompanying global convergence theorem. In fact, U.M. Ascher and M.R. Osborne in [10] constructed a simple example, which exhibits a 2–cycle in the Newton method when monitored by the natural level function. Details are left as Exercise 3.3. However, as shown in [33] by H.G. Bock, E.A. Kostina, and J.P. Schlöder, such 2–cycles can be generally avoided, if the theoretical optimal steplength $\overline{\lambda}_k$ is restricted such that $\lambda h_k \leq \eta < 1$. Details are left as Exercise 3.4. This restriction does not avoid m–cycles for $m > 2$—which still makes the derivation of a global convergence theorem *solely based on natural monotonicity* impossible. Numerical experience advises not to implement this kind of restriction—generically it would just increase the number of Newton iterations required.

Geometrical interpretation. This strategy has a nice geometrical interpretation, which is useful for a deeper understanding of the computational strategies to be developed in the sequel. Recalling the derivation in Section 3.3.1, the damped Newton method at some iterate x^k continues along the tangent of the Newton path $\overline{G}(x^k)$ with effective correction length

$$\|x^{k+1} - x^k\| = \overline{\lambda}_k \|\Delta x^k\| \leq \rho_k := 1/\omega_k\,.$$

Obviously, the radius ρ_k characterizes the *local trust region* of the linear Newton model around x^k. The situation is represented schematically in Figure 3.8.

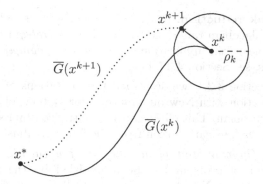

Fig. 3.8. Geometrical interpretation: Newton path $\overline{G}(x^k)$, trust region around x^k, and Newton step with locally optimal damping factor $\overline{\lambda}_k$.

Interpretation via Jacobian information. In terms of a relative change of the Jacobian matrix we may write

$$\left\| F'(x^k)^{-1}\big(F'(x^{k+1}) - F'(x^k)\big)\Delta x^k \right\| \Big/ \|\Delta x^k\| \le \overline{\lambda}_k h_k \le 1 .$$

This relation suggests the interpretation that Jacobian information at the center x^k of the trust region ball is valid along the Newton direction up to the surface of the ball, which is x^{k+1}. Of course, such an interpretation implicitly assumes that the maximum change actually occurs at the most distant point on the surface—this property certainly holds for the derived upper bounds. Beyond the trust region the Jacobian information from the center x^k is no longer useful, which then implies the construction of a *new* Newton path $\overline{G}(x^{k+1})$ and the subsequent continuation along the new tangent—see once again Figure 3.8.

Behavior near critical points. Finally, we want to discuss the expected behavior of the Newton method with damping in the presence of some close-by *critical point*, say $\hat{x}$ with *singular Jacobian* $F'(\hat{x})$. In this situation, *the Newton path and, accordingly, the Newton iteration with optimal damping will be attracted by $\hat{x}$.* Examples of such a behavior have been observed fairly often—in particular, when multiple solutions are separated by manifolds with singular Jacobian, compare, e.g., Figure 3.10. Nevertheless, even in such a situation, a structural advantage of the natural level function approach may play a role: whereas points $\hat{x}$ represent *local minima* of $T(x|I)$, which will attract iterative methods based on the residual monotonicity test, they show up as *local maxima* of the natural level functions since $T\big(\hat{x}|F'(\hat{x})^{-1}\big)$ is unbounded. For this reason, the above *theoretical* damped Newton method tends to avoid local minima of $T(x)$ whenever they correspond to locally isolated critical points.

Deliberate rank reduction. In rare emergency cases only, a deliberate reduction of the Jacobian rank (the so-called *rank-strategy*) turns out to be helpful—which means the application of *intermediate damped Gauss-Newton steps*. For details, see Section 4.3.5 below.

At the end of Section 3.3.1, we described the limitations of residual monotonicity in connection with Newton's method for systems of equations with ill–conditioned Jacobian. This effect can be neutralized by requiring natural monotonicity instead, as can be seen from the following illustrative example.

Example 3.1 *Optimal orbit plane change of a satellite around Mars.* This optimal control problem has been modeled by the space engineer E.D. Dickmanns [86] at NASA. The obtained ODE boundary value problem has been treated by multiple shooting techniques (see [71, Sect. 8.6.2.] and Section 7.1 below). This led to a system of $n = 72$ nonlinear equations with a Jacobian known to be ill–conditioned. The results given here are taken from the author's dissertation [59, 60]. The problem is a typical representative out of a large class of problems that 'look highly nonlinear', but are indeed essentially 'mildly nonlinear' as discussed at the end of Section 3.3.1.

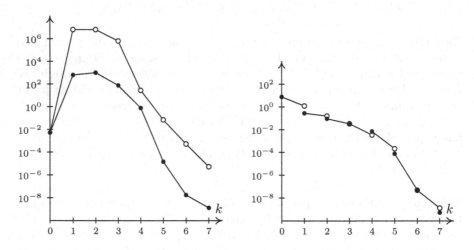

Fig. 3.9. Mars satellite orbit problem. *Left:* no convergence in residual norms $\|F(x^k)\|_2^2$ (o) or scaled residual norms $\|D_k^{-1}F(x^k)\|_2^2$ (•). *Right:* convergence in natural level function, ordinary Newton corrections $\|D_k^{-1}\Delta x^k\|_2^2$ (•) versus simplified Newton corrections $\|D_k^{-1}\overline{\Delta x}^{k+1}\|_2^2$ (o).

Figure 3.9 documents the comparative behavior of residual level functions (with and without diagonal scaling) and natural level functions. The Newton iteration has been controlled by scaled natural monotonicity tests

$$\|D_k^{-1}\overline{\Delta x}^{k+1}\|_2^2 \le \|D_k^{-1}\Delta x^k\|_2^2, \quad k = 0, 1, \dots,$$

as shown on the right; in passing, we note that the second Newton step has been performed using 'deliberate rank reduction' as described just above. On the left, the iterative values of the traditional residual level functions, both unscaled and scaled, are seen to increase drastically for the accepted Newton steps. Obviously, only the natural level functions reflect the 'approach' of the iterates x^k toward the solution x^*.

BIBLIOGRAPHICAL NOTE. The algorithmic concept of natural level functions has been suggested in 1972 by P. Deuflhard [59] for highly nonlinear problems with ill-conditioned Jacobian. In the same year, *linear preconditioning* has been suggested by O. Axelsson (see [11]) on a comparable geometrical basis, but for the purpose of speeding up the convergence of iterative solvers.

3.3.3 Adaptive trust region strategies

The above derived theoretical damping strategy (3.43) cannot be implemented directly, since the arising Kantorovich quantities h_k contain the computationally unavailable Lipschitz constants ω_k, which are defined over some domain D_0—in view of Figure 3.8, even a definition over some local trust region would be enough. The obtained theoretical results can nevertheless be exploited for the construction of computational strategies. Following our paradigm in Section 1.2.3 again, we determine damping factors in the course of the iteration *as close to the convergence analysis as possible* by introducing *computationally available estimates* $[\omega_k]$ and $[h_k] = [\omega_k]\|\Delta x^k\|$ for the unavailable theoretical quantities ω_k and $h_k = \omega_k\|\Delta x^k\|$.

Such estimates can only be obtained by *pointwise sampling* of the domain dependent Lipschitz constants, which immediately implies that

$$[\omega_k] \leq \omega_k \leq \omega_k(z) , \quad [h_k] \leq h_k \leq h_k(z) \tag{3.44}$$

compare Corollary 3.15. By definition, the estimates $[\cdot]$ will inherit the *affine covariant* structure. Suppose now that we have certain estimates $[h_k]$ at hand. Then associated estimates of the optimal damping factors may be naturally defined as

$$[\overline{\lambda}_k] := \min\left(1, 1/[h_k]\right) . \tag{3.45}$$

The above relation (3.44) gives rise to the equivalent relation

$$[\overline{\lambda}_k] \geq \overline{\lambda}_k .$$

Clearly, any computed estimated damping factors may be 'too large'—which, in turn, means that repeated reductions might be necessary. Therefore, any damping strategy to be derived will have to split into a *prediction strategy* and a *correction strategy*.

Bit counting lemma. The efficiency of such damping strategies will depend on the *required accuracy* of the computational estimates—a question, which is studied in the following lemma.

Lemma 3.16 *Notation as just introduced. Assume that the damped Newton method with damping factors as defined in* (3.45) *is realized. As for the accuracy of the computational estimates, let*

$$0 \le h_k - [h_k] < \sigma \max(1, [h_k]) \text{ for some } \sigma < 1. \tag{3.46}$$

Then the natural monotonicity test will yield

$$\|\overline{\Delta x}^{k+1}\| \le \left(1 - \tfrac{1}{2}(1 - \sigma)\lambda\right) \|\Delta x^k\|.$$

Proof. We reformulate the relation (3.46) as

$$[h_k] \le h_k < (1 + \sigma) \max(1, [h_k]).$$

For $[h_k] \ge 1$, the above notation directly leads to the estimation

$$\frac{\|\overline{\Delta x}^{k+1}\|}{\|\Delta x^k\|} \quad \le \quad \left[1 - \lambda + \tfrac{1}{2}\lambda^2 h_k\right]_{\lambda=[\overline{\lambda}_k]}$$

$$< \quad \left[1 - \lambda + \tfrac{1}{2}(1+\sigma)\lambda^2[h_k]\right]_{\lambda=[\overline{\lambda}_k]} \le 1 - \tfrac{1}{2}(1-\sigma)\overline{\lambda}_k.$$

The case $[h_k] < 1$ follows similarly.   □

The above lemma states that, for $\sigma < 1$, the computational estimates $[h_k]$ are just required to catch the *leading binary digit* of h_k, in order to assure natural monotonicity. For $\sigma \le \tfrac{1}{2}$, we arrive at the following *restricted natural monotonicity test*

$$\|\overline{\Delta x}^{k+1}\|_2 \le \left(1 - \tfrac{1}{4}\lambda\right) \|\Delta x^k\|_2, \tag{3.47}$$

which might be useful in actual computation to control the whole iterative process more closely—compare also the residual based restricted monotonicity test (3.32) and the Armijo strategy (3.19).

Correction strategy. After these abstract considerations, we now proceed to derive specific affine covariant computational estimates $[\cdot]$—preferably those, which are cheap to evaluate in the course of the damped Newton iteration. For this purpose, we first recall the interpretation of the damped Newton method as the tangent continuation along the Newton path as given in Section 3.1.4. Upon measuring the deviation in an affine covariant setting, we are led to the upper bound

$$\|\overline{\Delta x}^{k+1}(\lambda) - (1 - \lambda)\Delta x^k\| \le \tfrac{1}{2}\lambda^2\omega_k\|\Delta x^k\|^2,$$

which leads to estimates for the Kantorovich quantities

$$[h_k] = [\omega_k]\|\Delta x^k\| := \frac{2\|\overline{\Delta x}^{k+1}(\lambda) - (1-\lambda)\Delta x^k\|}{\lambda^2\|\Delta x^k\|} \leq h_k\,.$$

The evaluation of such an estimate requires at least one trial value λ_k^0 (or x^{k+1}, respectively). As a consequence, it can only be helpful in the design of a *correction strategy* for the damping factor :

$$\lambda_k^{j+1} := \min\left(\tfrac{1}{2}\lambda, 1/[h_k^j]\right)\big|_{\lambda=\lambda_k^j} \tag{3.48}$$

for repetition index $j = 0, 1, \dots$.

Prediction strategy. We are therefore still left with the task of constructing an efficient initial estimate λ_k^0. As it turns out, such an estimate can only be gained by switching from the above defined Lipschitz constant ω_k to some slightly different definition:

$$\|F'(x^k)^{-1}\big(F'(x) - F'(x^k)\big)v\| \leq \overline{\omega}_k\|x - x^k\|\|v\|\,,$$

wherein the direction v is understood to be 'not too far away from' the direction $x - x^k$ in order to mimic the above definition (3.42). With this modified Lipschitz condition we may proceed to derive the following bounds:

$$\|\overline{\Delta x}^k - \Delta x^k\| = \left\| \big(F'(x^{k-1})^{-1} - F'(x^k)^{-1}\big)F(x^k) \right\| =$$

$$= \left\| F'(x^k)^{-1}\big(F'(x^k) - F'(x^{k-1})\big)\overline{\Delta x}^k \right\| \leq \overline{\omega}_k\lambda_{k-1}\|\Delta x^{k-1}\| \cdot \|\overline{\Delta x}^k\|.$$

This bound inspires the local estimate

$$[\overline{\omega}_k] := \frac{\|\overline{\Delta x}^k - \Delta x^k\|}{\lambda_{k-1}\|\Delta x^{k-1}\| \cdot \|\overline{\Delta x}^k\|} \leq \overline{\omega}_k\,,$$

wherein, as required in the definition above, the direction $\overline{\Delta x}^k$ is 'not too far away from' the direction Δx^{k-1}. In any case, the above computational estimate exploits the 'newest' information that is available in the course of the algorithm just before deciding about the initial damping factor. We have thus constructed a *prediction strategy* for $k > 0$:

$$\lambda_k^0 := \min(1, \mu_k)\,, \quad \mu_k := \frac{\|\Delta x^{k-1}\|}{\|\overline{\Delta x}^k - \Delta x^k\|} \cdot \frac{\|\overline{\Delta x}^k\|}{\|\Delta x^k\|} \cdot \lambda_{k-1}\,. \tag{3.49}$$

The only empirical choice left to be made is the starting value λ_0^0. In the public domain code NLEQ1 (see [161]) this value is made an input parameter: if the user classifies the problem as 'mildly nonlinear', then $\lambda_0^0 = 1$ is set internally; otherwise the problem is regarded as 'highly nonlinear' and $\lambda_0^0 = \lambda_{\min} \ll 1$ is set internally.

Intermediate quasi-Newton steps. Whenever $\lambda_k = 1$ and the natural monotonicity test yields

$$\Theta_k = \frac{\|\overline{\Delta x}^{k+1}\|}{\|\Delta x^k\|} < \tfrac{1}{2},$$

then the error oriented quasi-Newton method of Section 2.1.4 may be applied in the present context—compare Theorem 2.9. In this case, Jacobian evaluations are replaced by *Broyden rank-1 updates*. As for a possible switch back from quasi-Newton steps to Newton steps just look into the details of the informal quasi-Newton algorithm QNERR.

Termination criterion. Instead of the termination criterion (2.14) we may here use its cheaper substitute

$$\|\overline{\Delta x}^{k+1}\| \leq \text{XTOL} .$$

Recall that then $x^{k+2} = x^{k+1} + \overline{\Delta x}^{k+1}$ is cheaply available with an accuracy of $O(\text{XTOL}^2)$.

The here described adaptive trust region strategy leads to the global Newton algorithm NLEQ-ERR, which is a slight modification of the quite popular code NLEQ1 [161].

Algorithm NLEQ-ERR. Set a required *error accuracy* ε sufficiently above the machine precision.

Guess an initial iterate x^0. Evaluate $F(x^0)$.

Set a damping factor either $\lambda_0 := 1$ or $\lambda_0 \ll 1$.

All norms of corrections below are understood to be scaled smooth norms such as $\|D^{-1} \cdot \|_2$, where D is a diagonal scaling matrix, which can be iteratively adapted together with the Jacobian matrix.

For iteration index $k = 0, 1, \ldots$ do:

1. **Step** k: Evaluate Jacobian matrix $F'(x^k)$. Solve linear system

$$F'(x^k)\Delta x^k = -F(x^k) .$$

Convergence test: If $\|\Delta x^k\| \leq \varepsilon$: **stop**. Solution found $x^* := x^k + \Delta x^k$.

For $k > 0$: compute a prediction value for the damping factor

$$\lambda_k := \min(1, \mu_k), \quad \mu_k := \frac{\|\Delta x^{k-1}\| \cdot \|\overline{\Delta x}^k\|}{\|\overline{\Delta x}^k - \Delta x^k\| \cdot \|\Delta x^k\|} \cdot \lambda_{k-1} .$$

Regularity test: If $\lambda_k < \lambda_{\min}$: **stop**. Convergence failure.

2. **Else**: compute the trial iterate $x^{k+1} := x^k + \lambda_k \Delta x^k$ and evaluate $F(x^{k+1})$. Solve linear system ('old' Jacobian, 'new' right hand side):

$$F'(x^k)\overline{\Delta x}^{k+1} = -F(x^{k+1}).$$

3. Compute the monitoring quantities

$$\Theta_k := \frac{\|\overline{\Delta x}^{k+1}\|}{\|\Delta x^k\|}, \quad \mu'_k := \frac{\frac{1}{2}\|\Delta x^k\| \cdot \lambda_k^2}{\|\overline{\Delta x}^{k+1} - (1-\lambda_k)\Delta x^k\|}.$$

If $\Theta_k \geq 1$ (or, if **restricted**: $\Theta_k > 1 - \lambda_k/4$):

> **then** replace λ_k by $\lambda'_k := \min(\mu'_k, \frac{1}{2}\lambda_k)$. **Go to** Regularity test.

Else: let $\lambda'_k := \min(1, \mu'_k)$.

If $\lambda'_k = \lambda_k = 1$:

> **If** $\|\overline{\Delta x}^{k+1}\| \leq \varepsilon$: **stop.**
>
> Solution found $x^* := x^{k+1} + \overline{\Delta x}^{k+1}$.
>
> **If** $\Theta_k < \frac{1}{2}$: switch to QNERR

Else: **If** $\lambda'_k \geq 4\lambda_k$: replace λ_k by λ'_k and **goto** 2.

Else: accept x^{k+1} as new iterate.
Goto 1 with $k \to k+1$.

In what follows we want to demonstrate the main feature of this algorithm at a rather simple, but very illustrative example for $n = 2$.

Example 3.2 [161]. The equations to be solved are

$$\exp(x^2 + y^2) - 3 = 0,$$
$$x + y - \sin(3(x+y)) = 0.$$

For this simple problem, *critical interfaces* with singular Jacobian can be calculated to be the straight line

$$y = x$$

and the family of parallels

$$y = -x \pm \tfrac{1}{3} \arccos\left(\tfrac{1}{3}\right) \pm \tfrac{2}{3}\pi \cdot j, \ j = 0, 1, 2 \ldots.$$

For illustration, the quadratic domain

$$-1.5 \leq x, \ y \leq 1.5$$

is picked out. This domain contains the six different solution points and five critical interfaces.

Computation of Newton paths. As derived in Section 3.1.4, the Newton path $\bar{x}(\lambda)$, $\lambda \in [0,1]$ may be defined either by the homotopy

$$F\big(\bar{x}(\lambda)\big) - (1-\lambda)F(x^k) = 0$$

or by the *initial value problem*

$$F'\,(\bar{x})\,\frac{d\bar{x}}{d\lambda} = -F(x^k)\,,\ \bar{x}(0) = x^k\,.$$

This implicit ordinary differential equation can be solved numerically, say, by implicit BDF codes [98] like DASSL [167] due to L. Petzold or by linearly implicit extrapolation codes like LIMEX [75, 79]. In any such discretization, linear subsystems of the kind

$$F'(\bar{x})\varDelta\bar{x} - \beta\varDelta\lambda F''(\bar{x})[F'(x^k)^{-1}F(x^k), \varDelta\bar{x}] = -\varDelta\lambda F(x^k)$$

must be solved. Apparently, this algorithmic approach involves *second order derivative* information in *tensor* form—to be compared with the above described global Newton methods, which involve second order derivative information only in *scalar* form (Lipschitz constant estimates $[\omega]$ entering into the adaptive trust region strategies). Note, however, that the Newton path should be understood as an underlying geometric concept rather than an object to be actually computed.

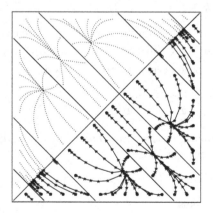

Fig. 3.10. Example 3.2: Newton paths ($\cdots$) versus Newton sequences (—).

Newton paths versus Newton sequences. Figure 3.10 shows the various Newton paths (left upper part) and sequences of Newton iterates (right lower part) as obtained by systematic variation of the initial guesses x_0—separated by the symmetry line $y = x$. The Newton paths have been integrated by

LIMEX, whereas the Newton sequences have been computed by NLEQ1 [161].
As predicted by theory—compare Section 3.1.4 and Figure 3.4 therein—each
Newton path either ends at a solution point or at a critical point with singular
Jacobian. The figure clearly documents that this same structure is mimicked
by the sequence of Newton iterates as selected by the error oriented trust
region strategy.

Attraction basins. An adjacent question of interest is the connectivity
structure of the different *attraction basins* for the global Newton iteration
around the different solution points. In order to visualize these structures, a
rectangular grid of starting points (with grid size $\Delta = 0.06$) has been defined
and the associated global Newton iteration performed.

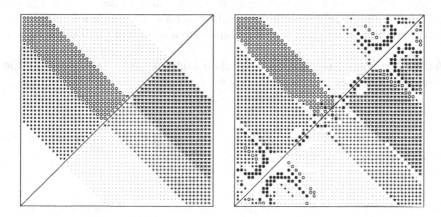

Fig. 3.11. **Example 3.2:** *attraction basins. Left:* Global Newton method, code
NLEQ1 [161]. *Right:* hybrid method, code HYBRJ1 [153]. Outliers are indicated as
bullets (•).

The results are represented in Figure 3.11: apart from very few exceptional
'corner points', the attraction basins nicely model the theoretical connectivity structure, which is essentially defined by the critical interfaces—a highly
satisfactory performance of the herein advocated global Newton algorithm
(code NLEQ1 due to [161]). For comparison, the attraction basins for a hybrid
method (code HYBRJ1 in MINPACK due to [153]) are given as well. There are
still some people who prefer the rather chaotic convergence pattern of such
algorithms. However, in most scientific and engineering problems, a *crossing
beyond critical interfaces* is undesirable, because this means an unnoticed
switching between different solutions—an important aspect especially in the
parameter dependent case.

3.3.4 Inexact Newton-ERR methods

Suppose that, instead of the *exact* Newton corrections Δx^k, we are only able to compute *inexact* Newton corrections δx^k from (dropping the inner iteration index i)

$$F'(x^k)\left(\delta x^k - \Delta x^k\right) = r^k, \quad x^{k+1} = x^k + \lambda_k \delta x^k, \quad 0 < \lambda_k \leq 1, \quad k = 0, 1, \dots .$$

As for local inexact Newton-ERR methods (Section 2.1.5), we characterize the inner iteration errors by the quantity

$$\delta_k = \frac{\|\delta x^k - \Delta x^k\|}{\|\delta x^k\|}. \tag{3.50}$$

Inner iterative solvers treated here are either CGNE or GBIT. As a guiding principle for global convergence, we will focus on *natural monotonicity* (3.41) subject to the perturbation coming from the truncation of the inner iteration.

Accuracy matching: inexact Newton corrections. First, we study the contraction factors

$$\Theta_k(\lambda) = \frac{\|\overline{\Delta x}^{k+1}\|}{\|\Delta x^k\|}$$

in terms of the exact Newton corrections Δx^k and the exact simplified Newton corrections $\overline{\Delta x}^{k+1}$ defined via

$$F'(x^k)\overline{\Delta x}^{k+1} = -F(x^k + \lambda \delta x^k).$$

Note that the *inexact* Newton correction arises in the argument on the right. Of course, none of the above exact Newton corrections will be actually computed.

Lemma 3.17 *We consider the inexact Newton iteration with CGNE or GBIT as inner iteration. Assume $\delta_k < \frac{1}{2}$. Then, with $h_k^\delta = \omega \|\delta x^k\|$, we obtain the estimate*

$$\Theta_k(\lambda) \leq 1 - \left(1 - \frac{\delta_k}{1 - \delta_k}\right)\lambda + \frac{1}{2}\lambda^2 \frac{h_k^\delta}{1 - \delta_k}. \tag{3.51}$$

The optimal damping factor is

$$\overline{\lambda}_k = \min\left(1, \frac{1 - 2\delta_k}{h_k^\delta}\right). \tag{3.52}$$

If we impose

$$\delta_k = \frac{\rho}{2}\lambda h_k^\delta, \quad \rho \leq 1, \tag{3.53}$$

we are led to the optimal damping factor

$$\overline{\lambda}_k = \min\left(1, \frac{1}{(1 + \rho)h_k^\delta}\right). \tag{3.54}$$

Proof. First, we derive the identity

$$\overline{\Delta x}^{k+1}(\lambda) = \Delta x^k - \lambda \delta x^k - F'(x^k)^{-1} \int_{s=0}^{\lambda} \left(F'(x^k + s\delta x^k) - F'(x^k) \right) \delta x^k \, ds.$$

Upon inserting definition (3.50) and using the triangle inequality

$$\frac{\|\delta x^k - \Delta x^k\|}{\|\Delta x^k\|} \leq \frac{\delta_k}{1 - \delta_k}, \qquad \|\delta x^k\| \leq \frac{\|\Delta x^k\|}{1 - \delta_k},$$

the above estimate (3.51) follows directly. The optimal damping factors $\overline{\lambda}_k$ in the two different forms arise by minimization of the upper bound parabola, as usual. □

Condition (3.53) is motivated by the idea that the $O(\lambda)$ perturbation due to the inner iteration should not dominate the $O(\lambda^2)$ term, which characterizes the nonlinearity of the problem.

Accuracy matching strategy. Upon inserting $\lambda = \overline{\lambda}_k$ into (3.53) and selecting some $\rho \leq 1$, we are led to

$$\delta_k \leq \overline{\delta} = \frac{\rho}{2(1 + \rho)} \leq \tfrac{1}{4} \quad \text{for} \quad \overline{\lambda}_k < 1 \tag{3.55}$$

and

$$\delta_k \leq \frac{\rho}{2} h_k^\delta \quad \text{for} \quad \overline{\lambda}_k = 1.$$

Of course, the realization of the latter rule will be done via computational Kantorovich estimates $[h_k^\delta] \leq h_k^\delta$ such that

$$\delta_k \leq \frac{\rho}{2} [h_k^\delta] \quad \text{for} \quad \overline{\lambda}_k = 1. \tag{3.56}$$

Obviously, the relation (3.55) reflects the 'fight for the first binary digit' as discussed in the preceding section; under this condition the optimal damping factors (3.52) and (3.54) are identical. In passing we note that requirement (3.56) nicely agrees with the 'quadratic convergence mode' (2.62) in the *local* Newton-ERR methods. (The slight difference reflects the different contraction factors in the local and the global case.) The condition (3.56) is a simple nonlinear scalar equation for an upper bound of δ_k.

As already mentioned at the beginning of this section, the exact natural monotonicity test cannot be directly implemented within our present algorithmic setting. We will, however, use this test and the corresponding optimal damping factor as a guideline.

Accuracy matching: inexact simplified Newton corrections. In order to construct an appropriate substitute for the nonrealizable Θ_k, we recur to the *inexact Newton path* $\widetilde{x}(\lambda), \lambda \in [0,1]$, from (3.34), which satisfies

$$F\left(\widetilde{x}(\lambda)\right) = (1 - \lambda)F(x^k) + \lambda r^k .$$

Recall that the local inexact Newton correction δx^k can be interpreted as the tangent direction $\dot{\widetilde{x}}(0)$ in x^k. On this background, we are led to define a perturbed (exact) simplified Newton correction via

$$F'(x^k)\widetilde{\Delta x}^{k+1} = -F(x^{k+1}) + r^k . \tag{3.57}$$

Lemma 3.18 *With the notation and definitions of this section the following estimate holds:*

$$\|\widetilde{\Delta x}^{k+1} - (1 - \lambda)\delta x^k\| \leq \tfrac{1}{2}\lambda^2 h_k^\delta\|\delta x^k\| . \tag{3.58}$$

Proof. It is easy to verify the identity

$$\widetilde{\Delta x}^{k+1}(\lambda) - (1 - \lambda)\delta x^k = -F'(x^k)^{-1} \int_{s=0}^{\lambda} \left(F'(x^k + s\delta x^k) - F'(x^k)\right) \delta x^k ds .$$

From this identity, the above estimate can be immediately derived in the usual manner. □

Of course, the linear equation (3.57) can only be solved iteratively. This means the computation of an *inexact simplified Newton correction* satisfying

$$F'(x^k)\widetilde{\delta x_i}^{k+1} = \left(-F(x^{k+1}) + r^k\right) + \widetilde{r}_i^{k+1}$$

for inner iteration index $i = 0, 1, \dots$. (In what follows, we will drop this index wherever convenient.)

Initial values for inner iterations. In view of (3.58) we set the initial value

$$\widetilde{\delta x_0}^{k+1} = (1 - \lambda)\delta x^k . \tag{3.59}$$

This means that the inner iteration has to recover only second order information. The same idea also supplies an initial guess for the inner iteration of the inexact ordinary Newton corrections:

$$\delta x_0^k = \widetilde{\delta x}^k . \tag{3.60}$$

This cross-over of initial values has proven to be really important in the realization of any Newton-ERR method.

In what follows, we replace the nonrealizable contraction factor $\Theta_k(\lambda)$ by its realizable inexact counterpart

$$\widetilde{\Theta}_k(\lambda) = \frac{\|\widetilde{\delta x}^{k+1}\|}{\|\delta x^k\|} \tag{3.61}$$

and study its dependence on the damping factor λ.

We start with CGNE as inner iterative solver.

Lemma 3.19 *Notation as just introduced. Assume that the inner CGNE iteration with initial guess (3.59) has been continued up to some iteration index $i > 0$ such that*

$$\widetilde{\rho}_i = \frac{\|\widetilde{\Delta x}^{k+1} - \widetilde{\delta x}_i^{k+1}\|}{\|\widetilde{\Delta x}^{k+1} - \widetilde{\delta x}_0^{k+1}\|} < 1 . \tag{3.62}$$

Then we obtain the estimate

$$\|\widetilde{\delta x}_i^{k+1} - (1 - \lambda)\delta x^k\| \le \tfrac{1}{2}\sqrt{1 - \widetilde{\rho}_i^2}\,\lambda^2 h_k^\delta\|\delta x^k\| . \tag{3.63}$$

Proof. In our context, the orthogonal decomposition (1.28) reads

$$\|\widetilde{\Delta x}^{k+1} - \widetilde{\delta x}_0^{k+1}\|^2 = \|\widetilde{\Delta x}^{k+1} - \widetilde{\delta x}_i^{k+1}\|^2 + \|\widetilde{\delta x}_i^{k+1} - \widetilde{\delta x}_0^{k+1}\|^2 . \tag{3.64}$$

Insertion of (3.62) then leads to

$$(1 - \widetilde{\rho}_i^2)\|\widetilde{\Delta x}^{k+1} - \widetilde{\delta x}_0^{k+1}\|^2 = \|\widetilde{\delta x}_i^{k+1} - \widetilde{\delta x}_0^{k+1}\|^2 . \tag{3.65}$$

With the insertion of (3.65) into (3.58) the proof is complete. $\square$

Observe that in CGNE the condition $\widetilde{\rho}_i < 1$ arises by construction. The parameter $\widetilde{\rho}_i$, however, is not directly computable from (3.62): the denominator cannot be evaluated, since we do not have $\widetilde{\Delta x}^{k+1}$, but for the numerator a rough estimate

$$\widetilde{\epsilon}_i \approx \|\widetilde{\Delta x}^{k+1} - \widetilde{\delta x}_i^{k+1}\|$$

is available (see Section 1.4.3). Therefore we may define the computable parameter

$$\overline{\rho}_i = \frac{\|\widetilde{\Delta x}^{k+1} - \widetilde{\delta x}_i^{k+1}\|}{\|\widetilde{\delta x}_i^{k+1} - \widetilde{\delta x}_0^{k+1}\|} \approx \frac{\widetilde{\epsilon}_i}{\|\widetilde{\delta x}_i^{k+1} - \widetilde{\delta x}_0^{k+1}\|} . \tag{3.66}$$

By means of (3.65), we then get

$$\|\widetilde{\Delta x}^{k+1} - \widetilde{\delta x}_i^{k+1}\| = \overline{\rho}_i\|\widetilde{\delta x}_i^{k+1} - \widetilde{\delta x}_0^{k+1}\| = \overline{\rho}_i\sqrt{1 - \widetilde{\rho}_i^2}\|\widetilde{\Delta x}^{k+1} - \widetilde{\delta x}_0^{k+1}\| .$$

This result can be compared with (3.62) to supply the identification

$$\bar{\rho}_i = \tilde{\rho}_i / \sqrt{1 - \tilde{\rho}_i^2} \quad \text{or} \quad \tilde{\rho}_i = \bar{\rho}_i / \sqrt{1 + \bar{\rho}_i^2} \,. \tag{3.67}$$

For GBIT as inner iterative solver, we also use $\bar{\rho}_i$ from (3.66), but in combination with a slightly different estimate.

Lemma 3.20 *Notation as in the preceding lemma. Let $\tilde{\rho}_i < 1$ according to (3.62). Then, for GBIT as inner iteration, we obtain*

$$\|\widetilde{\delta x}_i^{k+1} - (1 - \lambda)\delta x^k\| \le \frac{1 + \tilde{\rho}_i}{2} \lambda^2 h_k^\delta \|\delta x^k\| \,. \tag{3.68}$$

Proof. We drop the iteration index i. For GBIT, we cannot do better than apply the triangle inequality

$$\|\widetilde{\delta x}^{k+1} - (1 - \lambda)\delta x^k\| \le \|\widetilde{\Delta x}^{k+1} - (1 - \lambda)\delta x^k\| + \|\widetilde{\delta x}^{k+1} - \widetilde{\Delta x}^{k+1}\| \,.$$

With the requirement (3.62) we get

$$\|\widetilde{\delta x}^{k+1} - (1 - \lambda)\delta x^k\| \le (1 + \tilde{\rho})\|\widetilde{\Delta x}^{k+1} - (1 - \lambda)\delta x^k\| \,. \tag{3.69}$$

Application of Lemma 3.18 then directly verifies the estimate (3.68).   □

Note that in GBIT the condition $\tilde{\rho}_i < 1$ is not automatically fulfilled, but must be assured by the implementation. In order to actually estimate $\tilde{\rho}_i$, we again recur to $\bar{\rho}_i$ from (3.66). If we combine (3.66) with (3.69), we now arrive at

$$\|\widetilde{\Delta x}^{k+1} - \widetilde{\delta x}_i^{k+1}\| \le \bar{\rho}_i(1 + \tilde{\rho}_i)\|\widetilde{\Delta x}^{k+1} - \widetilde{\delta x}_0^{k+1}\| \,.$$

Comparison with (3.62) then supplies the identification

$$\bar{\rho}_i = \tilde{\rho}_i / (1 + \tilde{\rho}_i) \quad \text{or} \quad \tilde{\rho}_i = \bar{\rho}_i / (1 - \bar{\rho}_i) \text{ for } \bar{\rho}_i < 1 \,. \tag{3.70}$$

Accuracy matching strategy. On the basis of the presented convergence analysis, we might suggest to run the inner iteration until

$$\tilde{\rho}_i \le \tilde{\rho}_{\max} \quad \text{with} \quad \tilde{\rho}_{\max} \le \tfrac{1}{4}$$

for both CGNE and GBIT. By means of the transformations (3.67) or (3.70), respectively, this idea can be transferred to the realizable strategy

$$\bar{\rho}_i \le \bar{\rho}_{\max} \,. \tag{3.71}$$

Affine covariant Kantorovich estimates. Upon applying our algorithmic paradigm from Section 1.2.3, we will replace the optimal damping factor $\bar{\lambda}_k$ by computational estimates

$$[\overline{\lambda}_k] = \min\left(1, \frac{1 - 2\delta_k}{[h_k^\delta]}\right) = \min\left(1, \frac{1}{(1 + \rho)[h_k^\delta]}\right), \tag{3.72}$$

where $[h_k^\delta] = [\omega]\|\delta x^k\| \leq h_k^\delta$ are Kantorovich estimates to be carefully selected.

For CGNE, we will exploit (3.63) thus obtaining the *a-posteriori* estimates

$$[h_k^\delta]_i = \frac{2\|\widetilde{\delta x_i}^{k+1} - \widetilde{\delta x_0}^{k+1}\|_2}{\sqrt{1 - \widetilde{\rho}_i^2 \lambda^2 \|\delta x^k\|_2}} \leq h_k^\delta.$$

Note that (3.64) assures the saturation property

$$[h_k^\delta]_i \leq [h_k^\delta]_{i+1} \leq h_k^\delta.$$

Replacing $\widetilde{\rho}_i$ by $\overline{\rho}_i$ then gives rise to the computable a-posteriori estimate

$$[h_k^\delta]_i \approx \frac{2\sqrt{1 + \overline{\rho}_i^2}\|\widetilde{\delta x_i}^{k+1} - \widetilde{\delta x_0}^{k+1}\|_2}{\lambda^2 \|\delta x^k\|_2}. \tag{3.73}$$

For GBIT, we will exploit (3.68) and obtain the a-posteriori estimates

$$[h_k^\delta]_i = \frac{2\|\widetilde{\delta x_i}^{k+1} - \widetilde{\delta x_0}^{k+1}\|}{(1 + \widetilde{\rho}_i)\lambda^2 \|\delta x^k\|} \leq h_k^\delta.$$

Here a saturation property does not hold. Replacement of $\widetilde{\rho}_i$ by $\overline{\rho}_i$ leads to the computable a-posteriori estimates

$$[h_k^\delta]_i \approx \frac{2(1 - \overline{\rho}_i)\|\widetilde{\delta x_i}^{k+1} - \widetilde{\delta x_0}^{k+1}\|}{\lambda^2 \|\delta x^k\|}. \tag{3.74}$$

As for the construction of computational *a-priori* Kantorovich estimates, we suggest to simply go back to the definitions and realize the estimate

$$[h_{k+1}^\delta] = \frac{\|\delta x^{k+1}\|}{\|\delta x^k\|}[h_k^\delta]_*, \tag{3.75}$$

where $[h_k^\delta]_*$ denotes the final estimate $[h_k^\delta]_i$ from either (3.73) for CGNE or (3.74) for GBIT, i.e. the estimate obtained at the final inner iteration step $i = \widetilde{i}_k$ of the previous outer iteration step k.

Bit counting lemma. Once computational estimates $[h_k^\delta]$ are available, we may realize the damping strategy (3.72). In analogy to Lemma 3.16, we now study the influence of the accuracy of the Kantorovich estimates.

Lemma 3.21 *Notation as just introduced. Let an inexact Newton-ERR method with damping factors* $\lambda = [\bar{\lambda}_k]$ *due to (3.72) be realized. Assume that*

$$0 \le h_k^\delta - [h_k^\delta] < \sigma \max\left(\frac{1}{1+\rho}, [h_k^\delta]\right) \text{ for some } \sigma < 1. \tag{3.76}$$

Then the exact natural contraction factor satisfies

$$\Theta_k(\lambda) = \frac{\|\overline{\Delta x}^{k+1}\|}{\|\Delta x^k\|} < 1 - \frac{1}{2+\rho}(1-\sigma)\lambda .$$

For CGNE, *the inexact natural contraction factor is bounded by*

$$\widetilde{\Theta}_k = \frac{\|\widetilde{\delta x}^{k+1}\|}{\|\delta x^k\|} < 1 - \left(1 - \frac{1}{2}\frac{1+\sigma}{1+\rho}\right)\lambda .$$

For GBIT, *the inexact natural contraction factor is bounded by*

$$\widetilde{\Theta}_k = \frac{\|\widetilde{\delta x}^{k+1}\|}{\|\delta x^k\|} < 1 - \left(1 - \frac{1}{2}\frac{(1+\widetilde{\rho})(1+\sigma)}{1+\rho}\right)\lambda .$$

Proof. Throughout this proof, we will omit any results for $\lambda = 1$, since these can be directly verified by mere insertion. This means that we assume $\lambda = [\bar{\lambda}_k] < 1$ in the following.

For the exact natural monotonicity test we return to the inequality (3.51), which reads

$$\widetilde{\Theta}_k \le 1 - \frac{1-2\delta_k}{2(1-\delta_k)}\lambda + \frac{1}{2}\lambda^2\frac{h_k^\delta}{2(1-\delta_k)} .$$

Insertion of $\lambda = [\bar{\lambda}_k] < 1$ then yields

$$\lambda h_k^\delta = (1-2\delta_k)\frac{h_k^\delta}{[h_k^\delta]} < \frac{1+\sigma}{1+\rho} .$$

Inserting this into the above upper bound and switching from the parameter δ_k to ρ via (3.55) then verifies the first statement of the lemma.

For the inexact natural monotonicity test with CGNE as inner iterative solver, we go back to (3.63), which yields

$$\widetilde{\Theta}_k(\lambda) \le 1 - \lambda + \frac{1}{2}\sqrt{1-\widetilde{\rho}_i^2}\lambda^2 h_k^\delta < 1 - \lambda + \frac{1}{2}\lambda^2 h_k^\delta .$$

If we again insert the above upper bound λh_k^δ, we arrive at

$$\widetilde{\Theta}_k < 1 - \lambda + \frac{1}{2}\lambda\frac{1+\sigma}{1+\rho} ,$$

which is equivalent to the second statement of the lemma.

For the inexact natural monotonicity test with GBIT as inner iterative solver, we recur to (3.68), which yields

$$\widetilde{\Theta}_k(\lambda) \leq 1 - \lambda + \frac{1 + \widetilde{\rho}}{2} \lambda^2 h_k^\delta \, .$$

Following the same lines as for CGNE now supplies the upper bound

$$\widetilde{\Theta}_k < 1 - \lambda + \frac{1 + \widetilde{\rho}}{2} \lambda \frac{1 + \sigma}{1 + \rho} \, ,$$

which finally confirms the third statement of the lemma.   □

Inexact natural monotonicity tests. Suppose now that we require at least one binary digit in the Kantorovich estimate, i.e., $\sigma < 1$ in Lemma 3.21. In this case, exact natural monotonicity

$$\Theta_k(\lambda) = \frac{\|\overline{\Delta x}^{k+1}\|}{\|\Delta x^k\|} < 1$$

would hold—which, however, is not realizable in the present algorithmic setting.

For CGNE, a computable substitute is the inexact natural monotonicity test

$$\widetilde{\Theta}_k = \frac{\|\overline{\delta x}^{k+1}\|}{\|\delta x^k\|} < 1 - \frac{\rho}{1 + \rho} \lambda \, . \tag{3.77}$$

For GBIT, we similarly get

$$\widetilde{\Theta}_k = \frac{\|\widetilde{\delta x}^{k+1}\|}{\|\delta x^k\|} < 1 - \frac{\rho - \widetilde{\rho}}{1 + \rho} \lambda \, . \tag{3.78}$$

The latter result seems to suggest the setting $\widetilde{\rho} \leq \rho$ to assure $\widetilde{\Theta}_k < 1$; otherwise inexact natural monotonicity need not hold.

Correction strategy. This part of the adaptive trust region strategy applies, if inexact natural monotonicity, (3.77) for CGNE or (3.78) for GBIT, is violated. Then the damping strategy can be based on the *a-posteriori* Kantorovich estimates (3.73) for CGNE or (3.74) for GBIT, respectively, again written as $[h_k^\delta]_*$. In view of the exact correction strategy (3.48) with repetition index $j = 0, 1, \ldots$, we set

$$\lambda_k^{j+1} := \min \left(\tfrac{1}{2} \lambda, \frac{1}{(1 + \rho)[h_k^\delta]_*} \right) \Big|_{\lambda = \lambda_k^j} \, .$$

Prediction strategy. This part of the trust region strategy is based on the *a-priori* Kantorovich estimates (3.75). On the basis of information from outer iteration step $k - 1$, we obtain

$$\lambda_k^0 = \min\left(1, \frac{1}{(1 + \rho)[h_k^\delta]}\right), \quad k > 0.$$

For $k = 0$, we can only start with some prescribed initial value λ_0^0 to be chosen by the user.

If $\lambda < \lambda_{\min}$ for some threshold value $\lambda_{\min} \ll 1$ arises in the prediction or the correction strategy, then the outer iteration must be terminated—indicating some critical point with ill-conditioned Jacobian.

The described inexact Newton-ERR methods are realized in the programs GIANT-CGNE and GIANT-GBIT with error oriented adaptive trust region strategy and corresponding matching between inner and outer iteration. The realization of the local Newton part here is slightly different from the one suggested in the previous Section 2.1.5, since here we have the additional information of the inexact simplified Newton correction $\widetilde{\delta x}$ available.

Algorithms GIANT-CGNE and GIANT-GBIT.

1. **Step k** Evaluate $F'(x^k)$.
   Solve linear system

   $$F'(x^k)\delta x_i^k = -F(x^k) + r_i^k \quad \text{for} \quad i = 0, 1, \ldots, i_k$$

   iteratively by CGNE or GBIT. Control of i_k via accuracy matching strategy (3.55) or (3.56).

   If $\|\delta x^k\| \leq$ XTOL: **Solution:** $x^* = x^k + \delta x^k$.

   Else: For $k = 0$: select λ_0 ad hoc.
   For $k > 0$: determine $\lambda_k = \min\left(1, \frac{1}{(1 + \rho)[h_k^\delta]}\right)$ from the prediction strategy.

   Regularity test. If $\lambda_k < \lambda_{\min}$: **stop**

2. **Else:** compute trial iterate $x^{k+1} := x^k + \lambda_k \delta x^k$ and evaluate $F(x^{k+1})$.
   Solve linear system

   $$F'(x^k)\widetilde{\delta x}_i^{k+1} = \left(-F(x^{k+1}) + r^k\right) + \widetilde{r}_i^{k+1} \quad \text{for} \quad i = 0, 1, \ldots, \widetilde{i}_k$$

   iteratively by CGNE or GBIT. Control of $\widetilde{i}_k$ via accuracy matching strategy (3.71).

Computation of Kantorovich estimates $[h_k^\delta]$.

3. Evaluate the monitor

$$\widetilde{\Theta}_k := \frac{\|\widetilde{\delta x}^{k+1}\|_2}{\|\delta x^k\|_2} .$$

If monotonicity test (3.77) for CGNE or (3.78) for GBIT violated, **then** refine λ_k according to correction strategy and go to **regularity test**.

Else go to 1.

As soon as the global Newton-ERR method approaches the solution point, one may either directly switch to the local Newton-ERR methods presented in Section 2.1.5 or merge the Kantorovich estimates from here with the 'standard', 'linear', or 'quadratic' convergence mode as described there.

Remark 3.2   The *residual based* algorithm GIANT-GMRES as presented in Section 3.2.3 above seems to represent an easier implementable alternative to the here elaborated error oriented algorithms GIANT-CGNE and GIANT-GBIT. This is only true, if a 'good' *left* preconditioner C_L is available. Indeed, if spectral equivalence $C_L A \sim I$ holds, then the *preconditioned initial residual* satisfies $C_L r_0 \sim x^0 - x^*$. Otherwise, the here presented error oriented algorithms realize some *nonlinear preconditioning*.

A numerical comparison of GIANT-CGNE, GIANT-GBIT, and NLEQ-ERR in the context of discretized nonlinear PDEs is given in Section 8.2.1 below.

BIBLIOGRAPHICAL NOTE. The first affine covariant convergence proof for local inexact Newton methods has been given by T.J. Ypma [203]. A first error oriented global inexact Newton algorithm has been suggested by P. Deuflhard [67] on the basis of some slightly differing affine covariant convergence analysis. These suggestions led to the code GIANT by U. Nowak and coworkers [160], wherein the inner iteration has been realized by an earlier version of GBIT.

3.4 Convex Functional Descent

In the present section we want to minimize a general convex function f or, equivalently, solve the nonlinear system $F(x) = f'^T(x) = 0$ with $F'(x) = f''(x)$ symmetric positive definite. It is not at all clear whether for general functional the damped Newton method still is an efficient globalization. As the damped Newton method can be interpreted as a tangent continuation along the Newton path, we first study the behavior of an arbitrary convex functional f along the Newton path $\overline{x}(\lambda)$ as a function of λ.

Lemma 3.22 *Let $f \in C^2(D)$ denote some strictly convex functional to be minimized over some convex domain $D \in \mathbb{R}^n$. Let $F'(x) = f''(x)$ be symmetric positive definite in D and let $\overline{x} : [0, 1] \to D$ denote the Newton path starting at some iterate $\overline{x}(0) = x^k$ and ending at the solution point $\overline{x}(1) = x^*$ with $F(x^*) = f'^T(x^*) = 0$. Then $f(\overline{x}(\lambda))$ is a strictly monotone decreasing function of λ.*

Proof. In the usual way we just verify that

$$f(\overline{x}(\lambda)) - f(x^k) = \int_{\sigma=0}^{\lambda} \langle F(\overline{x}(\sigma)), \dot{\overline{x}}(\sigma) \rangle \, d\sigma \, .$$

Insertion of (3.22) and (3.24) then leads to

$$f(\overline{x}(\lambda)) - f(x^k) = - \int_{\sigma=0}^{\lambda} (1 - \sigma) \| F'(\overline{x}(\sigma))^{-1/2} F(x^k) \|_2^2 d\sigma$$

with a strictly positive definite integrand. Therefore, for $0 \le \lambda_2 < \lambda_1 \le 1$:

$$f(\overline{x}(\lambda_1)) - f(\overline{x}(\lambda_2)) = - \int_{\sigma=\lambda_2}^{\lambda_1} (1 - \sigma) \| F'(\overline{x}(\sigma))^{-1/2} F(x^k) \|_2^2 d\sigma < 0 \, .$$

$\square$

Obviously, this result is the desired generalization of the monotone level function decrease (3.23). We are now ready to analyze the *damped Newton iteration* $(k = 0, 1, \dots)$

$$F'(x^k)\Delta x^k = -F(x^k), \qquad x^{k+1} = x^k + \lambda_k \Delta x^k, \qquad \lambda_k \in]0, 1]$$

under the requirement of iterative *functional decrease* $f(x^{k+1}) < f(x^k)$.

3.4.1 Affine conjugate convergence analysis

As in the preceding sections, we first study the local reduction properties of the damped Newton method within one iterative step from iterate x^k to iterate x^{k+1}.

Theorem 3.23 *Let $f : D \to \mathbb{R}^1$ be a strictly convex C^2-functional to be minimized over some open convex domain $D \subset \mathbb{R}^n$. Let $F(x) = f'(x)^T$ and $F'(x) = f''(x)$ symmetric and strictly positive definite. For $x, y \in D$, assume the special affine conjugate Lipschitz condition*

$$\|F'(x)^{-1/2}(F'(y) - F'(x))(y - x)\| \le \omega\|F'(x)^{1/2}(y - x)\|^2 \qquad (3.79)$$

with $0 \le \omega < \infty$. *For some iterate* $x^k \in D$, *define the quantities*

$$\epsilon_k := \|F'(x^k)^{1/2}\Delta x^k\|_2^2\,, \quad h_k := \omega\|F'(x^k)^{1/2}\Delta x^k\|_2\,.$$

Moreover, let $x^k + \lambda\Delta x^k \in D$ *for* $0 \le \lambda \le \lambda_{\max}^k$ *with*

$$\lambda_{\max}^k := \frac{4}{1 + \sqrt{1 + 8h_k/3}} \le 2\,.$$

Then

$$f(x^k + \lambda\Delta x^k) \le f(x^k) - t_k(\lambda)\epsilon_k\,, \qquad (3.80)$$

where

$$t_k(\lambda) = \lambda - \tfrac{1}{2}\lambda^2 - \tfrac{1}{6}\lambda^3 h_k\,. \qquad (3.81)$$

The optimal choice of damping factor is

$$\overline{\lambda}_k = \frac{2}{1 + \sqrt{1 + 2h_k}} \le 1\,. \qquad (3.82)$$

Proof. Dropping the iteration index k, we apply the usual mean value theorem to obtain

$$f(x + \lambda\Delta x) - f(x) =$$
$$-\lambda\epsilon + \tfrac{1}{2}\lambda^2\epsilon + \lambda^2 \int\limits_{s=0}^{1} \int\limits_{t=0}^{1} s\,\langle \Delta x, (F'(x + st\lambda\Delta x) - F'(x))\Delta x\rangle\,dtds\,.$$

Upon recalling the Lipschitz condition (3.79), the Cauchy-Schwarz inequality yields

$$f(x + \lambda\Delta x) - f(x) + \left(\lambda - \tfrac{1}{2}\lambda^2\right)\epsilon$$
$$\le \lambda^3 \int\limits_{s=0}^{1} \int\limits_{t=0}^{1} s^2 t\|F'(x)^{1/2}\Delta x\|^3 dtds = \tfrac{1}{6}\lambda^3 h \cdot \epsilon\,, \qquad (3.83)$$

which confirms (3.80) and the cubic parabola (3.81). Maximization of t_k by $t_k' = 0$ and solving the arising quadratic equation then yields $\overline{\lambda}_k$ as in (3.84). Moreover, by observing that

$$t_k = \lambda\left(1 - \tfrac{1}{2}\lambda - \tfrac{1}{6}\lambda^2 h_k\right) = 0$$

has only one *positive* root $\lambda_{\max}^k$, the remaining statements are readily verified. $\qquad \square$

From these *local* results, we may easily proceed to obtain the following *global* convergence theorem.

Theorem 3.24 *General assumptions as before. Let the path-connected component of the level set $\mathcal{L}_0 := \{x \in D \mid f(x) \leq f(x^0)\}$ be compact. Let $F'(x) = f''(x)$ be symmetric positive definite for all $x \in \mathcal{L}_0$. Then the damped Newton iteration $(k = 0, 1, \ldots)$ with damping factors in the range*

$$\lambda_k \in [\varepsilon, \; \min(1, \lambda_{\max}^k - \varepsilon)]$$

and sufficiently small $\varepsilon > 0$, which depends on $\mathcal{L}_0$, converges to the solution point x^*.

Proof. The proof just applies the local reduction results of the preceding Theorem 3.23. The essential remaining task to show is that there is a common minimal reduction factor for all possible arguments $x^k \in \mathcal{L}_0$. For this purpose, just construct a polygonal upper bound for $t_k(\lambda)$ comparable to the polygon in Figure 3.5. We then merely have to select ε such that

$$\varepsilon < \min(\overline{\lambda}_k, \lambda_{\max}^k - \overline{\lambda}_k)$$

for all possible iterates x^k. Omitting the technical details, Figure 3.5 then directly helps to verify that

$$f(x^k + \lambda \Delta x^k) \leq (1 - \gamma \varepsilon) f(x^k)$$

for λ in the above indicated range with some global $\gamma > 0$, which yields the desired strict global reduction of the functional. □

Summarizing, we have thus established the *theoretical optimal damping strategy* (3.82) in terms of the computationally unavailable Kantorovich quantities h_k.

Remark 3.3 It may be worth noting that the above analysis is nicely connected with the *local* Newton methods (i.e., with $\lambda = 1$) as discussed in Section 2.3.1. If we require that

$$\lambda_{\max}^k = \frac{4}{1 + \sqrt{1 + 8h_k/3}} \geq 1,$$

then we arrive at the local contraction condition

$$h_k \leq 3.$$

This is exactly the condition that would have been obtained in the proof of Theorem 2.18, if the requirement $f(x^{k+1}) \leq f(x^k)$ had been made for the ordinary Newton method—just compare (2.94). However, just as in the framework of that section, the condition $h_{k+1} \leq h_k$ also cannot be guaranteed here, so that $\lambda_{\max}^{k+1} \geq 1$ is not assured. In order to assure such a condition, the more stringent assumption $h_k < 2$ as in (2.92) would be required.

3.4.2 Adaptive trust region strategies

Following our algorithmic paradigm from Section 1.2.3, we construct *computational* damping strategies on the basis of the above derived *theoretically optimal* damping strategy. This strategy contains the unavailable Kantorovich quantity h_k, which we want to replace by some computational estimate $[h_k] \leq h_k$ and, consequently, the theoretical damping factor $\overline{\lambda}_k$ defined in (3.82) by some computationally available value

$$[\overline{\lambda}_k] := \frac{2}{1 + \sqrt{1 + 2[h_k]}} \leq 1. \qquad (3.84)$$

Since $[h_k] \leq h_k$, we have

$$[\overline{\lambda}_k] \geq \overline{\lambda}_k$$

so that both a *prediction strategy* and a *correction strategy* need to be developed.

Bit counting lemma. As already observed in the comparable earlier cases, the efficiency of such strategies depends on the required accuracy of the computational estimate, which we now analyze.

Lemma 3.25 *Standard assumptions and notation of this section. Let*

$$0 \leq h_k - [h_k] \leq \sigma[h_k] \quad \text{for some} \quad \sigma < 1. \qquad (3.85)$$

Then, for $\lambda = [\overline{\lambda}_k]$, the following functional decrease is guaranteed:

$$f(x^k + \lambda \Delta x^k) \leq f(x^k) - \tfrac{1}{6}\lambda(\lambda + 2)\epsilon_k. \qquad (3.86)$$

Proof. With $h_k \leq (1 + \sigma)[h_k]$ and (3.83) we have (dropping the index k)

$$f(x + \lambda \Delta x) - f(x) \leq -t_k(\lambda)\epsilon_k = \left(-\lambda + \tfrac{1}{2}\lambda^2 + \tfrac{1}{6}\lambda^3 h_k\right)\epsilon_k$$

$$\leq \left(-\lambda + \tfrac{1}{2}\lambda^2 + \tfrac{1}{6}\lambda^3(1 + \sigma)[h_k]\right)\epsilon_k.$$

At this point, recall that $\overline{\lambda}_k$ is a root of $t'_k = 0$ so that $[\overline{\lambda}_k]$ is a root of

$$1 - \lambda - \tfrac{1}{2}\lambda^2[h_k] = 0.$$

Insertion of the above quadratic term into the estimate then yields

$$f(x + \lambda \Delta x) - f(x) \leq \left(-\lambda + \tfrac{1}{2}\lambda^2 + \tfrac{1}{3}\lambda(1 + \sigma)(1 - \lambda)\right)\epsilon_k. \qquad (3.87)$$

Upon using $\sigma < 1$ (3.86) is confirmed. □

The above *functional monotonicity test* (3.86) is suggested for use in actual computation. If we further impose $\sigma = 1/2$ in (3.85), i.e., if we require at least one exact binary digit in the Kantorovich quantity estimate, then (3.87) leads to the *restricted functional monotonicity test*

$$f(x^k + \lambda \Delta x^k) - f(x^k) \le -\tfrac{1}{2}\lambda \epsilon_k .$$

We are now ready to discuss specific computational estimates $[h_k]$ of the Kantorovich quantities h_k. Careful examination shows that we have three basic cheap options. From (3.83) we have the *third* order bound

$$E_3(\lambda) := f(x^k + \lambda \Delta x^k) - f(x^k) + \lambda \left(1 - \tfrac{1}{2}\lambda\right)\epsilon_k \le \tfrac{1}{6}\lambda^3 h_k \epsilon_k ,$$

which, in turn, naturally inspires the computational estimate

$$[h_k] := \frac{6|E_3(\lambda)|}{\lambda^3 \epsilon_k} \le h_k .$$

If $E_3(\lambda) < 0$, this means that the Newton method performs locally better than for the mere quadratic model of f (equivalent to $h_k = 0$). Therefore, we decide to set

$$[\overline{\lambda}_k] = 1 , \quad \text{if } E_3(\lambda) < 0 .$$

On the level of the first derivative we have the *second* order bound

$$E_2(\lambda) := \left\langle \Delta x^k , F(x^k + \lambda \Delta x^k) - (1 - \lambda)F(x^k) \right\rangle \le \tfrac{1}{2}\lambda^2 h_k \epsilon_k ,$$

which inspires the associated estimate

$$[h_k] := \frac{2|E_2(\lambda)|}{\lambda^2 \epsilon_k} \le h_k .$$

On the second derivative level we may derive the *first* order bound

$$E_1(\lambda) := \left\langle \Delta x^k , \left(F'(x^k + \lambda \Delta x^k) - F'(x^k)\right) \Delta x^k \right\rangle \le \lambda h_k \epsilon_k ,$$

which leads to the associated estimate

$$[h_k] := \frac{|E_1(\lambda)|}{\lambda \epsilon_k} \le h_k .$$

Cancellation of leading digits in the terms E_i, $i = 1, 2, 3$ should be carefully monitored—see Figure 3.12, where a snapshot at some iterate in a not further specified illustrative example is taken. Even though the third order expression is the most sensitive, it is also the most attractive one from the point of view of simplicity. Hence, one should first try E_3 and monitor rounding errors carefully.

In principle, any of the above three estimates can be inserted into (3.84) for $[\overline{\lambda}_k]$ requiring at least one trial value of λ (or, respectively, x^{k+1}). We have therefore only designed a possible *correction strategy*

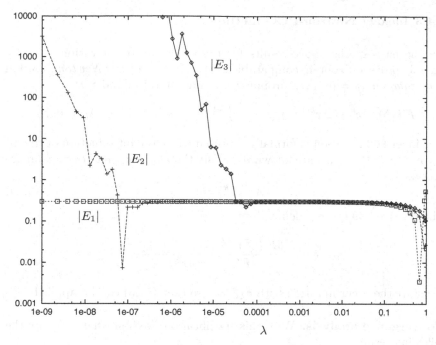

Fig. 3.12. Computational Kantorovich estimates $[h_k]$: cancellation of leading digits in $|E_3|$, $|E_2|$, $|E_1|$, respectively.

$$\lambda_k^{i+1} := \frac{2}{1 + \sqrt{1 + 2[h_k(\lambda)]}} \Big|_{\lambda = \lambda_k^i} . \tag{3.88}$$

In order to construct a theoretically backed initial estimate λ_k^0, we may recall that $h_{k+1} = \Theta_k h_k$, where

$$\Theta_k := \frac{\|F'(x^{k+1})^{-\frac{1}{2}} F(x^{k+1})\|_2}{\|F'(x^k)^{-\frac{1}{2}} F(x^k)\|_2} .$$

This relation directly inspires the estimate

$$[h_{k+1}^0] := \Theta_k [h_k^{i*},]$$

wherein i^* indicates the final computable index within estimate (3.88) for the previous iterative step k. Thus we are led to the following *prediction strategy* for $k \geq 0$:

$$\lambda_{k+1}^0 := \frac{2}{1 + \sqrt{1 + 2[h_{k+1}^0]}} \leq 1 . \tag{3.89}$$

As in the earlier discussed approaches, the starting value λ_0^0 needs to be set ad hoc—say, as $\lambda_0^0 = 1$ for 'mildly nonlinear' problems and as $\lambda_0^0 = \lambda_{\min} \ll 1$ for 'highly nonlinear' problems.

3.4.3 Inexact Newton-PCG method

On the basis of the above results for the exact Newton iteration, we may directly proceed to obtain comparable results for the *inexact Newton iteration with damping* ($k = 0, 1, \ldots$, dropping the inner iteration index i)

$$F'(x^k)\left(\delta x^k - \Delta x^k\right) = r^k, \qquad x^{k+1} = x^k + \lambda_k \delta x^k, \qquad \lambda_k \in \,]0, 1].$$

The inner PCG iteration is formally represented by the introduction of the inner residuals r^k, which are known to satisfy the *Galerkin condition* (compare Section 1.4).

$$\langle \delta x^k, r^k \rangle = 0. \tag{3.90}$$

The relative PCG error is denoted by

$$\delta_k := \frac{\|F'(x^k)^{1/2}(\Delta x^k - \delta x^k)\|}{\|F'(x^k)^{1/2}\delta x^k\|}.$$

We start the inner iteration with $\delta x_0^k = 0$ so that (1.26) can be applied.

Convergence analysis. With this specification, we immediately verify the following result.

Theorem 3.26 *The statements of Theorem* 3.23 *hold for the inexact Newton-PCG method as well, if only the exact Newton corrections* Δx^k *are replaced by the inexact Newton corrections* δx^k *and the quantities* ϵ_k, h_k *are replaced by*

$$\epsilon_k^\delta \quad := \quad \|F'(x^k)^{1/2}\delta x^k\|^2 = \frac{\epsilon_k}{1 + \delta_k^2},$$

$$h_k^\delta \quad := \quad \omega \|F'(x^k)^{1/2}\delta x^k\| = \frac{h_k}{\sqrt{1 + \delta_k^2}}.$$

Proof. Dropping the iteration index k, the first line of the proof of Theorem 3.23 may be rewritten as

$$f(x + \lambda \delta x) - f(x) =$$
$$-\lambda \epsilon^\delta + \tfrac{1}{2}\lambda^2 \epsilon^\delta + \lambda^2 \int\limits_{s=0}^{1} s \int\limits_{t=0}^{1} \left\langle \delta x, \left(F'(x + st\lambda \delta x) - F'(x)\right)\delta x \right\rangle dt\, ds + \langle \delta x, r \rangle,$$

wherein the last right hand term vanishes due to the Galerkin condition (3.90) so that merely the replacement of Δx by δx needs to be performed. $\square$

With these local results established, we are now ready to formulate the associated global convergence theorem.

Theorem 3.27 *General assumptions as Theorem* 3.23 *or Theorem* 3.26, *respectively (in the latter case δ_k is formally assumed to be bounded). Let the level set $\mathcal{L}_0 := \{x \in D \mid f(x) \le f(x^0)\}$ be closed and bounded. Let $F'(x) = f''(x)$ be symmetric strongly positive for all $x \in \mathcal{L}_0$. Then the damped (inexact) Newton iteration (for $k = 0, 1, \ldots$) with damping factors in the range*

$$\lambda_k \in [\varepsilon,\ \min(1, \lambda_{\max}^k - \varepsilon)]$$

and sufficiently small $\varepsilon > 0$ depending on $\mathcal{L}_0$ converges to the solution point x^*.

Proof. The proof just applies the local reduction results of the preceding Theorem 3.23 or Theorem 3.26. The essential remaining task to show is that there is a common minimal reduction factor for all possible arguments $x^k \in \mathcal{L}_0$. For this purpose, we simply construct a polygonal upper bound for $t_k(\lambda)$ such that (omitting technical details)

$$f(x^k + \lambda \Delta x^k) \le f(x^k) - \tfrac{1}{2}\varepsilon\epsilon_k$$

for λ in the above indicated range and all possible iterates x^k with some

$$\varepsilon < \min(\overline{\lambda}_k, \lambda_{\max}^k - \lambda_k).$$

This implies a strict reduction of the functional in each iterative step as long as $\epsilon_k > 0$ and therefore global convergence in the compact level set $\mathcal{L}_0$ towards the minimum point x^* with $\epsilon_* = 0$. □

Adaptive trust region strategy. The strategy as worked out in Section 3.4.2 can be directly copied, just replacing Δx^k by δx^k, ϵ_k by ϵ_k^δ, and h_k by h_k^δ; details are left to Exercise 3.5. In actual computation the orthogonality condition (3.90) may be perturbed by rounding errors from scalar products in PCG. Therefore the terms E_3 and E_2 should be evaluated in the special form

$$E_3(\lambda) := f(x^k + \lambda \delta x^k) - f(x^k) - \lambda\langle F(x^k), \delta x^k\rangle - \tfrac{1}{2}\lambda^2 \epsilon_k^\delta$$

and

$$E_2(\lambda) := \langle \delta x^k, F(x^k + \lambda \delta x^k) - F(x^k)\rangle - \lambda\epsilon_k^\delta$$

with the local energy computed as $\epsilon_k^\delta = \langle \delta x^k, F'(x^k)\delta x^k\rangle$.

As for the choice of accuracies δ_k arising from the inner PCG iteration, we once again require

$$\delta_k \le \tfrac{1}{4}$$

in the damping phase ($\lambda < 1$) and the appropriate settings as worked out in Section 2.3.3 for the local Newton-PCG ($\lambda = 1$)—merging either into the *linear* or the *quadratic* convergence mode.

The affine conjugate inexact Newton method with adaptive trust region method and corresponding matching of the inner PCG iteration and outer iteration is realized in the code GIANT-PCG.

BIBLIOGRAPHICAL NOTE. The first affine conjugate global Newton method has been derived and implemented by P. Deuflhard and M. Weiser [85], there even in the more complicated context of an adaptive multilevel finite element method for nonlinear elliptic PDEs—compare Section 8.3. The strategy presented here is just a finite dimensional analog of the strategy worked out there.

Exercises

Exercise 3.1 *Multipoint homotopy.* Let $F(x) = 0$ denote a system of nonlinear equations to be solved and x^* its solution. Let $x^0, \ldots, x^p$ be a sequence of iterates produced by some iterative process. Consider the homotopy $(\lambda \in \mathbb{R}^1)$

$$H_p(x, \lambda) := F(x) - \sum_{k=0}^{p-1} L_k(\lambda) F(x^k), \quad p \geq 1$$

with L_k being the fundamental Lagrangian polynomials defined over a set of nodes $0 = \lambda_0 < \lambda_1 < \cdots < \lambda_p = 1$.

a) Show that, under the standard assumptions of the implicit function theorem, there exists a homotopy path $\bar{x}(\lambda)$ such that

$$\bar{x}(\lambda_k) = x^k, \quad k = 0, \ldots, p-1, \quad \bar{x}(1) = x^*.$$

Derive the associated Davidenko differential equation.

b) Construct an iterative process for successively increasing $p = 1, 2, \ldots$ by appropriate discretization. What would be a reasonable assignment of the nodes $\lambda_1, \ldots, \lambda_{p-1}$? Consider the local convergence properties of such a process.

c) Write a program for $p = 1, 2$ and experiment with λ_1 over a set of test problems.

Exercise 3.2   An obstacle on the way toward a proof of global convergence for error oriented global Newton methods, controlled only by the natural monotonicity test

$$\|F'(x^k)^{-1}F(x^{k+1})\| \leq \|F'(x^k)^{-1}F(x^k)\|,$$

is the fact that a desirable property like

$$\|F'(x^{k+1})^{-1}F(x^{k+1})\| \leq \|F'(x^k)^{-1}F(x^k)\| \tag{3.91}$$

does *not* hold.

a) For $x^{k+1} = x^k + \lambda \Delta x^k$, upon applying Theorem 3.12, verify that

$$\|F'(x^{k+1})^{-1}F(x^{k+1})\| \leq \frac{1 - \lambda + \frac{1}{2}\lambda^2 h_k}{1 - \lambda h_k}\|F'(x^k)^{-1}F(x^k)\|\,.$$

b) Show that only under the Kantorovich-type assumption $h_k < 1$ the reduction (3.91) can be guaranteed for certain $\lambda > 0$.

Exercise 3.3 *2–cycle example* [10]. Consider a system $F(x) = 0$ of two equations in two unknowns. Let

$$F(0) = -\frac{1}{10}\begin{pmatrix} 4\sqrt{3} - 3 \\ -4\sqrt{3} - 3 \end{pmatrix} =: -a\,, \quad F'(0) = I\,,$$

$$F(a) = \frac{1}{5}\begin{pmatrix} 4 \\ -3 \end{pmatrix}\,, \quad F'(a) = \begin{pmatrix} 17\sqrt{3} & -1/\sqrt{3} \\ 1 & 1 \end{pmatrix}\,.$$

Starting a Newton method with $x^0 := 0$, we want to verify the occurrence of a 2–cycle, if only natural monotonicity is required.

a) Show that in the first Newton step $\lambda_0 = 1$ is acceptable, since the natural monotonicity criterion is passed, which leads to $x^1 = a$.

b) Show that in the second Newton step $\lambda_1 = 1$ is acceptable yielding $x^2 = x^0$.

Exercise 3.4 *Avoidance of 2–cycles* [33]. We study the possible occurrence of 2–cycles for a damped Newton iteration with *natural* monotonicity test and damping factor λ. A special example is given in Exercise 3.3. By definition, such a 2–cycle is characterized by the inequalities

$$\begin{aligned}
\|F'(x^k)^{-1}F(x^{k+1})\| &\leq \|F'(x^k)^{-1}F(x^k)\|\,, \\
\|F'(x^{k+1})^{-1}F(x^{k+2})\| &\leq \|F'(x^{k+1})^{-1}F(x^{k+1})\|
\end{aligned}$$

with $x^{k+2} = x^k$.

a) Upon applying Theorem 3.12 verify that

$$\|F'(x^{k+1})^{-1}F(x^{k+1})\| \leq \left(1 - \lambda + \frac{1}{2}\lambda^2 h_k \frac{1 + \lambda h_k}{1 - \lambda h_k}\right)\|F'(x^{k+1})^{-1}F(x^k)\|\,.$$

b) Show that under the restriction

$$\lambda h_k \leq \eta < \frac{1}{2}(\sqrt{17} - 3)$$

the occurrence of 2–cycles is impossible.

c) By a proper adaptation of the bit counting Lemma 3.16, modify the damping strategy (3.45) and the restricted monotonicity test (3.47) such that 2–cycles are also algorithmically excluded.

Exercise 3.5   Consider an inexact Newton method for convex optimization, where the inner iteration does *not* satisfy the Galerkin condition (3.90). The aim here is to prove an affine conjugate global convergence theorem as a substitute of Theorem 3.26. Define

$$\sigma_k = -\frac{\langle F(x^k), \delta x^k \rangle}{\epsilon_k^\delta} .$$

(a) Show that one obtains the upper bound

$$t_k(\lambda) = \sigma_k \lambda - \tfrac{1}{2}\lambda^2 - \tfrac{1}{6}\lambda^3 h_k^\delta ,$$

so that $\sigma_k > 0$ is required to assert functional decrease.

(b) On the basis of the optimal damping factor

$$\overline{\lambda}_k = \frac{2\sigma_k}{1 + \sqrt{1 + 2\sigma_k h_k^\delta}} \leq 1$$

prove a global convergence theorem.

How can this theorem also be exploited for the design of an adaptive inexact Newton method?

Exercise 3.6   Usual GMRES codes require the user to formulate the linear equation $Ay = b$ as $A\Delta y = r(y_0)$ with $\Delta y = y - y^0$ and $r(y_0) = b - Ay_0$ so that $\Delta y^0 = 0$. Reformulate the linear system

$$F'(x^k)\widetilde{\Delta x}^{k+1} = -F(x^{k+1}) + r^k$$

to be solved by the GMRES iteration for an initial guess

$$\widetilde{\delta x_0}^{k+1} = (1 - \lambda)\delta x^k .$$

4 Least Squares Problems: Gauss-Newton Methods

In many branches of science and engineering so-called *inverse problems* arise: given a series of system measurements and a conjectured model containing unknown parameters, determine these parameters in such a way that model and measurements 'match best possible'. In this section, let $x \in \mathbb{R}^n$ denote the *parameters* within a *model function* $\varphi(t, x)$ to be determined from a comparison with given *measurements* $(t_1, y_1), \ldots, (t_m, y_m)$. For $m > n$, a perfect match of model and data will not occur in general, caused by model deficiencies and/or measurement errors. Throughout this chapter, any vector norm $\| \cdot \|$ will be understood to be the Euclidean norm.

BIBLIOGRAPHICAL NOTE. This kind of problem has first been faced, formulated, and solved by Carl Friedrich Gauss in 1795. His nowadays so-called *maximum likelihood* method [96] has been published not earlier than 1809. For an extensive appreciation of the historical scientific context see the recent thorough treatise by A. Abdulle and G. Wanner [1]. The most elaborate recent survey about the numerical solution of least squares problems has been published by Å. Bjørck [26].

Following the arguments of Gauss, the deviations between model and data are required to satisfy a *least squares* condition of the type

$$\sum_{i=1}^{m} \left(\frac{y_i - \varphi(t_i, x)}{\delta y_i} \right)^2 = \min, \tag{4.1}$$

wherein the δy_i denote the *error tolerances* of the measurements y_i. Proper specification of the δy_i within this problem is of crucial importance to permit a reasonable statistical interpretation.

More generally, with

$$F = (f_i(x)) = \left(\frac{y_i - \varphi(t_i, x)}{\delta y_i} \right), \quad i = 1, 2, \ldots,$$

the above problem (4.1) appears as a special case of the following problem: Given a mapping $F : D \subseteq \mathbb{R}^n \rightarrow \mathbb{R}^m$ with $m > n$, find a solution $x^* \in D$ such that

$$\|F(x^*)\|_2 = \min_{x \in D} \|F(x)\|. \tag{4.2}$$

A nonlinear least squares problem is said to be *compatible* when

$$F(x^*) = 0 \,.$$

One way of attacking this problem is to reformulate it as a system of n nonlinear equations

$$G(x) := \tfrac{1}{2} \operatorname{grad} \|F(x)\|^2 = F'(x)^T F(x) = 0 \,.$$

Application of Newton's method to G would require the solution of a sequence of linear systems of the kind

$$\left[F'(x)^T F'(x) + F''(x)^T \circ F(x) \right] \Delta x = -F'(x)^T F(x) \,.$$

In a compatible problem, the second matrix term on the left will vanish at the solution point. With the vague idea that the deviation between model and data will not be 'too large', the tensor term F'' may be dropped to obtain the so-called *Gauss-Newton method*:

$$F'(x)^T F'(x) \Delta x = -F'(x)^T F(x) \,. \tag{4.3}$$

An alternative algorithmic approach directly starts from (4.2) using Taylor's expansion

$$\|F(x^*)\| = \|F(x^0) + F'(x^0)(x^* - x^0) + \cdots \| \,.$$

Dropping quadratic terms as in the algebraic derivation of the ordinary Newton method, one ends up with the iterative method ($k = 0, 1, \ldots$):

$$\|F'(x^k) \Delta x^k + F(x^k)\| = \min, \quad x^{k+1} := x^k + \Delta x^k \,.$$

It is an easy task to verify that the solution of this local minimization problem is again the Gauss-Newton method (4.3). Obviously, this method attacks the solution of the nonlinear least squares problem by solving a *sequence of linear least squares problems*.

As in Newton's method for nonlinear equations ($m = n$), a natural distinction between *local* and *global* methods arises: Local Gauss-Newton methods require 'sufficiently good' starting guesses x^0, whereas global Gauss-Newton methods are constructed to handle 'bad' guesses as well; in contrast to the nonlinear equation case ($m = n$), Gauss-Newton methods (for $m > n$) exhibit guaranteed convergence only for a subclass of nonlinear least squares problems. In what follows, local and global Gauss-Newton methods for *unconstrained* and *separable* nonlinear least squares problems will be elaborated in Sections 4.2 and 4.3, whereas the *constrained* case will only be treated in Section 4.3—for theoretical reasons to be explained at the end of Section 4.1.2. Affine invariance of both theory and algorithms will once again play a role, which here means *affine contravariance* and *affine covariance*.

Finally, in Section 4.4, we study *underdetermined* nonlinear systems, here only in the affine covariant setting. As it turns out, a *geodetic Gauss-Newton path* can be shown to exist generically and can be exploited to construct a quasi-Gauss-Newton algorithm and a corresponding adaptive trust region method.

Jacobian approximations. Before diving into details, a general warning concerning Jacobian approximations within Gauss-Newton methods seems to be in order. In contrast to Newton methods, Gauss-Newton methods may be seriously affected by 'too large' Jacobian errors: the mathematical reason for that is that they additionally carry information about *local projections*. Therefore, analytic expressions for the Jacobian or automated differentiation (see A. Griewank [112]) are strongly recommended.

4.1 Linear Least Squares Problems

In order to simplify the subsequent analysis of the nonlinear case, certain notations and relations are first introduced here for the linear special case.

BIBLIOGRAPHICAL NOTE. For more details about linear least squares problems the reader may refer, e.g., to the textbooks by M.Z. Nashed [157] or by A. Ben-Israel and T.N.E. Greville [24], wherein various generalized inverses are characterized. As for the numerical linear algebra, the textbook [107] by G.H. Golub and C.F. van Loan is still the classic. Moreover, the elaborate quite recent handbook article [26] due to Å. Bjørck is a rich source.

4.1.1 Unconstrained problems

For given rectangular matrix A with m rows and n columns and given m-vector y, we first consider the *unconstrained linear least squares problem*

$$\|Ax - y\|^2 = \min . \tag{4.4}$$

Recall that $m > n$ and, in view of the intended *data compression*, typically $m \gg n$. The solution structure of this problem can be described as follows.

Lemma 4.1 *Consider the linear least squares problem (4.4). Let*

$$p := \text{rank}(A) \leq n < m$$

denote the rank of the matrix A. Then the solution x^* is unique if and only if $p = n$. If $p < n$, then there exists an $(n - p)$-dimensional subspace X^* of solutions. In particular, let x^* denote the 'shortest' solution such that

$$\|x^*\| \leq \|x\| \quad \text{for all } x \in X^* .$$

Then the general solution can be written as

$$x = x^* + z$$

with arbitrary $z \in \mathcal{N}(A)$, the nullspace of A.

The proof can be found in any textbook on linear algebra (cf., e.g., [107]).

Moore-Penrose pseudo-inverse. For $0 \leq p \leq n$, the solution x^* is *unique* and can be formally written as

$$x^* = A^+ y$$

in terms of the Moore-Penrose pseudo-inverse A^+. This special generalized inverse is uniquely defined by the so-called *Penrose axioms:*

$$
\begin{aligned}
(A^+ A)^T &= A^+ A, \\
(A A^+)^T &= A A^+, \\
A^+ A A^+ &= A^+, \\
A A^+ A &= A.
\end{aligned}
$$

From these axioms define the following *orthogonal projectors* emerge:

$$
\begin{aligned}
P &:= A^+ A, & P^\perp &:= I_n - P, \\
\overline{P} &:= A A^+, & \overline{P}^\perp &:= I_m - \overline{P}.
\end{aligned}
$$

The orthogonality properties of the projectors follow directly as:

$$P^2 = P, \ P^T = P$$

and

$$\overline{P}^2 = \overline{P}, \ \overline{P}^T = \overline{P}.$$

Let $\mathcal{R}(A)$ denote the range of the matrix A. Then P projects onto $\mathcal{N}^\perp(A)$ and $\overline{P}$ projects onto $\mathcal{R}(A)$. Since $m > n$, the relation

$$\operatorname{rank}(A) = n \leq m \iff P = I_n, \ P^\perp = 0$$

holds.

Inner and outer inverses. Beyond the Moore-Penrose pseudo-inverse A^+, a variety of other generalized inverses appear in realistic problems. In Section 4.2 below so-called *inner inverses* A^- will play a role that satisfy only the fourth of the above Penrose axioms:

$$A A^- A = A. \tag{4.5}$$

Similarly, in Section 4.3, so-called *outer inverses* A^- will play a role that satisfy only the third of the above Penrose axioms:

$$A^- A A^- = A^-. \tag{4.6}$$

Of course, the Moore-Penrose pseudo-inverse is both an inner and an outer inverse.

Numerical solution. There are three basic approaches for the numerical solution of linear least squares problems:

- **QR-decomposition**: This approach uses the fact that $\|Ax - y\| = \|Q(Ax - y)\|$ is invariant under orthogonal transformation Q; one will then obtain

$$Q\,A = \begin{pmatrix} R \\ 0 \end{pmatrix}, \quad Q\,y = \begin{pmatrix} c \\ d \end{pmatrix} \tag{4.7}$$

with $R = (r_{ij})$ an upper triangular (n, n)-matrix. Formally speaking, whenever $\mathrm{rank}(A) = n$, then R is nonsingular. In the above introduced notation the residual norm of the solution is x^* then simply

$$\|d\| = \|Ax^* - y\| = \|\overline{P}^{\perp} y\|\,.$$

In a similar way, we have the relation

$$\|c\| = \|Ax^*\| = \|\overline{P}y\|\,. \tag{4.8}$$

- **Normal equations solution**: This approach is the one originally derived by Carl Friedrich Gauss who obtained

$$A^T A x = A^T y$$

with symmetric positive semi-definite (n, n)-matrix $A^T A$. Here (rational) Cholesky decomposition will be the method of choice.

- **Augmented system solution**: In this approach one explicitly defines the residual vector

$$r := y - Ax$$

as additional unknown, so that the normal equations can now be reformulated as

$$\begin{pmatrix} I & A \\ A^T & 0 \end{pmatrix} \begin{pmatrix} r \\ x \end{pmatrix} = \begin{pmatrix} y \\ 0 \end{pmatrix}\,.$$

This form is especially recommended for large sparse systems and for iterative refinement—see [25].

Numerical rank decision. In actual computation, a precise rank *determination* for a given matrix A is hard, if not impossible. Instead a numerical rank *decision* must be made, which depends on the selected linear *solver* as well as on *row and column scaling* of the matrix—which includes the effect of choosing the measurement tolerances δy_i in (4.1). Since both the normal equations and the augmented system approach essentially require full rank, we elaborate here only on a slight modification of the QR-decomposition that additionally applies column pivoting:

$$Q\,A\Pi = \begin{pmatrix} R \\ 0 \end{pmatrix}\,.$$

Once again $R = (r_{ij})$ is an upper triangular (n, n)-matrix. The matrix Π represents column permutations performed such that

$$|r_{11}| \geq |r_{22}| \geq \cdots \geq |r_{nn}|. \tag{4.9}$$

Let ε denote some reasonable input accuracy, then a *numerical rank* p may be defined by the maximum index such that

$$\varepsilon|r_{11}| < |r_{pp}|.$$

For $p = n$, the so-called *subcondition* number (cf. [83])

$$\mathrm{sc}(A) := \frac{|r_{11}|}{|r_{nn}|}$$

can be conveniently computed. Since

$$\mathrm{sc}(A) \leq \mathrm{cond}_2(A),$$

rank-deficiency will certainly occur whenever

$$\varepsilon \, \mathrm{sc}(A) \geq 1.$$

For $p < n$, a rank-deficient pseudo-inverse must be realized—e.g., by a QR-Cholesky decomposition due to [65] as presented in detail in Section 4.4.1 below.

Remark 4.1   An alternative, theoretically more satisfactory, but computationally more expensive numerical rank-decision is based on *singular value decomposition* (cf. [106]).

4.1.2 Equality constrained problems

Consider the equality constrained linear least squares problem given in the form

$$\|Bx - d\|^2 = \min \tag{4.10}$$

subject to

$$Ax - c = 0,$$

wherein A is an (m_1, n)-matrix, B an (m_2, n)-matrix, and c, d are vectors of appropriate length with

$$m_1 < n < m_1 + m_2 =: m.$$

There is a variety of well-known efficient methods to solve this type of problem numerically (see, e.g., the textbook [107]) so that there is no need to repeat this material. However, in view of our later treatment of the constrained nonlinear least squares problems, we nevertheless present here some derivation (brought to the knowledge of the author by C. Zenger) that will play a fruitful role in the nonlinear case as well.

Penalty method. To start with we replace the constrained problem by a *weighted* unconstrained problem of the kind

$$T_\mu(x) := \|Bx - d\|^2 + \mu^2 \|Ax - c\|^2 = \min,$$

which can also be written in the form

$$\left\| \begin{array}{c} \mu(Ax - c) \\ Bx - d \end{array} \right\|^2 = \min.$$

The idea is to set the penalty parameter μ 'sufficiently large': however, the associated *formal solution*

$$x_\mu^* := \left(\begin{array}{c} \mu A \\ B \end{array} \right)^+ \left(\begin{array}{c} \mu c \\ d \end{array} \right)$$

should not be computed directly (see [190]), since

$$\text{pseudo-rank} \left(\begin{array}{c} \mu A \\ B \end{array} \right) \longrightarrow \text{rank}(A) \text{ for sufficiently large } \mu.$$

Penalty limit method. Fortunately, there exists a stable numerical variant for

$$\mu \to \infty.$$

In order to see this, we apply simultaneous Householder transformations [43] both to the above matrices and the right-hand sides—with μ finite, for the time being. For ease of writing, only the first transformation is given: one has to compute the quantities

$$\sigma(\mu) := \left[\mu^2 \sum_{j=1}^{m_1} a_{j,1}^2 + \sum_{i=1}^{m_2} b_{i,1}^2 \right]^{\frac{1}{2}},$$

$$\beta(\mu) := \left[\sigma(\mu)\big(\sigma(\mu) + \mu|a_{1,1}|\big) \right]^{-1},$$

$$l = 2, \ldots, n: \quad y_l(\mu) := \beta(\mu) \left[\mu^2 \sum_{j=1}^{m_1} a_{j,1} a_{j,l} + \sum_{i=1}^{m_2} b_{i,1} b_{i,l} \right].$$

Then the effect of the Householder transformation on the matrices and vectors is

$$\begin{array}{lll}
\mu \cdot a_{jl} & \to & \mu\big(a_{jl} + a_{j1} \cdot y_l(\mu)\big), \quad j = 1, \ldots, m_1, \\
b_{il} & \to & b_{il} + b_{i1} \cdot y_l(\mu), \quad i = 1, \ldots, m_2, \\
\mu \cdot c_j & \to & \mu\big(c_j + a_{j1} \cdot y_1(\mu)\big), \quad j = 1, \ldots, m_1, \\
d_i & \to & d_i + b_{i,1} \cdot y_1(\mu), \quad i = 1, \ldots, m_2.
\end{array}$$

These relations show that the limiting process $\mu \to \infty$ can be performed yielding

$$\widehat{\sigma} := \lim_{\mu \to \infty} \frac{\sigma(\mu)}{\mu} = \left(\sum_{j=1}^{m_1} a_{j,1}^2 \right)^{\frac{1}{2}},$$

$$\widehat{\beta} := \lim_{\mu \to \infty} \mu^2 \beta(\mu) = \left[\widehat{\sigma}(\widehat{\sigma} + |a_{1,1}|) \right]^{-1},$$

$$l = 2, \ldots, n : \quad \widehat{y}_l = \lim_{\mu \to \infty} y_l(\mu) = \widehat{\beta} \sum_{j=1}^{m_1} a_{j,1} a_{j,l}$$

and eventually the transitions

$$a_{j,l} \quad \longrightarrow \quad a_{j,l} + a_{j,1} \cdot \widehat{y}_l, \quad j = 1, \ldots, m_1,$$

$$b_{il} \quad \longrightarrow \quad b_{il} + b_{i1} \cdot \widehat{y}_l, \quad i = 1, \ldots, m_2,$$

$$c_j \quad \longrightarrow \quad c_j + a_{j,1} \cdot \widehat{y}_1, \quad j = 1, \ldots, m_1,$$

$$d_i \quad \longrightarrow \quad d_i + a_{j,1} \cdot \widehat{y}_1, \quad i = 1, \ldots, m_2.$$

Note that the transformation produces vanishing subdiagonal entries in the matrices A and B. This type of transformation, however, is (column) orthogonal only with respect to A, but not with respect to B. Repeated application of the transformations may be represented by the following scheme

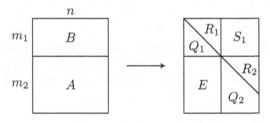

Here Q_1, Q_2 are (column) orthogonal matrices and E describes an elimination process producing zero entries in the associated part of the original matrix B. R_1, R_2 are upper triangular matrices with R_1 being nonsingular, whenever A is assumed to be of full rank m_1. Thus, if R_2 is also *nonsingular*, then x_∞^* is bounded, since it is the unique solution of an upper triangular system with triangular matrix

$$R := \begin{bmatrix} R_1 & S_1 \\ 0 & R_2 \end{bmatrix}.$$

Numerical rank decisions. If R_2 is *singular*, then the associated rank deficient Moore-Penrose pseudo-inverse must be taken. On the basis of this type of decomposition, the *two* sub-condition numbers

$$\mathrm{sc}(A) = \mathrm{sc}(R_1) \text{ and } \mathrm{sc}(R_2)$$

arise naturally. The first one, $\mathrm{sc}(R_1)$, monitors the *linear independence* of the equality constraints, whereas the second one, $\mathrm{sc}(R_2)$, represents the constrained least squares problem as such. If both sub-condition numbers are finite, then x_∞^* is certainly bounded. If $\mathrm{sc}(R_1) > 1/\mathrm{eps}$ for relative machine precision eps, then m_1 should be reduced. If $\mathrm{sc}(R_2) > 1/\mathrm{eps}$, then a best least squares solution can be aimed at. Again testing the condition number via singular values would be safer, but more costly. Moreover, such an approach would spoil the nice structure as presented above. Details are clarified by the subsequent lemma.

Lemma 4.2 *In the above introduced notation, let*

$$\mathrm{rank}(A) = m_1 \,.$$

Then,

$$x_\infty^* := \lim_{\mu \to \infty} \begin{pmatrix} \mu A \\ B \end{pmatrix}^+ \begin{pmatrix} \mu c \\ d \end{pmatrix}$$

is the (possibly shortest) solution of the above defined constrained linear least squares problem.

Proof. The existence of some bounded x_∞^* has been shown above by deriving an appropriate algorithm. It remains to be proven that x_∞^* is the solution of the constrained problem (4.10). For this purpose, let $\mu < \infty$ first. By definition of x_μ^*, one has

$$
\begin{aligned}
T_\mu(x_\mu^*) &= \|Bx_\mu^* - d\|^2 + \mu^2 \|Ax_\mu^* - c\|^2 \\
&\leq \|Bx - d\|^2 + \mu^2 \|Ax - c\|^2 = T_\mu(x)\,.
\end{aligned}
\tag{4.11}
$$

Now, let

$$S := \{ x \in \mathbb{R}^n \,|\, Ax - c = 0 \}\,.$$

Then, for any $x \in S$:

$$T_\mu(x_\mu^*) \leq \|Bx - d\|^2\,. \tag{4.12}$$

On the other hand, (4.11) implies

$$T_\mu(x_\mu^*) \geq \mu^2 \|Ax_\mu^* - c\|^2\,.$$

Combination of these two inequalities yields

$$\|Ax_\mu^* - c\|^2 \leq \frac{\|Bx - d\|^2}{\mu^2}\,,$$

which, for $\mu \to \infty$, shows that

$$x_\infty^* \in S\,.$$

Hence

$$T_\mu(x_\infty^*) = \|Bx_\infty^* - d\|^2\,,$$

which, together with (4.12) leads to

$$\|Bx_\infty^* - d\|^2 \leq \|Bx - d\|^2 \text{ for all } x \in S\,.$$

This confirms the statement of the lemma. ☐

Projection limits. For later purposes, we want to point out that, by columnwise application of Lemma 4.2, the domain space projector

$$P_\infty := \lim_{\mu \to \infty} \begin{pmatrix} \mu A \\ B \end{pmatrix}^+ \begin{pmatrix} \mu A \\ B \end{pmatrix}$$

exists, whereas the associated image space projector

$$\overline{P}_\mu := \begin{pmatrix} \mu A \\ B \end{pmatrix} \begin{pmatrix} \mu A \\ B \end{pmatrix}^+$$

would blow up in the limit $\mu \to \infty$. As a consequence, there exists a natural nonlinear extension of the just presented penalty limit approach within the *error oriented* framework (Section 4.3), but not within the *residual oriented* one (Section 4.2).

4.2 Residual Based Algorithms

Upon recalling that the objective function (4.2) for unconstrained nonlinear least squares problems is directly defined via the residual $F(x)$, an *affine contravariant* approach to Gauss-Newton methods seems to be certainly natural. This is the topic of the present section.

The class of iterations to be studied here is defined as follows: For given initial guess x^0, let the sequence $\{x^k\}$ be defined by

$$\Delta x^k := -F'(x^k)^- F(x^k), \quad x^{k+1} = x^k + \lambda_k \Delta x^k, \quad 0 < \lambda_k \leq 1, \qquad (4.13)$$

where $F : D \subseteq \mathbb{R}^n \to \mathbb{R}^m$ and $m > n$. For $\lambda_k = 1$ we have the *local* and for $0 < \lambda_k \leq 1$ the *global* Gauss-Newton method. The generalized inverse $F'(x)^-$ is merely assumed to be an *inner inverse* (compare (4.5)) satisfying

$$F'(x)F'(x)^- F'(x) = F'(x),$$

which, in terms of the projectors

$$\overline{P}(x) := F'(x)F'(x)^-, \quad \overline{P}^\perp(x) = I_m - \overline{P}(x)$$

implies that
$$\overline{P}(x)^2 = \overline{P}(x), \quad \overline{P}(x)F(x) = -F'(x)\Delta x.$$

Note that $\overline{P}$ is the projector onto $\mathcal{N}^\perp(F'(x))$, the orthogonal complement of the nullspace of the Jacobian. If, in addition, the projector is symmetric

$$\overline{P}(x)^T = \overline{P}(x),$$

then we may split the residual norm according to

$$\|F(x)\|^2 = \|\overline{P}(x)F(x)\|^2 + \|\overline{P}^\perp(x)F(x)\|^2 \,.$$

In any case, the minimum point x^* of the objective function can be characterized by

$$\overline{P}(x^*)F(x^*) = 0 \,.$$

Note that for *incompatible* nonlinear least squares problems

$$F(x^*) = \overline{P}^\perp(x^*)F(x^*) \neq 0 \,.$$

4.2.1 Local Gauss-Newton methods

This section presents a direct extension of the affine contravariant convergence theory for local Newton methods as given in Section 2.2.1. Again *affine contravariant* Lipschitz conditions of the type (1.8) will play an important role. For the sake of clarity we once more point out that any such residual based analysis cannot be expected to touch upon the question of local uniqueness of the solution.

Convergence analysis. In what follows we give a local convergence theorem which is the direct extension of the affine contravariant Newton-Mysovskikh theorem (Theorem 2.12) for Newton's method. The generalized Jacobian inverse is only assumed to be an *inner* inverse, which makes the analysis applicable to a wider class of algorithms—exemplified below for *unconstrained* and for *separable* Gauss-Newton algorithms.

Theorem 4.3 *Let $F : D \subset \mathbb{R}^n \to \mathbb{R}^m$ with $m > n$ be a differentiable mapping where $D \subset \mathbb{R}^n$ is open and convex. Let $F'(x)^-$ denote an inner inverse of the possibly rank-deficient Jacobian (m,n)-matrix and $\overline{P}(x) := F'(x)F'(x)^-$ a projector in the image space of F. At each iterate x^k define the quantities $h_k := \omega\|\overline{P}(x^k)F(x^k)\|$. Assume that the following affine contravariant Lipschitz condition holds:*

$$\left\|\big(F'(y) - F'(x)\big)(y - x)\right\| \leq \omega\|F'(x)(y - x)\|^2$$

for $y - x \in \mathcal{N}^\perp\left(F'(x)\right)$ and for $x, y \in D$.

Moreover, assume that

$$\|\overline{P}^\perp(y)F(y) - \overline{P}^\perp(x)F(x)\| \leq \rho(x)\|F'(x)(y - x)\|, \quad \rho(x) \leq \rho < 1 \tag{4.14}$$

for $y - x \in \mathcal{N}^\perp\left(F'(x)\right)$ and $x, y \in D$.

Define the open level set

$$\mathcal{L}_\omega = \Big\{x \in D\big|\ \|\overline{P}(x)F(x)\| < \frac{2(1 - \rho)}{\omega}\Big\}$$

and let $\bar{\mathcal{L}}_\omega \subset D$ be compact. For a given initial guess x^0 of an unknown solution x^* let

$$h_0 = \omega \|\overline{P}(x^0)F(x^0)\| < 2(1-\rho)\,, \text{ i.e., } x^0 \in \mathcal{L}_\omega\,. \qquad (4.15)$$

Then the Gauss-Newton iterates $\{x^k\}$ defined by (4.13) remain in $\mathcal{L}_\omega$ and converge to some solution point $x^* \in \mathcal{L}_\omega$ with $\overline{P}(x^*)F(x^*) = 0$. The iterative projected residuals $\{\overline{P}(x^k)F(x^k)\}$ converge to zero at an estimated rate

$$\|\overline{P}(x^{k+1})F(x^{k+1})\| \le (\rho + \tfrac{1}{2}h_k)\|\overline{P}(x^k)F(x^k)\|\,. \qquad (4.16)$$

Proof. For repeated induction, assume that $x^k \in \mathcal{L}_\omega$. The estimation to follow may conveniently start from the identity

$$F(x^{k+1}) = \overline{P}^\perp(x^k)F(x^k) + \int_{s=0}^{1} \left(F'(x^k + s\Delta x^k) - F'(x^k)\right)\Delta x^k\, ds\,.$$

From this, we may derive

$$\|\overline{P}(x^{k+1})F(x^{k+1})\| = \|F(x^{k+1}) - \overline{P}^\perp(x^{k+1})F(x^{k+1})\|$$

$$\le \|\overline{P}^\perp(x^{k+1})F(x^{k+1}) - \overline{P}^\perp(x^k)F(x^k)\|$$

$$+ \left\| \int_{s=0}^{1} \left(F'(x^k + s\Delta x^k) - F'(x^k)\right)\Delta x^k\, ds \right\|\,.$$

Since $x^{k+1} - x^k = \Delta x^k \in \mathcal{N}^\perp\left(F'(x^k)\right)$, the first right hand term can be estimated by means of assumption (4.14) to obtain

$$\|\overline{P}^\perp(x^{k+1})F(x^{k+1}) - \overline{P}^\perp(x^k)F(x^k)\| \le \rho\|\overline{P}(x^k)F(x^k)\|\,.$$

In a similar way the second term yields

$$\left\| \int_{s=0}^{1} \left(F'(x^k + s\Delta x^k) - F'(x^k)\right)\Delta x^k\, ds \right\| \le \tfrac{1}{2}h_k\|\overline{P}(x^k)F(x^k)\|\,.$$

Upon combining the two estimates, the result (4.16) is verified, which can be rewritten as

$$\frac{h_{k+1}}{h_k} \le \rho + \tfrac{1}{2}h_k\,. \qquad (4.17)$$

Now, for the purpose of the induction proof, define

$$\mathcal{L}_k := \left\{ x \in D \mid \|\overline{P}(x)F(x)\| \le \|\overline{P}(x^k)F(x^k)\| \right\}\,.$$

Due to assumption (4.15) we have $\mathcal{L}_0 \subset \mathcal{L}_\omega$. Now let $\mathcal{L}_k \subset \mathcal{L}_\omega$, then (4.17) implies that $h_{k+1} < h_k$ so that $\mathcal{L}_{k+1} \subseteq \mathcal{L}_k \subset \mathcal{L}_\omega$, which completes the induction. At this point we tacitly skip the usual contradiction argument to verify that $x^k + \lambda \Delta x^k \in \mathcal{L}_k$ for all $\lambda \in [0,1]$—just compare the similar proof of Theorem 2.12. As for the convergence to some (not necessarily unique) solution point x^*, arguments similar to those in the proof of Theorem 2.12 apply as well, mutatis mutandis, and are therefore omitted here.   □

Small residual problems. In the above convergence rate estimate (4.16) the *asymptotic convergence factor* ρ is seen to play a crucial role: obviously, $\rho < 1$ characterizes a class of nonlinear least squares problems that can be solved by the Gauss-Newton method. For lack of a better name, we will adopt the slightly misleading, but widespread name 'small residual' problems and call the factor ρ also *small residual factor*.

BIBLIOGRAPHICAL NOTE. It has been known for quite a while that the Gauss-Newton iteration fails to converge for so-called 'large residual' problems (compare, e.g., P.T. Boggs and J.E. Dennis [34] or V. Pereyra [165]). As for their definition or even computational recognition, there is a variety of options. In [101], P.E. Gill and W. Murray suggested to characterize such problems by comparison of the singular values of the Jacobian, i.e., in terms of *first order* derivative information. However, since linear problems converge within one iteration—independent of the size of the residual or the singular values of the Jacobian—the classification of a 'large residual' will necessarily require *second order* information about the true nonlinearity of the problem. The theory presented here asymptotically agrees with the beautiful geometric theory of P.-Å. Wedin [196], which also uses second order characterization. In fact, Theorem 4.3 implicitly contains the *necessary* condition

$$\rho(x^*) < 1.$$

Following [196] this condition can be interpreted as a stability restriction on the *curvature* of the manifold $F(x) = F(x^*)$ —details are left as Exercise 4.2.

Unconstrained Gauss-Newton algorithm. We now proceed to exploit the above convergence theory for the construction of a local Gauss-Newton algorithm to solve unconstrained nonlinear least squares problems (4.2). In this case we have the specification $F'(x)^- = F'(x)^+$, i.e., the Moore-Penrose inverse, so that the above theorem trivially applies.

Convergence monitor. Clearly, Theorem 4.3 suggests to introduce the contraction factors for the projected residuals

$$\Theta_k := \frac{\|\overline{P}(x^{k+1})F(x^{k+1})\|}{\|\overline{P}(x^k)F(x^k)\|} = \frac{\|F'(x^{k+1})\Delta x^{k+1}\|}{\|F'(x^k)\Delta x^k\|}.$$

Note that

$$\Theta_k \le \rho + \tfrac{1}{2}h_k < 1$$

characterizes the local convergence domain. Hence, *divergence* of the Gauss-Newton iteration is diagnosed, whenever $\Theta_k \ge 1$ arises. For *compatible* problems, quadratic convergence occurs. For *incompatible* problems, asymptotic linear convergence will be observed.

Numerical realization. Numerical techniques for the realization of each iterative step of the unconstrained Gauss-Newton method have been given in Section 4.1.1. Let Q_k denote the orthogonal matrix such that

$$Q_k F'(x^k) = \begin{pmatrix} R_k \\ 0 \end{pmatrix}, \quad Q_k F(x^k) = \begin{pmatrix} c_k \\ d_k \end{pmatrix} \tag{4.18}$$

with R_k an upper triangular (n, n)-matrix, assumed to be nonsingular here for ease of presentation. Then, according to (4.8), the computation of Θ_k can be realized in the simple form

$$\Theta_k := \frac{\|c_{k+1}\|}{\|c_k\|}.$$

In order to distinguish the linear convergence phase, one may additionally compute the modified contraction factor

$$\hat{\Theta}_k := \frac{\|\overline{P}(x^k)F(x^{k+1})\|}{\|\overline{P}(x^k)F(x^k)\|} = \frac{\|\hat{c}_{k+1}\|}{\|c_k\|}$$

via the decomposition

$$Q_k F(x^{k+1}) = \begin{pmatrix} \hat{c}_{k+1} \\ \hat{d}_{k+1} \end{pmatrix}. \tag{4.19}$$

It is an easy exercise to show that

$$\hat{\Theta}_k \le \tfrac{1}{2}h_k.$$

Termination criterion. On the basis of the above local convergence theory, the iteration should be terminated at some point $\hat{x}$ with

$$\|\overline{P}(\hat{x})F(\hat{x})\|^2 \le \text{FTOL},$$

where FTOL is a user prescribed *residual error tolerance*. An equivalent criterion is

$$|f(x^{k+1}) - f(x^k)| \le \text{FTOL},$$

where $f(x) = \|F(x)\|_2^2$ denotes the least squares functional—compare also the results of Exercise 4.1. The situation that the linear convergence behavior dominates the quadratic one, may be recognized by $\hat{\Theta}_k \ll \Theta_k$. There are good statistical reasons to terminate the iteration then.

Separable Gauss-Newton algorithms. In many applications of practical interest part of the parameters to be fitted arise *linearly*, whereas others arise *nonlinearly*. A typical representation would be

$$\|F(u,v)\| = \min,$$

where

$$F(u,v) := A(v)u - y,\tag{4.20}$$

$$A(v) : (m,s)\text{-matrix}, \ u \in \mathbb{R}^s, \ y \in \mathbb{R}^m, \ v \in \mathbb{R}^{n-s}.$$

Such problems are said to be *separable*. Typical examples occur in the analysis of spectra, where the model function is just a series of Gauss or Lorentz functions (interpreted as single spectral lines); then the variables u are the amplitudes, whereas the variables v contain information about the phases and widths of each of these bells.

Suppose now that the solution components v^* for the above problem were already given. Then the remaining components u^* would be

$$u^* = A(v^*)^+ y.$$

Hence

$$F(u^*,v^*) = -\overline{P}^{\perp}(v^*)y$$

in terms of the orthogonal projectors

$$\overline{P}(v) := A(v)A(v)^+, \qquad \overline{P}^{\perp}(v) := I_m - \overline{P}(v).$$

In this formulation, the parameters u have been totally eliminated. We might therefore aim at directly solving the substitute problem [105]

$$\|G(v)\| = \min$$

where $G : D \subseteq \mathbb{R}^{n-s} \to \mathbb{R}^m$ is defined as

$$G(v) := \overline{P}^{\perp}(v)y.$$

The associated Gauss-Newton method would then read

$$\Delta v^k := -G'(v^k)^+ G(v^k), \qquad v^{k+1} := v^k + \Delta v^k.\tag{4.21}$$

The convergence of this iteration is clearly covered by Theorem 4.3, since the Moore-Penrose pseudo-inverse is just a special *inner inverse*; in addition, the arising projector in the image space of G is orthogonal.

An interesting improvement [128] starts from the differentiation of the projector, which here specifies to

$$G'(v) = \frac{\partial}{\partial v}\overline{P}^{\perp}(v)y = -\overline{P}'(v)y.$$

Since $G(v) = \overline{P}^{\perp}(v)G(v)$, an intriguing idea is to replace

$$G'(v) \longrightarrow \overline{P}^{\perp}(v)G'(v) = \overline{P}^{\perp}(v)A'(v)A^{+}(v)y$$

in iteration (4.21). We then end up with the modified method ($k = 0, 1, \ldots$):

$$\Delta v^k := -\left(\overline{P}^{\perp}(v^k)G'(v^k)\right)^{+}G(v^k), \qquad v^{k+1} := v^k + \Delta v^k. \qquad (4.22)$$

Local convergence for this variant, too, is guaranteed, since here also a Moore-Penrose pseudo-inverse arises, which includes as well the orthogonality of the projector in the image space of G.

Numerical realization. Any of the separable Gauss-Newton methods just requires standard QR-decomposition as in the unconstrained case. The simultaneous computation of the iterates v^k and u^k can be simplified, if the Moore-Penrose pseudo-inverse $A^{+}(v)$ is replaced by some more general inner inverse $A^{-}(v)$ retaining the property that $\overline{P}(v) := A(v)A^{-}(v)$ remains an orthogonal projector.

Positivity constraints. Whenever the amplitudes $u = (u_1, \ldots u_s)$ must be positive, then a transformation of the kind $u_i = \hat{u}_i^2$ appears to be helpful. In this case, a special Gauss-Newton method for the $\hat{u}_i$ must be realized aside instead of just one linear least squares step.

Summarizing, numerical experiments clearly demonstrate that separable Gauss-Newton methods may well pay off compared to standard Gauss-Newton methods when applied to separable nonlinear least squares problems.

BIBLIOGRAPHICAL NOTE. The first idea about some separable Gauss-Newton method seems to date back to D. Braess [38] in 1970. His algorithm, however, turned out to be less efficient than the later suggestions due to G.H. Golub and V. Pereyra [105] from 1973, which, in turn, was superseded by the improved variant of L. Kaufman [128] in 1975. Theoretical and numerical comparisons have been performed by A. Ruhe and P.-Å. Wedin [179].

4.2.2 Global Gauss-Newton methods

We now turn to the convergence analysis and algorithm design for the global Gauss-Newton iteration as defined in (4.13) with damping parameter in the range $0 < \lambda \leq 1$. For ease of presentation, we concentrate on the special case that the projectors $\overline{P}(x)$ are *orthogonal*—which is the typical case for any unconstrained Gauss-Newton method.

As in the preceding theoretical treatment of Newton and Gauss-Newton methods, we first study the local descent from some x^k to the next iterate $x^{k+1} = x^k + \lambda \Delta x^k$ in terms of an appropriate level function and next prove the global convergence property on this basis. The first theorem is a confluence of Theorem 3.7 for the global Newton method and Theorem 4.3 for the local Gauss-Newton method.

Theorem 4.4 *Let $F : D \subset \mathbb{R}^n \rightarrow \mathbb{R}^m$ with $m > n$ be a differentiable mapping where $D \subset \mathbb{R}^n$ is open and convex. Let $F'(x)^-$ denote an inner inverse of the possibly rank-deficient Jacobian (m, n)-matrix and $\overline{P}(x) := F'(x)F'(x)^-$ a projector in the image space of F. Assume the affine contravariant Lipschitz condition*

$$\left\| (F'(y) - F'(x))(y - x) \right\| \le \omega \| F'(x)(y - x) \|^2$$
$$\text{for } y - x \in \mathcal{N}^\perp (F'(x)) \text{ and } x, y \in D.$$

together with

$$\left\| \overline{P}^\perp(y)F(y) - \overline{P}^\perp(x)F(x) \right\| \le \rho(x) \| F'(x)(y - x) \| \qquad (4.23)$$

$$\rho(x) \le \rho < 1 \text{ for } y - x \in \mathcal{N}^\perp (F'(x)) \text{ and } x, \ y \in D.$$

At iterate x^k define $h_k := \omega \| \overline{P}(x^k)F(x^k) \|$. Let $\mathcal{L}_k = \{ x \in D | \ \| \overline{P}(x)F(x) \| \le \| \overline{P}(x^k)F(x^k) \| \} \subset D$ denote a compact level set. Then the iterative projected residuals may be reduced at an estimated rate

$$\left\| \overline{P}(x^k + \lambda \Delta x^k)F(x^x + \lambda \Delta x^k) \right\| \le t_k(\lambda) \| \overline{P}(x^k)F(x^k) \|, \qquad (4.24)$$

where

$$t_k(\lambda) = 1 - (1 - \rho)\lambda + \tfrac{1}{2}\lambda^2 h_k.$$

The optimal choice of damping factor in terms of this local estimate is

$$\overline{\lambda}_k := \min \left(1, \frac{1 - \rho}{h_k} \right). \qquad (4.25)$$

Proof. Let $x^{k+1} = x^k + \lambda \Delta x^k$. For the purpose of estimation we may conveniently start from the identity

$$F(x^{k+1}) = \overline{P}^\perp(x^k)F(x^k) + (1 - \lambda)\overline{P}(x^k)F(x^k)$$
$$+ \int_{s=0}^{\lambda} \left(F'(x^k + s\Delta x^k) - F'(x^k) \right) \Delta x^k ds. \qquad (4.26)$$

From this, we may derive

$$\left\| \overline{P}(x^{k+1})F(x^{k+1}) \right\| = \| F(x^{k+1}) - \overline{P}^\perp(x^{k+1})F(x^{k+1}) \|$$

$$\le \| \overline{P}^\perp(x^{k+1})F(x^{k+1}) - \overline{P}^\perp(x^k)F(x^k) \| + (1 - \lambda)\| \overline{P}(x^k)F(x^k) \|$$

$$+ \left\| \int_{s=0}^{\lambda} \left(F'(x^k + s\Delta x^k) - F'(x^k) \right) \Delta x^k ds \right\|.$$

Note that $x^{k+1} - x^k \in \mathcal{N}^\perp (F'(x^k))$. Hence, the first right hand term can be estimated by means of assumption (4.23) to obtain

$$\|\overline{P}^\perp(x^{k+1})F(x^{k+1}) - \overline{P}^\perp(x^k)F(x^k)\| \le \rho\lambda\|\overline{P}(x^k)F(x^k)\|\,.$$

The second term is trivial and the third term yields from the Lipschitz condition

$$\left\| \int\limits_{s=0}^{\lambda} \left(F'(x^k + s\Delta x^k) - F'(x^k)\right)\Delta x^k ds\right\| \le \tfrac{1}{2}\lambda^2 h_k\|\overline{P}(x^k)F(x^k)\|\,.$$

Upon combining these estimates, the result (4.24) is verified. Finally, the optimal damping factor $\bar{\lambda}_k$ can be seen to minimize the parabola $t_k(\lambda)$ subject to $\lambda \le 1$ for the local convergence case.   □

On the basis of the local Theorem 4.4 we are now ready to derive a global convergence theorem.

Theorem 4.5 *Notation and assumptions as in the preceding Theorem 4.4. In addition, let $D_0 \subseteq D$ denote the path-connected component of the level set $\mathcal{L}_0$ in x^0 assumed to be compact. Then the damped Gauss-Newton iteration $(k = 0, 1, \ldots)$ with damping factors in the range*

$$\lambda_k \in [\varepsilon,\; 2\bar{\lambda}_k - \varepsilon]$$

and sufficiently small $\varepsilon > 0$, which depends on D_0, converges to some solution point x^* with $\overline{P}(x^*)F(x^*) = 0$.

Proof. The proof is by induction using the local results of the preceding theorem. It is a direct copy of the proof of Theorem 3.8 for the global Newton method—only replace the residuals by the projected residuals and the associated level sets.   □

4.2.3 Adaptive trust region strategy

The clear message from the above global convergence theory is that *monotonicity* of the iterates should *not* be required directly in terms of the nonlinear residual norm $\|F(x)\|$ but in terms of the projected residual norm $\|\overline{P}(x)F(x)\|$. Let QR-decomposition be used for the computation of the (unconstrained) Gauss-Newton corrections. Then, in the notation of Section 4.2.1, (4.18), we have to assure monotonicity as

$$\Theta_k(\lambda) = \frac{\|\overline{P}(x^k + \lambda x^k)F(x^k + \lambda\Delta x^k)\|}{\|\overline{P}(x^k)F(x^k)\|} = \frac{\|c_{k+1}(\lambda)\|}{\|c_k\|} < 1\,,$$

wherein $c_{k+1}(0) = c_k$ holds. Unfortunately, the monotonicity criterion would require a Jacobian evaluation also for rejected trial iterates $x^k + \lambda\Delta x^k$. In order to avoid this unwanted computational amount, we suggest to replace the evaluation of Θ_k by the much cheaper evaluation of

$$\widehat{\Theta}_k(\lambda) := \frac{\|\overline{P}(x^k)F(x^k + \lambda\Delta x^k)\|}{\|\overline{P}(x^k)F(x^k)\|} = \frac{\|\hat{c}_{k+1}(\lambda)\|}{\|c_k\|}, \tag{4.27}$$

where $\hat{c}_{k+1}(0) = c_k$ also holds. From the above Theorem 4.4 we know that

$$\Theta_k \leq 1 - (1 - \rho)\lambda + \tfrac{1}{2}\lambda^2 h_k, \tag{4.28}$$

whereas $\widehat{\Theta}_k$ can be shown to satisfy

$$\widehat{\Theta}_k \leq 1 - \lambda + \tfrac{1}{2}\lambda^2 h_k. \tag{4.29}$$

Colloquially speaking, $\widehat{\Theta}_k$ does not 'see' the small residual factor ρ. Therefore, rather than just requiring $\widehat{\Theta}_k < 1$, we will have to 'mimic' $\Theta_k < 1$ by virtue of some modified test in terms of $\widehat{\Theta}_k$.

Bit counting lemma. Following our paradigm in Section 1.2.3, we replace the optimal damping factor $\overline{\lambda}_k$ from (4.25) by

$$[\overline{\lambda}_k] := \min\left(1, \frac{1 - [\rho]}{[h_k]}\right)$$

in terms of computational estimates $[\rho]$ and $[h_k]$ replacing the unavailable small residual factor ρ and the Kantorovich quantities h_k. Then a modified monotonicity test with $\widehat{\Theta}_k$ can be derived by means of the following lemma.

Lemma 4.6 *Notation as just introduced. Let $\rho = [\rho] < 1$ for simplicity. For some $0 < \sigma < 1$ assume that*

$$0 \leq h_k - [h_k] < \sigma\max\left(1 - \rho, [h_k]\right) ESp.$$

Then for $\lambda = [\overline{\lambda}_k]$ the following monotonicity results hold:

$$\Theta_k \leq 1 - \tfrac{1}{2}(1 - \rho)(1 - \sigma)\lambda < 1, \tag{4.30}$$

$$\widehat{\Theta}_k \leq 1 - \left(1 - \tfrac{1}{2}(1 - \rho)(1 + \sigma)\right)\lambda < 1 - \rho\lambda. \tag{4.31}$$

Proof. We insert the relation

$$[h_k] \leq h_k < (1 + \sigma)\max(1 - \rho, [h_k])$$

first into (4.28) to obtain

$$\Theta_k \leq 1 - (1 - \rho)\lambda + \tfrac{1}{2}\lambda(1 - \rho)(1 + \sigma),$$

which verifies (4.30), and second into (4.29) to obtain

$$\widehat{\Theta}_k \leq 1 - \lambda + \tfrac{1}{2}\lambda(1 - \rho)(1 + \sigma),$$

which leads to (4.31). Finally, insert $\sigma = 1$ into Θ_k and $\widehat{\Theta}_k$. $\square$

Residual monotonicity tests. On this basis, we recommend to mimic the condition $\Theta_k < 1$ by the substitute condition

$$\widehat{\Theta}_k < 1 - \rho\lambda \tag{4.32}$$

with $\widehat{\Theta}_k$ evaluated as indicated in (4.27). Alternatively, one might also first use condition (4.32) as *a-priori monotonicity test* and second, after the computation of the new correction, use the condition $\Theta_k < 1$ in terms of the above defined contraction factor Θ_k, as *a-posteriori monotonicity test*. Furthermore, if we insert $\sigma = \frac{1}{2}$ in (4.31), then we arrive at the *restricted a-priori monotonicity test*

$$\widehat{\Theta}_k < 1 - \tfrac{1}{4}(1 + 3\rho)\lambda. \tag{4.33}$$

In order to design an adaptive trust region strategy, we are only left with the task of constructing affine contravariant computational estimates $[\rho]$ and $[h_k]$.

Correction strategy. From (4.26) we may directly obtain

$$\|F(x^k + \lambda\Delta x^k) - (1 - \lambda)\overline{P}(x^k)F(x^k) - \overline{P}^{\perp}(x^k)F(x^k)\| \le \tfrac{1}{2}\lambda^2 h_k\|\overline{P}(x^k)F(x^k)\|.$$

Note that the left hand term can be easily evaluated using (4.18) and (4.19). We thus obtain the cheaply computable *a-posteriori estimate*

$$[h_k] := \frac{2\left(\|\widehat{c}_{k+1} - (1 - \lambda)c_k\|^2 + \|\widehat{d}_{k+1} - d_k\|^2\right)^{1/2}}{\lambda^2\|c_k\|} \le h_k, \tag{4.34}$$

which can be exploited for the following correction strategy $(i = 0, 1, \ldots)$:

$$\lambda_k^{i+1} := \min\left(\tfrac{1}{2}\lambda_k^i, \ \frac{1 - [\rho]}{[h_k^{i+1}]}\right). \tag{4.35}$$

Prediction strategy. Next, for an *a-priori estimate* we may simply use

$$[h_{k+1}^0] = \Theta_k(\lambda_k)[h_k^{i_*}],$$

wherein i_* indicates the final computed index within estimate (4.34) from the previous iterative step k. With this estimate we are now able to define the prediction strategy

$$\lambda_{k+1}^0 := \min\left(1, \ \frac{1 - [\rho]}{[h_{k+1}^0]}\right), \tag{4.36}$$

Small residual factor estimate. Both of the above strategies require an estimate of the factor ρ. For this we may just insert available quantities into the definition (4.23). The newest available information at iterate x^{k+1} is

$$\rho_{k+1} = \frac{\|\overline{P}^\perp(x^{k+1})F(x^{k+1}) - \overline{P}^\perp(x^k)F(x^k)\|}{\lambda_k\|\overline{P}(x^k)F(x^k)\|} \leq \rho.$$

The above numerator requires both orthogonal transformations Q_k, Q_{k+1} based on Jacobian evaluations $F'(x^k)$, $F'(x^{k+1})$. A quite economic version of evaluating ρ_{k+1} computes first

$$Q_{k+1}\overline{P}^\perp(x^k)F(x^k) = Q_{k+1}Q_k^T \left(\begin{array}{c} 0 \\ d_k \end{array} \right) = \left(\begin{array}{c} \tilde{c}_k \\ \tilde{d}_k \end{array} \right)$$

and then

$$\rho_{k+1} = \frac{\sqrt{\|\tilde{c}_k\|^2 + \|d_{k+1} - \tilde{d}_k\|^2}}{\lambda_k\|c_k\|}.$$

Consequently, we will insert

$$[\rho] = \rho_k \quad \text{into (4.35), (4.32) and (4.33)}$$

and

$$[\rho] = \rho_{k+1} \quad \text{into (4.36)}.$$

Of course, the iteration is terminated whenever $\rho_k \geq 1$ occurs at some iterate x^k. In this case, a 'large residual' problem has been identified that cannot be solved by a Gauss-Newton method.

The adaptive trust region method worked out here on an affine contravariant basis will be implemented in the code NLSQ-RES.

BIBLIOGRAPHICAL NOTE. Most of the published convergence studies about Gauss-Newton methods for nonlinear least squares problems focus on the residual convergence aspect, but do not care too much about affine contravariance. The same is true for the implemented algorithms—see, e.g., the algorithm by P. Lindström and P.-Å. Wedin [144]. As a matter of fact, pure Gauss-Newton algorithms are definitely less popular than Levenberg-Marquardt algorithms—see, e.g., [118, 152, 54]. Note, however, that when more than one solution exists, Levenberg-Marquardt methods may supply 'a numerical solution' that cannot be interpreted in terms of the underlying model and data—which is a most undesirable occurrence in scientific computing.

4.3 Error Oriented Algorithms

This section deals with extensions of the error oriented Newton methods for nonlinear equations (see Section 2.1) to the case of nonlinear least squares problems. Attention focuses on a special class of Gauss-Newton algorithms to be characterized as follows: For given initial guess x^0, let the sequence $\{x^k\}$ be defined by

$$\Delta x^k := -F'(x^k)^- F(x^k), \quad x^{k+1} = x^k + \lambda_k \Delta x^k, \quad 0 < \lambda_k \leq 1,$$

where $F : D \subseteq \mathbb{R}^n \to \mathbb{R}^m$ and $m > n$. Herein $F'(x)^-$ denotes an *outer* inverse of the Jacobian (m, n)-matrix $F'(x)$—recall property (4.6), which implies that

$$F'(x)^- F'(x) F'(x)^- F(x) = F'(x)^- F(x)$$

or, in terms of the projectors

$$P(x) := F'(x)^- F'(x), \quad P^\perp = I_n - P$$

that

$$P(x)^2 = P(x), \quad P(x)\Delta x = \Delta x. \tag{4.37}$$

The presentation covers both local Gauss-Newton methods ($\lambda = 1$) and global Gauss-Newton methods ($0 < \lambda \leq 1$). At first glance, affine covariance in the domain space does not seem to be a valid concept here; however, as will be shown below, there still exists a *hidden affine covariance structure* even in the case of nonlinear least squares problems, which will be exploited next.

4.3.1 Local convergence results

This section is a direct extension of Section 2.1. We start with a local convergence theorem of Newton-Mysovskikh type, which may be compared with Theorem 2.2, the affine covariant Newton-Mysovskikh theorem for the ordinary Newton method.

Theorem 4.7 *Let* $F : D \subseteq \mathbb{R}^n \to \mathbb{R}^m$ *with* $D \in \mathbb{R}^n$ *open convex denote a continuously differentiable mapping. Let* $F'(\cdot)^-$ *denote an outer inverse of the Jacobian matrix with possibly deficient rank. Assume that one can find a starting point* $x^0 \in D$, *a mapping* $\kappa : D \to \mathbb{R}^+$ *and constants* $\alpha, \omega, \overline{\kappa} \geq 0$ *such that*

$$\|\Delta x^0\| \leq \alpha,$$

$$\|F'(z)^- (F'(y) - F'(x))(y - x)\| \leq \omega \|y - x\|^2,$$
$$\text{for all } x, y, z \in D \text{ collinear, } y - x \in \mathcal{R}(F'(x)^-),$$

$$\left\| F'(y)^- \overline{P}^\perp(x) F(x) \right\| \leq \kappa(x) \|y - x\| \text{ for all } x, y \in D, \tag{4.38}$$
$$\kappa(x) \leq \overline{\kappa} < 1 \text{ for all } x \in D,$$

$$h := \alpha \omega < 2(1 - \overline{\kappa}),$$

$$\overline{S}(x^0, \rho) \subset D \text{ with } \rho := \alpha / (1 - \overline{\kappa} - \tfrac{1}{2} h).$$

Then:

(I) *The sequence $\{x^k\}$ of local Gauss-Newton iterates (with $\lambda_k = 1$) is well-defined, remains in $\overline{S}(x^0, \rho)$ and converges to some $x^* \in \overline{S}(x^0, \rho)$ with*

$$F'(x^*)^- F(x^*) = 0.$$

(II) *The convergence rate can be estimated according to*

$$\|x^{k+1} - x^k\| \leq \left(\kappa(x^{k-1}) + \tfrac{1}{2}\omega\|x^k - x^{k-1}\|\right)\|x^k - x^{k-1}\|. \qquad (4.39)$$

Proof. Let $x^{k-1},\ x^k \in D$ for $k \geq 1$. Then the following estimates hold:

$$\|x^{k+1} - x^k\| = \|F'(x^k)^- F(x^k)\|$$

$$\leq \|F'(x^k)^-\left(F(x^k) - F(x^{k-1}) - F'(x^{k-1})(x^k - x^{k-1})\right)\|$$

$$+\|F'(x^k)^-\left(I - F'(x^{k-1})F'(x^{k-1})^-\right)F(x^{k-1})\|$$

$$\leq \tfrac{1}{2}\omega\|x^k - x^{k-1}\|^2 + \kappa(x^{k-1})\|x^k - x^{k-1}\|.$$

This confirms (4.39).

The rest of the induction proof follows along the usual lines of such convergence proofs. Finally, continuity of $F'(x^*)^- F(x^k)$ and boundedness of $F'(x^*)^- F'(x^k)$ for $x^k \to x^*$ help to complete the proof. □

Note that the above condition (4.38) may be equivalently written in the form

$$\left\|F'(y)^-\left(\overline{P}^{\perp}(y)F(y) - \overline{P}^{\perp}(x)F(x)\right)\right\| \leq \kappa(x)\|y - x\|, \qquad (4.40)$$

which can be more directly compared with the small residual condition (4.14). The above convergence theorem states *existence* of a solution x^* in the sense that $F'(x^*)^- F(x^*) = 0$ even in the situation of *deficient* Jacobian rank, which may even vary throughout D as long as $\omega < \infty$ and $\kappa < 1$ are assured. In order to show *uniqueness*, *full* rank will be required to extend Theorem 2.3, as will be shown next.

Theorem 4.8 *Under the assumptions of Theorem 4.7, let a solution x^* with $F'(x^*)^- F(x^*) = 0$ exist. Let*

$$\sigma := \|x^0 - x^*\| < \overline{\sigma} := 2\left(1 - \kappa(x^*)\right) / \omega$$

and assume full Jacobian rank such that

$$P(x) = F'(x)^- F'(x) = I_n \text{ for all } x \in D.$$

Then the following results hold:

(I) *For any starting point $x^0 \in S(x^*, \overline{\sigma})$, the Gauss-Newton iterates remain in $\overline{S}(x^*, \sigma)$ and converge to x^* at the estimated rate*

$$\|x^{k+1} - x^*\| \leq \left(\kappa(x^*) + \tfrac{1}{2}\omega\|x^k - x^*\|\right)\|x^k - x^*\|. \tag{4.41}$$

(II) *The solution x^* is unique in the open ball $S(x^*, \overline{\sigma})$.*

(III) *The following error estimate holds*

$$\|x^k - x^*\| \leq \frac{\|x^{k+1} - x^k\|}{1 - \overline{\kappa}} - \tfrac{1}{2}\omega\|x^{k+1} - x^k\|. \tag{4.42}$$

Proof. First, we compare the assumptions of Theorem 4.7 and 4.8. With $\kappa(x^*) \leq \overline{\kappa} < 1$, one immediately obtains

$$\alpha < \frac{2}{\omega}(1 - \overline{\kappa}) \leq \frac{2}{\omega}(1 - \kappa(x^*)) = \overline{\sigma}.$$

Therefore, the assumption here

$$\|x^0 - x^*\| = \sigma < \overline{\sigma}$$

is sharper than the corresponding result of Theorem 4.7, since there

$$\|x^0 - x^*\| \leq \rho = \frac{\alpha}{1 - \overline{\kappa} - \tfrac{1}{2}h} < \frac{\overline{\sigma}}{1 - \overline{\kappa} - \tfrac{1}{2}h}.$$

On this basis we may therefore proceed to estimate the convergence rate as

$$\|x^{k+1} - x^*\| = \|x^k - x^* - F'(x^k)^- F(x^k)\|$$
$$\leq \|F'(x^k)^- \left(F(x^*) - F(x^k) - F'(x^k)(x^* - x^k)\right)\| + \|F'(x^k)^- F(x^*)\|$$

The first term vanishes as $x^k \longrightarrow x^*$. The same is true for the second term, since $F'(x^*)^- F(x^*) = 0$ implies

$$\|F'(x^k)^- F(x^*)\| = \|F'(x^k)^-(I - F'(x^*)F'(x^*)^-)F(x^*)\|$$
$$\leq \kappa(x^*)\|x^k - x^*\|.$$

Thus one confirms (4.41). The proof of contraction follows inductively from

$$\tfrac{1}{2}\omega\|x^k - x^*\| + \kappa(x^*) \leq \tfrac{1}{2}\omega\|x^0 - x^*\| + \kappa(x^*) < \tfrac{1}{2}\omega\overline{\sigma} + \kappa(x^*) < 1.$$

The *error estimate* (4.42) follows directly by shifting the index from $k = 0$ to general k within the above assumptions.

Finally, we prove *uniqueness* by contradiction: let $x^{**} \neq x^*$ denote a different solution, then (4.41) and the above assumptions with $x^0 := x^{**}$ would imply that

$$\|x^{**} - x^*\| \leq \left(\kappa(x^*) + \tfrac{1}{2}\omega\|x^{**} - x^*\|\right)\|x^{**} - x^*\| < \|x^{**} - x^*\|.$$

Hence, $x^{**} = x^*$, which completes the proof.   □

4.3.2 Local Gauss-Newton algorithms

The above local convergence theorems will now be illustrated for three types of methods—*unconstrained, separable,* and *constrained* Gauss-Newton methods. On this basis, details of the realization of the algorithms will be worked out.

Unconstrained Gauss-Newton method. Consider once again the unconstrained nonlinear least squares problem (4.2). The corresponding Gauss-Newton method is then

$$\Delta x^k := -F'(x^k)^+ F(x^k), \ x^{k+1} = x^k + \Delta x^k, \tag{4.43}$$

with the specification of the Moore-Penrose pseudo-inverse:

$$F'(x)^- = F'(x)^+.$$

At the solution point x^*, the condition

$$F'(x^*)^+ F(x^*) = 0$$

will hold, which is equivalent to

$$F'(x^*)^T F(x^*) = 0,$$

since in finite dimensions

$$\mathcal{N}\big(F'(x)^T\big) = \mathcal{N}\big(F'(x)^+\big).$$

Numerical realization. Each iterative step of the Gauss-Newton method (4.43) simply realizes any of the numerical techniques for linear least squares problems—such as the QR-decomposition (4.7) described in Section 4.1.1, which, in the realistic case $m \gg n$, requires a computational amount of $\sim 2mn^2$.

Warning. We want to mention explicitly that any 'simplified' Gauss-Newton method with a specification of the kind

$$F'(x)^- = F'(x^0)^+$$

will usually converge to some point $\hat{x} \neq x^*$ depending on x^0 and therefore solve the 'wrong' nonlinear least squares problem

$$F'(x^0)^+ F(\hat{x}) = 0. \tag{4.44}$$

A similar warning applies to most 'Gauss-Newton-like' or 'quasi-Gauss-Newton' methods (as long as $m > n$). In the optimization literature such an occurrence is well-known as 'caging' of the iterates depending on the initial guess.

Incompatibility factor. Upon comparing the convergence theorems for Newton and Gauss-Newton methods, the essential new item seems to be the condition $\kappa(x) < 1$, which restricts the class of problems that are successfully tractable by a local Gauss-Newton method. In order to gain some insight into this condition, it is now studied for several cases.

(I) In purely *linear least squares problems*, one has

$$F'(x)^- = F'(x)^+\,, \ \ F'(x) = F'(y)\,,$$

which directly implies the best possible choices

$$\omega = 0\,, \ \kappa(x) \equiv \overline{\kappa} = 0\,.$$

As a consequence, one iteration step produces $x^* = x^1$ independent of x^0.

(II) For general *systems of nonlinear equations* $(m = n)$, let $F'(x)$ be nonsingular for all $x \in D$. Then

$$F'(x)^- = F'(x)^{-1} \Rightarrow \kappa(x) \equiv 0\,,$$

which guarantees the usual *quadratic* convergence of the ordinary Newton method.

(III) For general *nonlinear least squares problems* $(m > n)$ Theorem 4.7, which is independent of the actual Jacobian rank, states that

$$\lim_{k \to \infty} \frac{\|x^{k+1} - x^k\|}{\|x^k - x^{k-1}\|} \le \kappa(x^*)\,,$$

showing that $\kappa(x^*)$ is the *asymptotic convergence rate*.

(IV) An interesting reformulation of condition (4.38) at $x = x^*$ is

$$\kappa(x^*) := \sup_{x \in D} \frac{\|F'(x)^+ F(x^*)\|}{\|x - x^*\|} < 1\,.$$

Obviously, this condition rules out problems with 'too large' residual $F(x^*)$ in terms of *second order* derivative information.

For *compatible* problems especially, the definition of κ directly implies that

$$\kappa(x^*) = 0\,.$$

In this interpretation κ measures the incompatibility and is therefore called *incompatibility factor* herein. Rather than speaking of 'small residual' problems, we want to coin a different wording here:

Definition: *Adequate nonlinear least squares problems* are characterized by the condition

$$\kappa(x^*) < 1\,.$$

Assume that condition (4.38) permits the definition of a *continuous* mapping κ in a neighborhood $D_0 \subseteq D$ of x^*, where we assume D_0 to be restricted such that $\kappa(x) \le \overline{\kappa} < 1$. In short, Theorem 4.7 may then be summarized as:

The Gauss-Newton method converges locally linearly for adequate *and quadratically for* compatible *nonlinear least squares problems.*

Note that a *necessary* condition for the continuity of κ is that $F'(x)$ has *constant* rank in D_0, which need *not* be *full* rank! If $F'(x)$ has *full* rank, then $P(x) = I_n$ guarantees even *local uniqueness* of the solution x^*.

Convergence monitor. The convergence of the local Gauss-Newton method will be monitored by the contraction condition

$$\Theta_k := \frac{\|\Delta x^{k+1}\|}{\|\Delta x^k\|} < 1$$

in terms of the Gauss-Newton corrections. From Theorem 4.7 we know that

$$\Theta_k \le \kappa(x^k) + \tfrac{1}{2}h_k,$$

where the $h_k = \omega\|\Delta x^k\|$ denote the associated Kantorovich quantities. In order to separate the effect of the two right hand terms above, it is advisable to additionally compute the simplified Gauss-Newton corrections

$$\overline{\Delta x}^{k+1} = -F'(x^k)^+ F(x^{k+1}),$$

which can be easily shown to satisfy

$$\overline{\Theta}_k := \frac{\|\overline{\Delta x}^{k+1}\|}{\|\Delta x^k\|} \le \tfrac{1}{2}h_k.$$

For *compatible* nonlinear least squares problems, the behavior of the two contraction factors will be roughly the same, whereas for *incompatible* problems a rather different behavior will show up: as soon as

$$h_k \ll \kappa,$$

we will observe that

$$\|\Delta x^{k+1}\| \approx \kappa\|\Delta x^k\|, \; \|\overline{\Delta x}^{k+1}\| \approx \kappa^2\|\overline{\Delta x}^k\|. \tag{4.45}$$

This situation is illustrated schematically in Figure 4.1.

Termination criteria. On the basis of the statistical interpretation of the Gauss-Newton method, we suggest to terminate the iteration as soon as linear convergence dominates the iteration process: iteration beyond that point would just lead to an accuracy far below reasonable in view of model and data accuracy. A convenient criterion to detect this point will be

$$\overline{\Theta}_k \ll \Theta_k .$$

As an alternative, we also use the modified termination criterion

$$\|\overline{\Delta x}^{k+1}\| \leq \text{XTOL} \tag{4.46}$$

in terms of some user prescribed error tolerance XTOL. After termination in the linear convergence phase, the *finally achieved accuracy* will be

$$\|x^{k+1} - x^*\| \approx \epsilon_{k+1} := \frac{\Theta_k}{1 - \Theta_k}\|\Delta x^k\| . \tag{4.47}$$

Herein the norms are understood to be *scaled* norms, of course, to permit such an interpretation. The above criteria also prevent codes from getting inefficient, if unaware users require a too stringent error tolerance parameter XTOL. A feature of this kind is important within any nonlinear least squares algorithm, since already in the purely linear case iterative refinement may fail to converge—compare the detailed rounding error analysis of G.H. Golub and J.H. Wilkinson [108] and of Å. Bjørck [25].

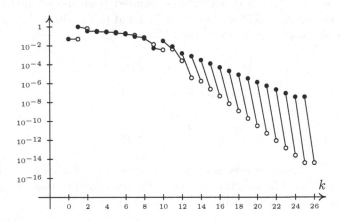

Fig. 4.1. Enzyme reaction problem. Linear convergence pattern of ordinary Gauss-Newton corrections $\|\Delta x^k\|^2 (\bullet)$ and of simplified Gauss-Newton corrections $\|\overline{\Delta x}^{k+1}\|^2 (\circ)$ for $k \geq 12$.

Example 4.1 *Enzyme reaction problem.* In order to illustrate the convergence pattern for incompatible problems, the following unconstrained problem due to J. Kowalik and M.R. Osborne [136] is selected ($m = 11, n = 4$):

$$\|F(x)\|^2 = \sum_{i=1}^{11} \big(y_i - \varphi(t_i, x)\big)^2 = \min .$$

With starting point $x^0 = (0.25, 0.39, 0.415, 0.39)$, the *local* Gauss-Newton method presented here would fail to converge, whereas the *global* Gauss-Newton method to be presented in Section 4.3.4 converged after $k = 16$ iterations with an estimated contraction factor $\Theta_{16} \approx \kappa \approx 0.65$. The iterative behavior of the ordinary and the simplified Gauss-Newton corrections is represented graphically in Figure 4.1. In the linear convergence phase a *scissors* is seen to open between the iterative behavior of the ordinary and the simplified corrections—just as described in (4.45). In this case, the Gauss-Newton iteration has been deliberately continued below any reasonable required accuracy. For XTOL $= 10^{-7}$ the termination criterion (4.46) would have been activated at $k = 16$. According to (4.47) this would have led to a finally achieved accuracy

$$\|x^{17} - x^*\| \approx \epsilon_{17} = 0.9 \cdot 10^{-2}.$$

The final value of the least squares objective function appeared to be

$$\|F(x^*)\|^2 = 3.1045 \cdot 10^{-4}.$$

A-posteriori perturbation analysis. The above incompatibility factor still has a further interpretation, this time in terms of *statistics*. In such problems, the minimization formulation is only reasonable, if small perturbations of the measured data lead to acceptable perturbations of the parameters, i.e., if the solution x^* is *stable* under small perturbations. Let $\delta F(x^*)$ denote such a perturbation. Then standard *linear* perturbation analysis yields the parameter perturbations

$$\delta x_L^* := -F'(x^*)^+ \delta F(x^*).$$

For a *nonlinear* perturbation analysis, Theorem 4.7 can be applied with

$$x^0 := x_{\text{old}}^*, \qquad x_{\text{new}}^* := x_{\text{old}}^* + \delta x_{\text{NL}}^*$$

to obtain the more appropriate estimate

$$\|\delta x_{\text{NL}}^*\| \stackrel{\cdot}{\leq} \frac{\|\delta x_L^*\|}{1 - \kappa(x^*)}.$$

As a consequence, stability of the underlying statistical model can only be guaranteed for *adequate* least squares problems. In [32] H.G. Bock even proved that for $\kappa(x^*) \geq 1$ the point x^* is no longer a local minimum of $\|F(x)\|$. This insight leads to the following rules:

Whenever the ordinary Gauss-Newton method with full rank *Jacobian fails to converge, then one should rather improve the model than just turn to a different iterative solver. In the convergent, but* rank-deficient *case, one should try to get more information by acquiring better data.*

BIBLIOGRAPHICAL NOTE. These rules are backed by careful theoretical considerations in terms of the underlying statistics—for details see, e.g., the early paper [123] by R.L. Jennrich and the more recent, computationally oriented study [164] by M.R. Osborne. In the statistics community the Gauss–Newton method is usually named as *scoring method*.

Separable Gauss-Newton methods. Separable nonlinear least squares problems have been defined in (4.20) above as

$$\|F(u,v)\| = \|A(v)u - y\| = \min,$$

wherein the matrix

$$A(v) : (m,s)\text{-matrix}, \ u \in \mathbb{R}^s, \ y \in \mathbb{R}^m, \ v \in \mathbb{R}^{n-s}$$

carries the nonlinearity of the model parameters.

Following G.H. Golub and V. Pereyra [105], one aims at solving the substitute problem

$$\|G(v)\| = \|\overline{P}^{\perp}(v)y\| = \min$$

in terms of the projectors

$$\overline{P}(v) = A(v)A(v)^+ , \ \overline{P}^{\perp}(v) := I_m - \overline{P}(v).$$

The convergence of the associated Gauss-Newton method is covered by Theorem 4.7, since the Moore-Penrose pseudo-inverse is just a special *outer inverse*.

The simpler variant due to L. Kaufman [128] is characterized by

$$G'(v)^- = \left(\overline{P}^{\perp}(v)G'(v)\right)^+ ,$$

which obviously is again a special *outer inverse* so that local convergence for this variant is also guaranteed by Theorem 4.7. There is some theoretical evidence that the Kaufman variant, which is anyway cheaper to implement, may also have a possibly *larger convergence domain* than the Golub-Pereyra suggestion—compare Exercise 4.8.

Constrained Gauss-Newton method. Consider the nonlinear least squares problem

$$\|G(x)\|^2 = \min, \\ G : D_2 \subseteq \mathbb{R}^n \to \mathbb{R}^{m_2} \tag{4.48}$$

subject to the nonlinear equality constraints

$$H(x) = 0, \\ H : D_1 \subset \mathbb{R}^n \to \mathbb{R}^{m_1}, \tag{4.49}$$

where $D := D_1 \cap D_2 \neq \emptyset$, and

$$m_1 < n < m_1 + m_2 =: m \, .$$

In order to treat the truly *nonlinear* case (in both the least squares functional and the constraints), we apply the same kind of idea as for the linear case in Section 4.1.2. Instead of (4.48) subject to (4.49) we consider the unconstrained *penalty* problem

$$\|G(x)\|^2 + \mu^2 \|H(x)\|^2 = \min \, .$$

With the notation

$$F_\mu(x) := \left(\begin{array}{c} \mu H(x) \\ G(x) \end{array} \right), \quad F'_\mu(x) := \left(\begin{array}{c} \mu H'(x) \\ G'(x) \end{array} \right)$$

the above problem is equivalent to

$$\|F_\mu(x)\|^2 = \min$$

and, for $\mu < \infty$, the associated local Gauss-Newton method is defined via the corrections

$$\Delta x_\mu^k := F'_\mu(x^k)^+ F_\mu(x^k) \, .$$

As in the linear case, the penalty limit $\mu \to \infty$ is of interest.

Lemma 4.9 *Notation as just introduced. Let*

$$\mathrm{rank}\big(H'(x)\big) = m_1 \, , \ x \in D \subseteq \mathbb{R}^n \, .$$

Then, for $x, y \in D$, the following forms exist:

$$A(x, y) \quad := \quad - \lim_{\mu \to \infty} F'_\mu(x)^+ F_\mu(y) \, , \tag{4.50}$$

$$P(x, y) \quad := \quad \lim_{\mu \to \infty} F'_\mu(x)^+ F'_\mu(y) \, , \tag{4.51}$$

which satisfy the projection properties:

$$P^T(x, x) = P^2(x, x) = P(x, x) \, ,$$
$$P(x, x) A(x, x) = A(x, x) \, .$$

Proof. Columnwise application of Lemma 4.2 for the linear case.   □

In the just introduced notation, the local Gauss-Newton method for the constrained problem—in short: *constrained Gauss-Newton method*—can be written as:

$$x^{k+1} := x^k + A(x^k, x^k) \quad k = 0, 1, \dots \, . \tag{4.52}$$

The convergence of this iteration is easily seen by simply rewriting Theorem 4.7 in terms of the forms P and A, which leads to the following local convergence theorem.

Theorem 4.10 *Notation as just introduced in this section. Let $D \subseteq \mathbb{R}^n$ denote some open convex domain of the differentiable mappings G, H. Let $P(\cdot, \cdot)$, $\Delta(\cdot, \cdot)$ denote the forms introduced in Lemma 4.9. Assume that one can find a starting point $x^0 \in D$, a mapping $\kappa : D \to \mathbb{R}^+$ and constants $\alpha, \omega, \overline{\kappa} \geq 0$ such that*

$$\|\Delta(x^0, x^0)\| \leq \alpha,$$

$$\| (P(z, y) - P(z, x)) (y - x)\| \leq \omega \|y - x\|^2$$
$$\text{for all } x, y, z \in D \text{ collinear, } y - x \in \mathcal{R}\big(P(x, x)\big),$$

$$\|\Delta(y, x) - P(y, x)\Delta(x, x)\| \leq \kappa(x)\|y - x\| \text{ for all } x, y \in D,$$

$$\kappa(x) \leq \overline{\kappa} < 1 \text{ for all } x \in D,$$

$$h := \alpha\omega < 2(1 - \overline{\kappa}),$$

$$\overline{S}(x^0, \rho) \subset D \text{ with } \rho := \alpha/(1 - \overline{\kappa} - \tfrac{1}{2}h).$$

Then:

(I) *The sequence $\{x^k\}$ of constrained Gauss-Newton iterates (4.52) is well-defined, remains in $\overline{S}(x^0, \rho)$ and converges to some $x^* \in \overline{S}(x^0, \rho)$ with*

$$\Delta(x^*, x^*) = 0.$$

(II) *The convergence rate can be estimated according to*

$$\|x^{k+1} - x^k\| \leq \big(\kappa(x^{k-1}) + \tfrac{1}{2}\omega\|x^k - x^{k-1}\|\big)\|x^k - x^{k-1}\|.$$

Numerical realization. Each iterative step of the constrained Gauss-Newton method (4.52) realizes the penalty limit method described in Section 4.1.2— or, of course, any other numerical technique for constrained linear least squares problems (see, e.g., the textbook [107]).

Incompatibility factor. As in the unconstrained case, this factor has an interpretation as *asymptotic convergence* factor—see Theorem 4.10. Also its *statistical* interpretation carries over—with the natural modification that now statistical perturbations are only allowed to produce some $\delta G \neq 0$, but must preserve the equality constraints so that $\delta H = 0$. Similar statements hold for the *a-posteriori perturbation analysis*.

Convergence criteria. The monitoring of convergence as well as the termination criterion carry over—just replace the unconstrained corrections by the appropriate constrained $\Delta(\cdot, \cdot)$.

Inequality constraints. In this frequently occurring case a modification of this equality constrained Gauss-Newton method is possible. The essential additionally required technique is a so-called *active set strategy*: active inequality constraints are included into the set of equality constraints. The detection criterion for an active inequality is usually based on the sign of the corresponding *Lagrange multiplier*. For this reason the decomposition technique due to J. Stoer [186] may be preferable here, which permits an easy and cheap evaluation of the Lagrange multipliers. Whenever the set of active inequalities remains constant for $x \in D$, then the above convergence results apply directly. If, however, the active set changes from iterate to iterate, then even local convergence cannot be guaranteed. For practical purposes, such a situation is usually avoided by deliberate suppression of active set *loops*. However, proceeding like that might sometimes lead to solving the 'wrong' problem.

BIBLIOGRAPHICAL NOTE. A first error oriented convergence theorem for rather general Gauss-Newton methods has been given in 1978 by P. Deuflhard and G. Heindl [76, Theorem 4]. In parallel, the application to equality constrained Gauss-Newton methods via the penalty limit has been worked out by P. Deuflhard and V. Apostolescu [69]. Since 1981, theorems of this kind have also been derived and used for the construction of algorithms for parameter identification in ODEs by H.G. Bock [29, 31, 32].

4.3.3 Global convergence results

In this section, we will study global convergence of the damped Gauss-Newton method in the error oriented framework. In order to include both *constrained* and *unconstrained* nonlinear least squares problems in a common treatment, we will use the forms $P(\cdot, \cdot)$ and $\Delta(\cdot, \cdot)$ introduced in Lemma 4.9 for both cases, meaning the limiting definitions (4.51) and (4.50) in the constrained case, but the definitions

$$P(x, y) := F'(x)^+ \, F'(y) \,, \Delta(x, y) := -F'(x)^+ F(y)$$

in the unconstrained case. For simplicity, all derivations will be made in the unconstrained framework.

Global versus local Gauss-Newton path. At first glance, (unconstrained) nonlinear least squares problems do not seem to exhibit any affine covariance property apart from the *trivial class* $O(m)$ containing the orthogonal (m, m)-matrices. However, upon careful examination a nontrivial invariance class for this kind of problem can be detected, which is usually overlooked in theoretical treatments. For this purpose, just observe that the nonlinear least squares problem is equivalent to the system of n nonlinear equations

$$F'(x)^+ F(x) = 0 \,.$$

Application of Newton's method to this formulation, however, would require second-order tensor information beyond mere Jacobian information—which is of no interest in the present context. Rather, we will restrict our attention to the following more special system of equations

$$F'(x^*)^+ F(x) = 0, \tag{4.53}$$

wherein the Jacobian is fixed at the solution point. This formulation immediately gives rise to the *affine covariance class*

$$\mathcal{A}(x^*) := \left\{ A = BF'(x^*)^+ | B \in \mathrm{GL}(n) \right\},$$

which means that equation (4.53) is equivalent to any of the equations

$$AF(x) = 0 \quad \text{with} \quad A \in \mathcal{A}(x^*).$$

On this basis, we may monitor the Gauss-Newton iteration by means of any of the *test functions*

$$T(x|A) := \tfrac{1}{2} \|AF(x)\|^2, \quad A \in \mathcal{A}(x^*).$$

The case $A = I$ has been discussed in the preceding Section 4.2 in the affine contravariant setting. Here we are interested in an extension to the affine covariant setting. As in the derivation of global Newton methods in Section 3.1.4, we define the *level sets*

$$G(z|A) := \{x \in D | T(x|A) \le T(z|A)\}$$

and study their intersection

$$\overline{G}_*(x^0) := \bigcap_{A \in \mathcal{A}(x^*)} G(x^0|A) \tag{4.54}$$

for a given starting point x^0.

Theorem 4.11 *Let $F \in C^1(D)$, $D \subseteq \mathbb{R}^n$ and $P(x^*, x)$ nonsingular for all $x \in D$. For some $\widehat{A} \in \mathcal{A}(x^*)$ let the path-connected component D_0 of $G(x^0|\widehat{A})$ in x^0 be compact and $D_0 \subseteq D$. Then the path-connected component of $\overline{G}_*(x^0)$ in (4.54) defines a topological path $\overline{x}_* : [0, 2] \longrightarrow \mathbb{R}^n$, which satisfies:*

$$\Delta(x^*, \overline{x}_*(\lambda)) = (1 - \lambda)\Delta(x^*, x^0), \tag{4.55}$$

$$T(\overline{x}_*(\lambda)|\widehat{A}) = (1 - \lambda)^2 T(x^0|\widehat{A}),$$

$$P(x^*, \overline{x}_*) \frac{d\overline{x}_*}{d\lambda} = \Delta(x^*, x^0),$$

$$\overline{x}_*(0) = x^0, \quad \overline{x}_*(1) = x^*,$$

$$\frac{d\overline{x}_*}{d\lambda} \Big|_{\lambda=0} = P(x^*, x^0)^{-1} \Delta(x^*, x^0). \tag{4.56}$$

Proof. One proceeds as in the proof of Theorem 3.6, but here for the mapping $F'(x^*)^+F$. Intersection of level sets in the image space of that mapping leads to the *homotopy*

$$\Phi(x, \lambda) := F'(x^*)^+F(x) - (1 - \lambda)F'(x^*)^+F(x^0) = 0$$

for $\lambda \in [0, 2]$. One easily verifies that

$$\Phi_x = F'(x^*)^+F'(x) = P(x^*, x), \Phi_\lambda = F'(x^*)^+F(x^0) = -\Delta(x^*, x^0).$$

The process of local continuation to construct the path $\overline{x}_*$ will naturally start at $\overline{x}_*(0) = x^0$. As $\overline{G}_*(x^0) \subset D_0 \subseteq D$, the nonsingularity of $P(x^*, \overline{x}_*(\lambda))$ is assured, which permits continuation up to $\overline{x}_*(1) = x^*$ and $\overline{x}_*(2)$. From this, the results (4.55) up to (4.56) follow directly.   □

Remark 4.2   The somewhat obscure assumption '$P(x^*, x)$ nonsingular' cannot be replaced by just 'rank$(F'(x)) = n$'. Under the latter assumption, one concludes from

$$\mathcal{N}(P(x^*, x)) = \mathcal{N}(F'(x^*)^+) \cap \mathcal{R}(F'(x))$$

that

$$\dim \mathcal{N}(F'(x^*)^+) \;=\; m - n,$$
$$\dim \mathcal{R}(F'(x)) \;=\; n.$$

Hence, even though the *generic* case will be that

$$\dim \mathcal{N}(P(x^*, x)) = 0,$$

the case

$$\dim \mathcal{N}(P(x^*, x)) > 0$$

cannot be generally excluded. On the other hand, if the Jacobian $F'(x)$ is *rank-deficient*, then $P(x^*, x)$ is certainly *singular*, which violates the corresponding assumption.

The crucial result of the above theorem is that the local tangent direction of $\overline{x}_*$ in x^0 requires *global* information via $F'(x^*)$—which means that the tangent cannot be realized within any Gauss-Newton method. If we replace

$$A(x^*) \longrightarrow A(x^0),$$

we end up with a result similar to the one before.

Theorem 4.12 *Under assumptions that are the natural modifications of those in Theorem 4.11, the intersection*

$$\overline{G}_0(x^0) := \bigcap_{A \in \mathcal{A}(x^0)} G(x^0 | A)$$

defines a topological path $\overline{x}_0 : [0,2] \longrightarrow \mathbb{R}^n$, *which satisfies*

$$\Delta\big(x^0, \overline{x}_0(\lambda)\big) = (1-\lambda)\Delta(x^0, x^0),$$

$$P(x^0, \overline{x}_0)\frac{d\overline{x}_0}{d\lambda} = \Delta(x^0, x^0),$$

$$\overline{x}_0(0) = x^0, \quad \overline{x}_0(1) \neq x^* \text{ in general,}$$

$$\frac{d\overline{x}_0}{d\lambda}\bigg|_{\lambda=0} = \Delta(x^0, x^0).$$

The proof can be omitted. The most important result is that the tangent direction of the path $\overline{x}_0$ in x^0 is now the Gauss-Newton direction. Comparing the two theorems, we will call the path $\overline{x}_*$ *global Gauss-Newton path* and the path $\overline{x}_0$ *local Gauss-Newton path*. The situation is represented graphically in Figure 4.2. Each new iterate induces a new 'wrong' problem

$$\Delta(x^0, x) = 0$$

via the new local Gauss-Newton path—compare also (4.44). The whole procedure may nevertheless approach the 'true' problem under the natural assumption that the tangent directions $\dot{\overline{x}}_*(0)$ and $\dot{\overline{x}}_0(0)$ do not differ 'too much'—a more precise definition of this term will be given below.

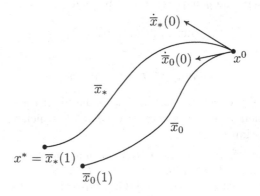

Fig. 4.2. Geometrical scheme: local and global Gauss-Newton path.

Remark 4.3   For $m = n = \mathrm{rank}(F'(x))$, which is the nonlinear equation case, the two paths turn out to be identical—the Newton path treated in Section 3.1.4.

Convergence analysis. On the basis of the geometrical concept of the Gauss-Newton paths, we are now ready to study the convergence of the global Gauss-Newton method including damping.

Theorem 4.13 *Notation and assumptions as before. Let the damped Gauss-Newton iteration be defined as*

$$x^{k+1} := x^k + \lambda \Delta(x^k, x^k).$$

Let $x, y, x^k, x^{k+1} \in D_0$, for some convex $D_0 \subseteq D$. Define a Lipschitz constant ω_* by

$$\left\| (P(x^*, y) - P(x^*, x))(y - x) \right\| \leq \omega_* \|y - x\|^2.$$

Moreover, assume that

$$\left\| \Delta(x^*, x) - P(x^*, x)\Delta(x, x) \right\| \leq \delta_*(x) \|\Delta(x^*, x)\|, \delta_*(x) \leq \overline{\delta} < 1. \quad (4.57)$$

Then, with the convenient notation

$$h_k^* := \omega_* (1 + \delta_*(x^k)) \|P(x^*, x^k)^{-1}\| \cdot \|\Delta(x^k, x^k)\|, \quad (4.58)$$

we obtain

$$\left\| \Delta(x^*, x^k + \lambda \Delta(x^k, x^k)) \right\| \leq t_k^*(\lambda) \|\Delta(x^*, x^k)\|, \quad (4.59)$$

wherein

$$t_k^*(\lambda) := 1 - (1 - \delta_*(x^k))\lambda + \tfrac{1}{2}\lambda^2 h_k^*.$$

The globally optimal damping factor is

$$\lambda_k^* := \min(1, (1 - \delta_*(x^k))/h_k^*). \quad (4.60)$$

Proof. Using standard tools we arrive at the following estimate:

$$\|F'(x^*)^+ F(x^{k+1}))\| = \left\| F'(x^*)^+ \int\limits_{s=0}^{\lambda} (F'(x^k + s\Delta x^k) - F'(x^k))\Delta x^k ds \right.$$

$$\left. + (1 - \lambda)F'(x^*)^+ F(x^k) + \lambda F'(x^*)^+ (I - F'(x^k)F'(x^k)^+)F(x^k) \right\|.$$

Assumption (4.57) directly leads to

$$\left\| \Delta(x^*, x^k + \lambda \Delta(x^k, x^k)) \right\|$$

$$\leq [1 - \lambda(1 - \delta_*(x^k))] \|\Delta(x^*, x^k)\| + \tfrac{1}{2}\lambda^2 \omega_* \|\Delta(x^k, x^k)\|^2.$$

The second order term needs additional treatment:

$$\|\Delta(x^k, x^k)\| \leq \|P(x^*, x^k)^{-1}\| \cdot \|P(x^*, x^k)\Delta(x^k, x^k)\|$$

$$\leq \|P(x^*, x^k)^{-1}\|(1 + \delta_*(x^k))\|\Delta(x^*, x^k)\|.$$

Upon combining these estimates and using the notation for h_k^*, one obtains the result (4.59) and the optimal damping factor (4.60).   □

At iterate x^k, the above assumption (4.57) is equivalent to

$$\frac{\left\| P(x^*, x^k)\left(\dot{\overline{x}}_*(0) - \dot{\overline{x}}_k(0)\right)\right\|}{\|P(x^*, x^k)\dot{\overline{x}}_*(0)\|} \le \delta_*(x^k) \le \overline{\delta} < 1\,,$$

which has a nice *geometric interpretation* in terms of Figure 4.2: in this form, the condition is seen to restrict the 'directional discrepancy' of the local Gauss-Newton path $\overline{x}_k$ and the global Gauss-Newton path $\overline{x}_*$ starting at x^k. The assumption may also be characterized in *algebraic* terms (for the unconstrained case):

$$\|F'(x^*)^+ \overline{P}^{\perp}(x)F(x)\| \le \delta_*(x)\|F'(x^*)^+ F(x)\|\,. \tag{4.61}$$

With these preparations, we are now ready to state global convergence.

Theorem 4.14 *Notations and assumptions as before in Theorem* 4.13. *Let* $P(x, x) = I$ *and* $P(x^*, x)$ *be nonsingular for all* $x \in D$. *Denote the path-connected component of* $G\big(x^0 | F'(x^*)^+\big)$ *in* x^0 *by* D_0 *and assume* $D_0 \subseteq D$, D_0 *compact. Then the Gauss-Newton iteration with damping factors*

$$\lambda_k \in [\varepsilon,\ 2\lambda_k^* - \varepsilon]$$

converges globally to x^*. *In addition, there exists an index* $k_0 \ge 0$ *such that*

$$\lambda_k^* = 1 \ \text{for all}\ \ k \ge k_0\,.$$

In comparison with Theorem 4.7 *the asymptotic convergence factors show the relation*

$$\lim_{x \to x^*} \delta_*(x) = \kappa(x^*) \tag{4.62}$$

assuming best possible estimates for $\delta_*(x)$ *and* $\kappa(x)$.

Proof. The first part of the proof is just along the lines of earlier proofs of global convergence made so far—and is therefore omitted. A new piece of proof comes up with the verification of the above *merging property*. For this purpose we merely need to study the limit for $x^k \to x^*$, since global convergence is already guaranteed. For best possible estimates and $D_0 \to D_k \to D_* = \{x^*\}$ one obtains

$$
\begin{aligned}
\delta_*(x) &\longrightarrow \delta_*(x^*) \le \overline{\delta} < 1\,,\\
P(x^*, x^k) &\longrightarrow P(x^*, x^*) = I_n\,,\\
\|P(x^*, x^k)^{-1}\| &\longrightarrow 1\,,\\
h_k^* &\longrightarrow 0\,.
\end{aligned}
$$

Hence, if x^k is close enough to x^*, then h_k^* decreases monotonically.

Finally, the connection property (4.62) can be shown rather elementarily. Just recall condition (4.40) with $y = x^*$ and use Taylor's expansion for the right hand side in (4.61) to obtain

$$F'(x^*)^+ \left(F(x^*) + F'(x^*)(x - x^*) + \cdots \right) = P(x^*, x^*)(x - x^*) + \mathcal{O}(\|x - x^*\|^2).$$

Hence, with $P(x^*, x^*) = I$ and $x^k \to x^*$, we have verified (4.62), which completes the proof. $\square$

Instead of the unavailable solution point x^* we will certainly insert the best available iterate x^k. This leads to the following theorem.

Theorem 4.15 *Notation and assumptions as in Theorem 4.13 before. Let $x, y, x^k, x^{k+1} \in D_0 \subseteq D$. Define Lipschitz constants ω_k by*

$$\left\| \left(P(x^k, y) - P(x^k, x) \right)(y - x) \right\| \le \omega_k \|y - x\|^2. \tag{4.63}$$

Then, with the notation

$$h_k := \omega_k \|\Delta(x^k, x^k)\|,$$

we obtain

$$\left\| \Delta\left(x^k, x^k + \lambda \Delta(x^k, x^k) \right) \right\| \le \left(1 - \lambda + \tfrac{1}{2}\lambda^2 h_k \right) \|\Delta(x^k, x^k)\|. \tag{4.64}$$

Compared with (4.58) we obtain

$$h_k \le h_k^*. \tag{4.65}$$

Proof. As in the derivations before, we may obtain the following estimate:

$$\|F'(x^k)^+ F(x^{k+1}))\| =$$

$$\left\| F'(x^k)^+ \int_{s=0}^{\lambda} \left(F'(x^k + s\Delta x^k) - F'(x^k) \right) \Delta x^k \, ds + (1 - \lambda) F'(x^k)^+ F(x^k) \right\|,$$

which yields (4.64). Finally, using best possible estimates for ω_k, ω_* as defined in the two above theorems, one may verify that

$$\omega_k \le \omega_* \|P(x^*, x^k)^{-1}\| \left(1 + \delta_*(x^k) \right),$$

which implies (4.65). $\square$

4.3.4 Adaptive trust region strategies

As for the global convergence of the damped Gauss-Newton method, the situation can be described as follows: in order to *guarantee* global convergence, *global Jacobian information* at x^* would be required. Since such information is computationally unavailable, all global information must be replaced by *local* information at the newest iterate x^k—at the expense that then there is no guarantee of convergence. For the time being, assume *full rank* $P(x, x) = I_n$ for all arguments x in question.

Natural level function. First, the global level function

$$T(x|F'(x^*)^+) = \tfrac{1}{2}\|\Delta(x^*, x)\|^2$$

is replaced by its local counterpart

$$T(x|F'(x^k)^+) = \tfrac{1}{2}\|\Delta(x^k, x)\|^2,$$

which is called natural level function in view of the detailed investigations for nonlinear equations in Section 3.3.2. Even in the extended case here, this level function has intriguing properties:

(I) Application of the Penrose axioms directly leads one to

$$\Delta(x^k, x^k) = -\operatorname{grad} T(x|F'(x^k)^+)\big|_{x=x^k},$$

for *arbitrary Jacobian rank*. This means that the Gauss-Newton direction is the direction of *steepest descent* with respect to the natural level function.

(II) For $F \in C^2(D)$, one easily verifies that

$$\|\Delta(x^*, x)\| = \|P(x^*, x^*)(x - x^*)\| + O(\|x - x^*\|^2),$$

which then implies

$$T(x|F'(x^*)^+) = \tfrac{1}{2}\|x - x^*\|^2 + O(\|x - x^*\|^3).$$

Obviously, for $x^k \to x^*$, the natural level function is an *asymptotic distance function*.

Convergence criteria. As a consequence of Theorem 4.13, the iterates might be tested via the contraction factor

$$\Theta_k^* := \frac{\|\Delta(x^*, x^{k+1})\|}{\|\Delta(x^*, x^k)\|} \le (1 - (1 - \bar{\delta})\lambda + \tfrac{1}{2}\lambda^2 h_k^*),$$

which, however, contains computationally unavailable information. Therefore we again suggest to 'mimic' the situation by means of a substitute test based on Theorem 4.15. We thus arrive at the computationally available contraction factor

$$\Theta_k := \frac{\|\varDelta(x^k, x^{k+1})\|}{\|\varDelta(x^k, x^k)\|} \leq \left(1 - \lambda + \tfrac{1}{2}\lambda^2 h_k\right)$$

containing the *simplified* Gauss-Newton correction, which in the unconstrained case reads

$$\varDelta(x^k, x^{k+1}) = -F'(x^k)^+ F(x^{k+1}).$$

Colloquially speaking, the contraction factor Θ_k does not 'see' the incompatibility factor κ.

Bit counting lemma. In order to derive the desired substitute test, we may exploit the following lemma.

Lemma 4.16 *Notation as just introduced. Let $\bar{\delta} = \lfloor\bar{\delta}\rfloor < 1$ for simplicity. For some $0 < \sigma < 1$ assume that*

$$0 \leq h_k - \lfloor h_k \rfloor < \sigma \max\left(1 - \bar{\delta}, \lfloor h_k \rfloor\right).$$

Then for the damping factor

$$\lambda = \lfloor\bar{\lambda}_k\rfloor = \min\left(1, \frac{1 - \bar{\delta}}{\lfloor h_k \rfloor}\right)$$

the following monotonicity results hold:

$$\Theta_k^* \leq 1 - \tfrac{1}{2}(1 - \bar{\delta})(1 - \sigma)\lambda < 1, \tag{4.66}$$

$$\Theta_k \leq 1 - \left(1 - \tfrac{1}{2}(1 - \bar{\delta})(1 + \sigma)\right)\lambda < 1 - \bar{\delta}\lambda. \tag{4.67}$$

Proof. As in the proof of Lemma 4.6 we just insert the relation

$$\lfloor h_k \rfloor \leq h_k < (1 + \sigma)\max(1 - \bar{\delta}, \lfloor h_k \rfloor)$$

to obtain (4.66) and (4.67) and set $\sigma = 1$.   □

Natural monotonicity test. On this basis, we now recommend to replace the condition $\Theta_k^* < 1$ by the test

$$\Theta_k < 1 - \bar{\delta}\lambda.$$

If we insert $\sigma = \tfrac{1}{2}$ in (4.67), then we arrive at its *restricted* counterpart

$$\Theta_k < 1 - \tfrac{1}{4}(1 + 3\bar{\delta})\lambda.$$

Correction strategy. In order to design an adaptive trust region strategy, we are left with the task of constructing cheap computational estimates $[\bar{\delta}]$ and $[h_k]$. Note that due to (4.65) the estimates $[h_k]$ will automatically be also estimates $[h_k^*]$. From Theorem 4.14 the a-posteriori estimate

$$[h_k] := \frac{2 \left\| \Delta\left(x^k, x^k + \lambda \Delta(x^k, x^k)\right) - (1 - \lambda)\Delta(x^k, x^k) \right\|}{\lambda^2 \|\Delta(x^k, x^k)\|} \leq h_k \qquad (4.68)$$

can be obtained, which supplies the correction strategy $(i = 0, 1, \ldots)$:

$$\lambda_k^{i+1} := \min \left(\tfrac{1}{2}\lambda, \frac{1 - \bar{\delta}}{[h_k(\lambda)]} \right) \Big|_{\lambda = \lambda_k^i} . \qquad (4.69)$$

Prediction strategy. For its derivation we will proceed as in the nonlinear equation case and slightly modify the Lipschitz condition (4.63). Let some Lipschitz constant $\bar{\omega}_k$ be defined by

$$\left\| \left(P(x^k, y) - P(x^k, x) \right) v \right\| \leq \bar{\omega}_k \|y - x\| \|v\| .$$

By construction, we have $\bar{\omega}_k \geq \omega_k$. Upon specification of the above arguments, we may exploit the relation

$$\left\| \left(P(x^k, x^k) - P(x^k, x^{k-1}) \right) \Delta(x^{k-1}, x^k) \right\| \leq \bar{\omega}_k \|x^k - x^{k-1}\| \|\Delta(x^{k-1}, x^k)\|$$

to obtain the *a-priori estimate*

$$[\bar{\omega}_k] := \frac{\|\Delta(x^{k-1}, x^k) - \Delta(x^k, x^k) + \overline{\Delta}(x^k, x^{k-1})\|}{\lambda_{k-1} \|\Delta(x^{k-1}, x^{k-1})\| \|\Delta(x^{k-1}, x^k)\|} \leq \bar{\omega}_k , \qquad (4.70)$$

wherein in the unconstrained case

$$\overline{\Delta}(x^k, x^{k-1}) := -F'(x^k)^+ \left(F(x^k) + F'(x^{k-1})\Delta(x^{k-1}, x^k) \right) .$$

This establishes the *prediction strategy* for $k > 0$ as

$$\lambda_k^0 := \min \left(1, \frac{1 - \bar{\delta}}{[\bar{\omega}_k] \|\Delta(x^k, x^k)\|} \right) . \qquad (4.71)$$

Finally, we are still left with the specification of the quantity $\bar{\delta}$ in the choice of the damping factors. Once again, we will replace the condition $\delta_*(x) \leq \bar{\delta} < 1$ from (4.57) by some substitute local condition

$$\delta_k(x) := \frac{\|\Delta(x^k, x) - P(x^k, x)\Delta(x, x)\|}{\|\Delta(x^k, x)\|} \leq \bar{\delta} < 1$$

for appropriate arguments x. For $x = x^k$, the Penrose axioms trivially yield $\delta_k(x^k) = 0$. For $x = x^{k-1}$, $k > 0$ the condition $\delta_k(x^{k-1}) < 1$ may naturally be imposed. Hence, either $[\bar{\delta}] := 0$ or $[\bar{\delta}] = \delta_k(x^{k-1})$ might be used in actual computation. As soon as the iterates approach the solution point (i.e. at least $\lambda_k = 1$), we switch to the local option

$$[\bar{\delta}] = \frac{\|\Delta(x^{k+1}, x^{k+1})\|}{\|\Delta(x^k, x^k)\|} \approx \kappa(x^k)$$

based on the local contraction result (4.39) and the asymptotic result (4.62).

Remark 4.4   For *inadequate* nonlinear least squares problems, the adaptive damping strategy will typically supply values

$$\lambda_k \approx 1/\kappa < 1.$$

Vice versa, this effect can be conveniently taken as an indication of the inadequacy of an inverse problem under consideration.

In most realizations of global Gauss-Newton methods, the choice $\bar{\delta}_k := 0$ is made ad-hoc. As in the simpler Newton case, numerical experience has shown that mostly λ_k^0 is successful, whereas λ_k^1 has turned out to be sufficient in nearly all remaining cases. The whole strategy appeared to be extremely efficient even in sensitive nonlinear least squares problems up to large scale—despite the lack of a really satisfactory global convergence proof.

4.3.5 Adaptive rank strategies

In contrast to the preceding section, the full rank assumption for $P(x, x)$ is now dropped. For ease of writing, we will introduce the short hand notation $J(x) := F'(x)$. In the unconstrained case, the situation of possible rank deficiency

$$q(x) := \mathrm{rank}(J(x)) \leq n, P(x, x) \neq I_n \text{ for } q < n$$

may arise either if the Jacobian at some iterate turns out to be *ill-conditioned* or if a *deliberate rank reduction* has been performed, say, by replacing

$$J(x) \longrightarrow J'(x) \text{ with } q' := \mathrm{rank}(J'(x)) < q.$$

In the rank-deficient case, the corrections $\Delta(x^0, x^0)$ are confined to the cokernel of $J(x^0)$, since

$$P(x^0, x^0)\Delta(x^0, x^0) = \Delta(x^0, x^0).$$

As an immediate consequence, the Gauss-Newton method starting at x^0 is locally solving either the equations

$$J(x^0)^+ F\big(x^0 + P(x^0, x^0)z\big) = 0$$

or the equations

$$J'(x^0)^+ F\big(x^0 + P'(x^0, x^0)z\big) = 0\,,$$

where the intuitive notation $P'(x^0, x^0)$ just denotes the associated rank-deficient projector. On this basis, a Gauss-Newton method with damping strategy may be constructed in the rank-deficient case as well. In order to do so, the following feature of the associated Lipschitz constants is of crucial importance.

Lemma 4.17 *Associated with the Jacobian matrices $J(x^0)$ and $J'(x^0)$, respectively, let $\omega_0 \le \overline{\omega}_0$, $\omega'_0 \le \overline{\omega}'_0$ denote the best possible Lipschitz constants in the notation of Section 4.3.4. Then rank reduction implies that*

$$\omega'_0 \le \omega_0\,, \quad \overline{\omega}'_0 \le \overline{\omega}'_0\,. \tag{4.72}$$

Proof. For simplicity, the result is only shown for ω'_0 in the unconstrained case—the remaining part for $\overline{\omega}'_0$ is immediate. One starts from the definitions of best possible estimates in the form

$$\omega_0 := \sup \frac{\big\| J(x^0)^+ \big(J(u+v) - J(u)\big)v \big\|}{\|v\|^2}\,,$$

$$\omega'_0 := \sup \frac{\big\| J'(x^0)^+ \big(J(u + s\cdot v) - J(u)\big)v \big\|}{s\|v\|^2}\,.$$

The extremal property of the Moore-Penrose pseudoinverse implies that

$$\big\| J'(x^0)^+ \big(J(u+v) - J(u)\big)v \big\| \le \big\| J(x^0)^+ \big(J(u+v) - J(u)\big)v \big\|\,,$$

which directly implies $\omega'_0 \le \omega_0$. $\qquad\qquad\square$

Next, since

$$\big\| \Delta'(x^0, x^0) \big\| \le \big\| \Delta(x^0, x^0) \big\|\,, \tag{4.73}$$

an immediate consequence of the preceding lemma is that the associated Kantorovich quantities

$$h'_0 := \omega'_0 \big\| \Delta'(x^0, x^0) \big\|\,, \quad h_0 := \omega_0 \big\| \Delta(x^0, x^0) \big\|$$

satisfy the comparison property

$$h'_0 \le h_0\,.$$

Hence, by *deliberate* rank reduction, larger damping factors should be possible. In order to proceed to a theoretically backed choice of the damping factors, computational estimates of the associated rank-deficient Lipschitz constants are needed. Upon carefully revisiting the above full rank analysis we may end up with the estimates:

$$[\overline{\omega}_k] := \frac{\left(\|\Delta(x^{k-1},x^k) - \Delta(x^k,x^k) + \overline{\Delta}(x^k,x^{k-1})\|^2 - \|\Delta(x^{k-1},x^k)\|^2\right)^{\frac{1}{2}}}{\lambda_{k-1}\|\Delta(x^{k-1},x^{k-1})\|\|\Delta(x^{k-1},x^k)\|}$$

$$(4.74)$$

wherein, for the unconstrained case, the quantities

$$\overline{\Delta}(x^k,x^{k-1}) := -J(x^k)^+\left(F(x^k) + J(x^{k-1})\Delta(x^{k-1},x^k)\right),$$

$$\Delta(x^{k-1},x^k) := P^\perp(x^k,x^k)\Delta(x^{k-1},x^k)$$

can be cheaply computed. Moreover, with either $[\overline{\delta}] := 0$ or

$$[\overline{\delta}] := \frac{\|J(x^k)^+ r(x^{k-1})\|}{\|J(x^k)^+ F(x^{k-1})\|} < 1,$$

the *trust region* strategy for the rank-deficient case is complete: both the prediction strategy (4.71) and the correction strategy (4.69) remain unchanged—with the proper identification of terms such as $[\overline{\omega}_k]$ by (4.74) in the prediction case.

An interesting question is, whether the above Lipschitz constant *estimates* also inherit property (4.72).

Lemma 4.18 Let $[\omega_k]$, $[\omega'_k]$ and $[\overline{\omega}_k]$, $[\overline{\omega}'_k]$ denote the computational estimates as defined above, associated with the Jacobian matrices $J(x^k)$ and $J'(x^k)$, respectively. Then these quantities satisfy

$$[\omega'_k] \le [\omega_k]\,, \quad [\overline{\omega}'_k] \le [\overline{\omega}_k]\,,$$

where equality only holds, if the residual component dropped in the rank reduction process vanishes.

Proof. First, one observes that, in the unconstrained case

$$[\overline{\omega}_k] = \left\|J(x^k)^+\left(J(x^k) - J(x^{k-1})\right)\Delta(x^{k-1},x^k)\right\| / d$$

with d a short hand notation for the denominator in (4.70). After rank reduction, the corresponding expression is

$$[\overline{\omega}'_k] = \left\|J'(x^k)^+\left(J'(x^k) - J(x^{k-1})\right)\Delta(x^{k-1},x^k)\right\| / d.$$

By means of the usual extremal property of the Moore-Penrose pseudoinverse, one would immediately have that

$$\left\|J'(x^k)^+\left(J(x^k) - J(x^{k-1})\right)\Delta(x^{k-1},x^k)\right\|$$
$$\le \left\|J(x^k)^+\left(J(x^k) - J(x^{k-1})\right)\Delta(x^{k-1},x^k)\right\|.$$

Hence, in order to connect the numerators in $[\overline{\omega}_k]$ and $[\overline{\omega}'_k]$, one needs to show that the vectors

$$z := J'^+ J y , \quad z' := J'^+ J' y$$

are identical for arbitrary y. For this purpose, consider the QR-decompositions

$$J = Q^T \begin{bmatrix} R_{11} & R_{12} \\ 0 & R_{22} \\ 0 & 0 \end{bmatrix}, \quad J' = Q'^T \begin{bmatrix} R_{11} & R_{12} \\ 0 & 0 \\ 0 & 0 \end{bmatrix} .$$

Note that, as long as only the Euclidean norm arises, QR-decomposition formally includes SVD-decomposition (they just differ by orthogonal transformations). Therefore the above representation models our above rank reduction strategy. Since the difference of Q and Q' only acts on R_{22}, one may also write:

$$J' = Q^T \begin{bmatrix} R_{11} & R_{12} \\ 0 & 0 \\ 0 & 0 \end{bmatrix} .$$

Hence,

$$\begin{aligned} z &= \begin{pmatrix} z_1 \\ z_2 \end{pmatrix} = \begin{bmatrix} R_{11} & R_{12} \\ 0 & 0 \end{bmatrix}^+ Q Q^T \begin{bmatrix} R_{11} & R_{12} \\ 0 & R_{22} \\ 0 & 0 \end{bmatrix} \begin{pmatrix} x_1 \\ x_2 \end{pmatrix} \\ &= \begin{bmatrix} R_{11} & R_{12} \\ 0 & 0 \end{bmatrix}^+ \begin{pmatrix} R_{11} x_1 + R_{12} x_2 \\ R_{22} x_2 \end{pmatrix} =: \begin{bmatrix} R_{11} & R_{12} \\ 0 & 0 \end{bmatrix}^+ \begin{pmatrix} u \\ v \end{pmatrix} . \end{aligned}$$

The associated linear least squares problem is

$$\| R_{11} z_1 + R_{12} z_2 - u \|^2 + \| v \|^2 = \min$$

subject to

$$\| z \| = \min .$$

This must be compared with the linear least squares problem defining z':

$$\| R_{11} z_1' + R_{12} z_2' - u \|^2 = \min$$

subject to

$$\| z' \| = \min .$$

From this, one concludes that

$$z = z' .$$

One is now ready to further treat the numerator in $[\overline{\omega}_k']$ by virtue of

$$\begin{aligned} & \| J'(x^k)^+ \big(J'(x^k) - J(x^{k-1}) \big) \Delta(x^{k-1}, x^k) \| \\ &= \| J'(x^k)^+ \big(J(x^k) - J(x^{k-1}) \big) \Delta(x^{k-1}, x^k) \| . \end{aligned}$$

Upon collecting all results, we finally have proven that

$$[\overline{\omega}_k'] \leq [\overline{\omega}_k] ,$$

where equality only holds, if the dropped residual component of the vector

$$\left(J(x^k) - J(x^{k-1})\right)\Delta(x^{k-1}, x^k)$$

actually vanishes. The same kind of argument also verifies that

$$[\omega_k'] \le [\omega_k]\,,$$

which completes the proof of the lemma.   □

In view of (4.72) and (4.73) together with the (technical) assumption $\overline{\delta}' := \overline{\delta}$, the estimated damping factors

$$\overline{\lambda}_k' := \min\left(1\,, \ \frac{1 - \overline{\delta}'}{[\omega_k']\|\Delta'(x^k, x^k)\|}\right)$$

should increase

$$\overline{\lambda}_k' \ge \overline{\lambda}_k\,,$$

a phenomenon that has been observed in actual computation.

Summarizing, the above theoretical considerations nicely back the observations made in practical applications of such a *rank strategy*. Numerical experience, however, indicates that only rather restricted use of this device should be made in realistic examples, since otherwise the iteration tends to converge to critical points with a rank-deficient Jacobian or to *trivial solutions* that are often undesirable in scientific applications. For this reason, a deliberate rank reduction is recommended only, if

$$\overline{\lambda}_k < \lambda_{\min} \ll 1$$

for some prescribed default parameter $\lambda_{\min}$ (say $\lambda_{\min} = 0.01$). In such a situation, deliberate rank reduction may often lead to damping factors

$$\overline{\lambda}_k' \ge \lambda_{\min}$$

at some intermediate iteration step and allow one to continue the iteration with greater rank afterwards.

A convergence analysis comparable to the preceding sections would be extremely technical and less satisfactory than in the simpler cases treated before. As it turns out, the whole rank reduction device is rarely activated, since the refined damping strategies anyway tend to avoid critical points with rank-deficient Jacobian. The device *does* help, however, in *inverse problems* whenever the identification of some unknown model parameters is impossible from the comparison between assumed model and given data. In these cases, at least *local* convergence is guaranteed by Theorem 4.7.

BIBLIOGRAPHICAL NOTE. The possible enlargement of the convergence domain by means of deliberate rank reduction has been pointed out by the author in 1972 in his dissertation [59, 60]. In 1980, the effect was partially redetected by J. Blue [28].

The described adaptive trust region and rank strategies have been implemented in the codes NLSQ-ERR for *unconstrained* and NLSCON for equality *constrained* nonlinear least squares problems.

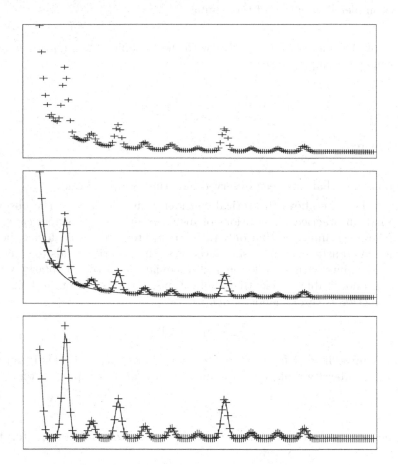

Fig. 4.3. Parameter fit: X-ray spectrum. The background signal (primary beam) of unknown shape gives rise to a rank-deficiency 2. *Top:* measurements, *center:* fit to measurements, *bottom:* fit after subtraction of primary beam.

Example 4.2 *X-ray spectrum.* In order to illustrate the convergence of the rank-deficient Gauss-Newton method, a rather typical example from X-ray spectroscopy is given now. In this example, measurements of small angle diffraction on the crystalline regions of collagen fibers are compared with a model function representing a finite series of Gaussian peaks. The peaks of actual interest are contaminated by a background signal ('primary beam') of unknown shape.

Assessing a combination of Lorentz peaks, Gaussian peaks and a constant to the background signal, *Jacobian rank-deficiency* naturally arises. The data $\{y_i\}$ were obtained from an electron multiplier with a noise level characterized by absolute measurement tolerances

$$\delta y_i = \sqrt{y_i}\,, \ i = 1, \ldots, m\,.$$

Without explicit setting of these error tolerances, this problem had turned out to be not solvable.

In Figure 4.3, the fit of $n = 23$ model parameters to given $m = 178$ measurement values is represented. In a neighborhood of the solution, a constant rank deficiency of 2 occurred. Nevertheless, the Gauss-Newton method succeeded in supplying a satisfactory solution.

It may be worth mentioning that standard *filtering techniques*, which just separate high and low frequencies, are bound to fail in this example, since the unwanted background signal contains both low and high frequencies.

4.4 Underdetermined Systems of Equations

In this section we discuss the iterative solution of *underdetermined* systems of nonlinear equations

$$F(x) = 0\,, \tag{4.75}$$

where $F : D \subset \mathbb{R}^n \to \mathbb{R}^m$ with $m < n$. First, in Section 4.4.1, error oriented *local Gauss-Newton algorithms* are studied in close connection with the treatment of *overdetermined* systems in Section 4.3. In particular, a *local quasi-Gauss-Newton method* is worked out including an appropriate extension of Broyden's 'good' update technique—compare Section 2.1.4 above. Next, in Section 4.4.2, a globalization of these methods is derived.

4.4.1 Local quasi-Gauss-Newton method

This section deals with *local* Gauss-Newton methods, which will require 'sufficiently good' starting points.

Linear underdetermined systems. We start with a discussion of the linear special case

$$Ax = b\,,$$

where A is an (m,n)-matrix and x, b are vectors of corresponding length. As for the solution structure, we may adapt Lemma 4.1 of Section 4.1.1 to the present situation: For $p := \operatorname{rank}(A) \leq m < n$, there exists some $(n-p)$-dimensional solution subspace X^*. The general solution may be decomposed as

$$x = x^* + z$$

with arbitrary $z \in \ker(A)$ and x^* the 'shortest' solution in X^* defined by

$$x^* = A^+ b \qquad (4.76)$$

in terms of the Moore-Penrose pseudo-inverse of A. As in Section 4.1.1 we define the orthogonal projectors $P = A^+ A, \overline{P} = AA^+$. In particular, we have

$$\operatorname{rank}(A) = m < n \Longleftrightarrow \overline{P} = I_m, \quad \overline{P}^\perp = 0.$$

As in the overdetermined case, the actual computation of (4.76) is done via a QR-decomposition with column pivoting

$$QA\Pi = [R, S], \qquad (4.77)$$

where R is a nonsingular upper triangular (m, m)-matrix with diagonal entries ordered according to modulus as in (4.9) and S an $(m, n-m)$-matrix. If $m = n + 1$, then the permutation Π naturally defines some *external parameter*, say ξ, as the variable corresponding to the *last* column S.

Let

$$x = A^+ b = \Pi^T \begin{pmatrix} v_1 \\ v_2 \end{pmatrix}$$

with partitioning $v_1 \in \mathbb{R}^m$, $v_2 \in \mathbb{R}^{n-m}$. For $n - m \ll m$, which is the usual case treated here, the QR-*Cholesky decomposition* [65] is recommended:

Backsubstitutions:	$Rw = S$,	$R\overline{v}_1 = b$,
Inner products:	$M := I_{n-m} + w^T w$,	$\overline{v}_2 := w^T \overline{v}_1$,
Cholesky decomposition:	$M = LL^T$,	$Mv_2 = \overline{v}_2$,
Projection:	$v_1 = \overline{v}_1 - w^T v_2$.	

The total computational amount is $n-m$ square root evaluations and roughly $\frac{1}{2} m^2 (n-m)$ multiplications. If the row rank p must be reduced successively within some adaptive rank strategy (compare Section 4.3.5), then some additional $\sim m(n-m)$ array storage places are needed.

Local Gauss-Newton method. We are now ready to treat the truly nonlinear case. Linearization of (4.75) leads to the underdetermined system

$$F'(x)\Delta x = -F(x).$$

From this we arrive at the *local Gauss-Newton method* $(k = 0, 1, \ldots)$:

$$\Delta x^k = -F'(x^k)^+ F(x^k)\,, \quad x^{k+1} := x^k + \Delta x^k\,. \tag{4.78}$$

Again we define the orthogonal projectors P and $\overline{P}$ (assuming full row rank) via

$$\overline{P}(x) := F'(x)F'(x)^+ = I_m\,, \qquad P(x) := F'(x)^+ F'(x) \neq I_n\,. \tag{4.79}$$

As for local convergence, a specialization of Theorem 4.7 will apply. Since $\overline{P}^{\perp}(x) = 0$ from (4.79), we may directly conclude that $\kappa(x) \equiv 0$ and therefore drop assumption (4.38). Moreover, uniqueness of the solution now longer holds. The associated convergence theorem for the underdetermined case then reads:

Theorem 4.19 *Let* $F : D \subseteq \mathbb{R}^n \to \mathbb{R}^m$, $m < n$, *with* D *open, convex denote a continuously-differentiable mapping. Consider the Gauss-Newton method (4.78). Assume that one can find a starting point* $x^0 \in D$, *and constants* α, $\omega > 0$ *such that*

$$\|F'(x^0)^+ F(x^0)\| \leq \alpha\,,$$

$$\|F'(x)^+ \big(F'(y) - F'(x)\big)(y - x)\| \leq \omega \|y - x\|^2$$

for all $x, y \in D$ *collinear,* $y - x \in \mathcal{R}\big(F'(x)^+\big)$.

Moreover, let

$$h := \alpha\omega < 2\,, \qquad \overline{S}(x^0, \rho) \subset D \text{ with } \rho := \alpha/(1 - \tfrac{1}{2}h)\,.$$

Then:

(I) *The sequence* $\{x^k\}$ *of Gauss-Newton iterates is well-defined, remains in* $\overline{S}(x^0, \rho)$ *and converges to some* $x^* \in \overline{S}(x^0, \rho)$ *with*

$$F'(x^*)^+ F(x^*) = 0\,.$$

(II) *Quadratic convergence can be estimated according to*

$$\|x^{k+1} - x^k\| \leq \tfrac{1}{2}\omega \|x^k - x^{k-1}\|^2\,.$$

In passing, we want to emphasize that the local quadratic convergence property for the underdetermined case also implies that *Jacobian approximation errors* up to considerable size are *self-corrected*—just as in the ordinary Newton method.

Local quasi-Gauss-Newton method. For the underdetermined case a variant of the *error oriented* Jacobian rank-1 update technique, the so-called 'good' Broyden method can be developed (cf. Section 2.1.4). In analogy to formula (2.26), let $J_0 := F'(x^0)$ and define the update formula ($k = 0, 1, \ldots$)

$$J_{k+1} := J_k + F(x^{k+1})\frac{(\delta x^k)^T}{\|\delta x^k\|_2^2} \qquad (4.80)$$

together with the iteration

$$\delta x^k := -J_k^+ F(x^k), \ x^{k+1} := x^k + \delta x^k . \qquad (4.81)$$

This *quasi-Gauss-Newton* method has a nice *geometric interpretation* to be derived now.

Lemma 4.20 *Let $\mathcal{N}(J_k)$ denote the nullspace of the Jacobian updates J_k, $k = 0, 1, \ldots$ with $J_0 := F'(x^0)$. Then the quasi-Gauss-Newton method (4.81) is equivalent to the good Broyden method within the m-dimensional hyperplane*

$$H := x^0 \oplus \mathcal{N}^\perp\big(F'(x^0)\big) = \{x \in \mathbb{R}^n | x = x^0 + P(x^0, x^0)z, \ z \in \mathbb{R}^n\} .$$

Proof. From the definition (4.81), we may directly verify that

$$\delta x^k \perp \mathcal{N}^\perp(J_k),$$

which with the recursion (4.80) implies that

$$\mathcal{N}(J_{k+1}) = \mathcal{N}(J_k).$$

Hence, for $k > 0$

$$x^k - x^0 \in \mathcal{N}^\perp(J_0),$$

which confirms the basic statement of the lemma.

The situation depicted in Figure 4.4 inspires an *orthogonal* coordinate transformation including some *shift* such that x^0 is the new origin and H spans $\mathbb{R}^m$. In this coordinate frame, the *quasi-Gauss-Newton correction δx^k* may be interpreted as the *quasi-Newton correction* in H. At the same time, the update formula (4.80) in $\mathbb{R}^n$ degenerates to the good Broyden update (2.26) in $\mathbb{R}^m$. This completes the proof of the Lemma.   □

From the above lemma we may directly conclude that, under reasonable assumptions (cf. Theorem 2.9), the quasi-Gauss-Newton iteration converges *superlinearly* to some solution point x^*, which is the intersection of the hyperplane H with the solution manifold—see also Figure 4.4. Moreover, the quasi-Gauss-Newton iteration can be cheaply computed by a variant of algorithm QNERR of Section 2.1.4: just replace each solution of the linear system

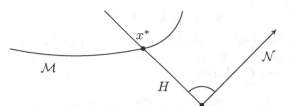

Fig. 4.4. Quasi-Gauss-Newton algorithm: convergence toward some x^* at intersection of solution manifold $\mathcal{M}$ and hyperplane H orthogonal to Jacobian nullspace $\mathcal{N}$.

$$J_0 v = -F$$

by the solution of the underdetermined linear system

$$v = -J_0^+ F\,,$$

which can be realized via the QR-Cholesky-decomposition.

This algorithm requires only the usual array storage for the decomposition of J_0 and, in addition, $(k_{\max} + 1)n$ extra storage places for $\delta x^0, \ldots, \delta x^k$.

Convergence monitor. We need a convergence criterion to assure 'sufficiently good' starting points. On the basis of Lemma 4.20 we can directly adopt the criteria from the simplified Newton and the quasi-Newton method. On this theoretical basis, we merely compute the usual *contraction factors* Θ_k in terms of the quasi-Newton corrections. The iteration is terminated whenever

$$\Theta_k = \frac{\|\delta x^{k+1}\|_2}{\|\delta x^k\|_2} > \tfrac{1}{2}\,.$$

4.4.2 Global Gauss-Newton method

This section again deals with *global* Gauss-Newton methods, which (under the usual mild assumptions) have no principal restriction on the starting points. We proceed as in the case of the global Newton method and try first to construct some *Gauss-Newton path* from an *affine covariant* level concept and second, on this geometrical basis, an *adaptive trust region strategy* to be realized via some damping of the Gauss-Newton corrections.

Geodetic Gauss-Newton path. The underdetermined nonlinear system $F(x) = 0$ generically describes some $(n - m)$-dimensional solution manifold $\mathcal{M}_*$ in $\mathbb{R}^n$. As in the case $m = n$, we argue that this systems is equivalent to any system of the kind

$$AF(x) = 0, \quad A \in \mathrm{GL}(m),$$

which means that *affine covariance* is again a natural requirement for both theory and algorithms. Therefore, as in Section 3.1.4, definition (3.20), we introduce *level sets* $G(x^0|A)$ at the initial guess x^0. Intersection of these level sets according to

$$\overline{G}(x^0) := \bigcap_{A \in \mathrm{GL}(m)} G(x^0|A)$$

generates the λ-family of underdetermined mappings

$$\Phi(x, \lambda) := F(x) - (1 - \lambda)F(x^0) = 0, \lambda \in [0, 1],$$

which generically will define a family of manifolds $\mathcal{M}(\lambda)$, $\lambda \in [0, 1]$, such that $\mathcal{M}_* = \mathcal{M}(1)$. Let $\overline{y}(\lambda)$ denote any path out of a continuum $\overline{Y}$ of paths, all of which start at x^0 and satisfy

$$\Phi(\overline{y}(\lambda), \lambda) \equiv 0, \ \lambda \in [0, 1].$$

Differentiation with respect to λ then yields an underdetermined linear system for the direction fields

$$F'(\overline{y})\dot{\overline{y}} = -F(x^0),$$

which is equivalent to

$$P(\overline{y})\dot{\overline{y}} = \Delta(\overline{y}, x^0), \tag{4.82}$$

wherein P denotes the already introduced orthogonal projector and Δ is defined as

$$\Delta(y, x) := -F'(y)^+ F(x).$$

In order to construct a locally *unique* path $\overline{x}(\lambda)$, we will naturally require the additional *local orthogonality* condition

$$\dot{\overline{x}}(\lambda) \perp \mathcal{N}\Big(F'\big(\overline{x}(\lambda)\big)\Big),$$

which is equivalent to

$$P^\perp(\overline{x})\dot{\overline{x}} = 0. \tag{4.83}$$

Upon adding (4.82) and (4.83), we end up with the unique representation

$$\dot{\overline{x}} = \Delta(\overline{x}, x^0), \ \overline{x}(0) = x^0.$$

of some path $\overline{x}$. Its tangent direction in the starting point x^0 is seen to be

$$\dot{\overline{x}}(0) = \Delta(x^0, x^0) = -F'(x^0)^+ F(x^0),$$

i.e., just the ordinary Gauss-Newton correction. A geometric picture of the situation is given in Figure 4.5.

As can be verified without much technicalities, the path $\overline{x}(\lambda)$ is defined uniquely: it starts at the given point $x^0 = \overline{x}(0)$ and continues up to some solution point $x_* = \overline{x}(1) \in \mathcal{M}_*$. Among all possible paths $\overline{y} \in \overline{Y}$ the path $\overline{x}$ exhibits an intriguing *geodetic property* to be shown in the following lemma.

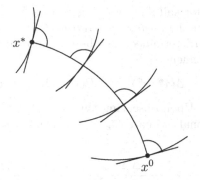

Fig. 4.5. Manifold family $\mathcal{M}(\lambda)$ and geodetic Gauss-Newton path $\bar{x}$.

Lemma 4.21 *Let $\overline{Y}$ denote the continuum of all C^1-paths $\bar{y}$ satisfying*

$$F\big(\bar{y}(\lambda)\big) - (1-\lambda)F(x^0) = 0\,,\ F'\big(\bar{y}(\lambda)\big)\dot{\bar{y}} = -F(x^0)\,.$$

Then the Gauss-Newton path $\bar{x} \in \overline{Y}$ satisfies the minimal property

$$\int\limits_{\lambda=0}^{1} \|\dot{\bar{x}}(\lambda)\|_2^2 d\lambda = \min_{\bar{y}\in\overline{Y}} \int\limits_{\lambda=0}^{1} \|\dot{\bar{y}}(\lambda)\|_2^2 d\lambda\,.$$

Proof. The minimal property can be shown to hold starting from the pointwise relation

$$\begin{aligned}
\|\dot{\bar{y}}(\lambda)\|_2^2 &= \|P(\bar{y})\dot{\bar{y}}(\lambda)\|_2^2 + \|P^\perp(\bar{y})\dot{\bar{y}}(\lambda)\|_2^2 \\[2mm]
&= \|\Delta(\bar{y},x^0)\|_2^2 + \|P^\perp(\bar{y})\dot{\bar{y}}(\lambda)\|_2^2 \geq \|\dot{\bar{x}}(\lambda)\|_2^2\,.
\end{aligned}$$

Moreover, since $\bar{x} \in \overline{Y}$ and $P^\perp(\bar{x})\dot{\bar{x}} = 0$ determines $\bar{x}$ uniquely, the above result is essentially verified. □

Because of the above geodetic minimal property, the path $\bar{x}$ will be called *geodetic Gauss-Newton path* herein.

Adaptive trust region strategy. On the basis of the above geometrical insight, a global Gauss-Newton method with *affine covariant damping strategy*

$$\Delta(x^k,x^k) = -F'(x^k)^+F(x^k)\,,\qquad x^{k+1} = x^k + \lambda_k\Delta(x^k,x^k)$$

can be constructed. An extended *natural monotonicity test*

$$\|\Delta(x^k,x^{k+1})\|_2 \leq \|\Delta(x^k,x^k)\|_2\,.$$

is obtained to test for suitable damping factors λ_k. Within formula (4.74), which is the basis for the *prediction strategy* (4.71) in the rank deficient case (including the underdetermined case here), the term $\overline{\Delta}(x^k, x^{k-1})$ vanishes. However, the replacement

$$[\Delta(x^{k-1}, x^k) - \Delta(x^k, x^k)] \longrightarrow P(x^k)[\Delta(x^{k-1}, x^k) - \Delta(x^k, x^k)]$$

must be performed. Upon observing that $P^\perp(x^k)\Delta(x^k, x^k) = 0$, we end up with the computational estimate $[\overline{\omega}_k] \leq \overline{\omega}_k$ of the local Lipschitz constant $\overline{\omega}_k$ as

$$[\overline{\omega}_k] := \frac{\left[\|\Delta(x^{k-1}, x^k) - \Delta(x^k, x^k)\|^2 - \|P^\perp(x^k)\Delta(x^{k-1}, x^k)\|^2\right]^{1/2}}{\lambda_{k-1}\|\Delta(x^{k-1}, x^{k-1})\|\|\Delta(x^{k-1}, x^k)\|}$$

and, consequently, with the first trial value for the damping factor as

$$\lambda_k^0 := \min\left(1, 1/([\overline{\omega}_k]\|\Delta(x^k, x^k)\|)\right).$$

The *correction strategy* for possibly refined trials λ_k^i, $i = 1, 2, \ldots$ remains unchanged as in (4.69) based on the estimate (4.68).

The thus defined adaptive damping strategies work surprisingly well in practice. Unfortunately, as in the simpler case of nonlinear equations ($m = n$), an associated *global convergence* theorem cannot be proved here: each new Gauss-Newton iterate x^k induces a new geodetic Gauss-Newton path with a corresponding solution point $x^* \in \mathcal{M}_*$—a structure that is prohibitive to proving convergence. Fortunately, *local quadratic* convergence is guaranteed by the above Theorem 4.19.

Exercises

Exercise 4.1   We consider the iterative behavior of the unconstrained nonlinear least squares functional

$$f(x) = \|F(x)\|_2^2$$

during the Gauss-Newton iteration.

a) For the local Gauss-Newton iteration prove that

$$f(x^{k+1}) \leq \left(\|\overline{P}^\perp(x^k)F(x^k)\| + \tfrac{1}{2}h_k\|\overline{P}(x^k)F(x^k)\|\right)^2$$

in the notation of Section 4.2.1.

b) For the global Gauss-Newton iteration (notation of Section 4.2.2) prove that

$$f(x^{k+1}) \leq \left(\|\overline{P}^\perp(x^k)F(x^k)\| + \left(1 - \tfrac{1}{2}\lambda(1 + \rho)\right)\|\overline{P}(x^k)F(x^k)\|\right)^2,$$

if the optimal damping factor $\lambda = \overline{\lambda}_k$ from (4.25) is used.

Exercise 4.2   Revisit assumption (4.14) for the local convergence of Gauss-Newton methods in the residual framework. In particular, consider the small residual factor $\rho(x)$ in the limiting case $x \to x^*$.

a) Interpret this term as local curvature of the C^2-manifold $F(x) = F(x^*)$.

b) Compare the condition $\rho(x^*) < 1$ with the stability condition for nonlinear least squares problems as imposed by P.-Å. Wedin (see [196, 144]).

Exercise 4.3   In Newton's method, sufficiently small Jacobian perturbations are known to still keep *superlinear* convergence—this is the so-called self-correction property. Consider the ordinary Gauss-Newton method for *compatible* nonlinear least squares problems, which is known to converge superlinearly: here Jacobian perturbations, even if they are 'sufficiently small', deteriorate the iterative behavior to *linear* convergence. Analyze this convergence pattern in terms of theory.

Exercise 4.4   For unconstrained nonlinear least squares problems

$$\|F(x)\|_2 = \min,$$

we consider a modification of Broyden's rank-1 update technique. Starting from some given $J_0 = J(x^0)$, Jacobian updates are obtained by

$$J_{k+1} := J_k - J_k \overline{\Delta x}_{k+1} \frac{\Delta x_k^T}{\|\Delta x_k\|_2^2},$$

where $x^{k+1} = x^k + \Delta x_k$ with

$$\Delta x_k := -J_k^+ F(x^k), \quad \overline{\Delta x}_{k+1} := -J_k^+ F(x^{k+1}).$$

a) Derive the kind of secant condition that is satisfied by this update.

b) Show that such a *quasi-Gauss-Newton* iteration solves the 'wrong' nonlinear least squares problem

$$\|\overline{P}_0 F(x)\|_2 = \min,$$

where $\overline{P}_0 = J_0 J_0^+$ is an orthogonal projector.

Exercise 4.5   Consider a general *steepest descent* method based on the corrections

$$\Delta x(A) = -\operatorname{grad}(T(x|A)$$

in terms of the general level function

$$T(x|A) := \tfrac{1}{2}\|AF(x)\|_2^2 \,.$$

Let the Jacobian J be nonsingular, say, of the form

$$J = \left[\begin{array}{c|c} R & S \\ \hline 0 & 0 \end{array}\right] , \quad R \text{ nonsingular } (p,p)\text{-matrix} \,.$$

Let $v := R^{-1}S \in \mathbb{R}^{n-p}$, and define the projectors

$$P := \left(\begin{array}{cc} I_p & 0 \\ 0 & 0 \end{array} \right) , \quad P^\perp = I_n - P \,.$$

a) Show that, for *all* nonsingular matrices A, the corrections $\Delta x(A)$ are confined to the $(n-p)$-dimensional hyperplane

$$(H_p) \quad P^\perp \Delta x - v^T P \Delta x = 0 \,.$$

b) Show that the associated Levenberg-Marquardt correction (3.17) also lies in H_p.

c) Appropriate projection of Newton's method yields

$$(PN) \qquad P\Delta x + vP^\perp \Delta x = -R^{-1}PF(x) \,,$$

$$(P^\perp N) \qquad\qquad\qquad 0 = -P^\perp F(x) \,.$$

In general, the relation $(P^\perp N)$ is a contradiction. Verify, however, that the intersection

$$(H_p) \cap (PN)$$

leads to the Gauss-Newton correction

$$\Delta x = -J^+ F(x) \,.$$

Hint: Compare the QR-Cholesky decomposition for the rank-deficient Moore-Penrose pseudo-inverse.

Exercise 4.6   We again consider the (possibly shortest) solution of the weighted linear least squares problem

$$\|Bx - d\|_2^2 + \mu^2\|Ax - c\|_2^2 = \min \,,$$

defined by some penalty parameter μ. In Section 4.1.2 we had presented the treatment using Householder transformations. Here we study the numerical solution via orthogonalization using *fast Givens rotations*—for details see, e.g., [107].

Perform the penalty limiting process $\mu \to \infty$ for this case. What kind of elimination method arises for the equality constraints? What kind of pivoting is required? Which two sub-condition numbers arise naturally? How can these be interpreted? Treatment of rank-deficiency in either case?

Exercise 4.7   Let
$$A^- := LB^+R$$
denote a generalized inverse, defined via the Moore-Penrose pseudoinverse B^+ with two nonsingular matrices L, R. Derive a set of four axioms which uniquely defines A^-.

Hint: Start from the four Penrose axioms for B.

Exercise 4.8   *Separable Gauss-Newton method.* We compare the two variants suggested by Golub-Pereyra [105] and by Kaufman [128]. Let $\alpha_{GP}, \omega_{GP}, \kappa_{GP}$ denote the best possible theoretical estimates in Theorem 4.7 when applied to the Golub-Pereyra suggestion (4.21) and $\alpha_K, \omega_K, \kappa_K$ the corresponding quantities for the Kaufman variant (4.22).

Assuming best possible estimates, prove that

$$\alpha_K \le \alpha_{GP}, \quad \omega_K \le \omega_{GP}, \quad \kappa_K(v^*) \le \kappa_{GP}(v^*).$$

Exercise 4.9   Show that the results of Theorem 4.7 also hold, if assumption (4.38) is replaced by

$$\left\| \left(F'(x+v)^- - F'(x)^- \right) F(x) \right\| \le \kappa_1(x)\|v\|, \quad x \in D.$$

Note that in this case the projection property (4.37) is not required to hold.

Exercise 4.10   Consider underdetermined nonlinear systems

$$F(x) = 0, \quad F : D \subset \mathbb{R}^n \longrightarrow \mathbb{R}^m, \quad n > m$$

with generically nonunique solution x^*. Verify that the residual based affine contravariant Newton-Mysovskikh theorem (Theorem 2.12) still holds, if only the iterative corrections Δx^k solve the underdetermined linearized systems

$$F'(x^k)\Delta x^k = -F(x^k).$$

Interpret this result in terms of the uniqueness of the residual $F(x^*)$.

5 Parameter Dependent Systems: Continuation Methods

In typical scientific and engineering problems not only a single isolated non-linear system is to be solved, but a family of problems depending on one or more *parameters* $\lambda \in \mathbb{R}^p, p \geq 1$. The subsequent presentation will be mainly restricted to the case $p = 1$ (with the exception of Section 4.4). In fact, parameter dependent *systems of nonlinear equations*

$$F(x, \lambda) = 0, \quad x \in D \subseteq \mathbb{R}^n, \quad \lambda \in [0, L] \tag{5.1}$$

are the basis for *parameter studies* in *systems analysis* and *systems design*, but can also be deliberately exploited for the *globalization of local Newton or Gauss-Newton methods*, if only poor initial guesses are available.

In order to understand the structure of this type of problem, assume that (5.1) has a locally unique solution $(x^*, \lambda^*) \in D \times [0, L]$. Let the (n, n)-matrix $F_x(x, \lambda)$ be *regular* in some neighborhood of this point. Then, by the implicit function theorem, there exists a unique *homotopy path* $\overline{x}$ defined by virtue of the *homotopy*

$$F\left(\overline{x}(\lambda), \lambda\right) \equiv 0, \; \lambda \in [0, L]$$

or, equivalently, by the linearly implicit ODE, often called the *Davidenko differential equation* (in memory of the early paper [48] by D. Davidenko),

$$F_x \dot{\overline{x}} + F_\lambda = 0 \tag{5.2}$$

with a selected solution x^* on the homotopy path as *initial value*, say

$$\overline{x}(\lambda^*) := x^* \,.$$

Note that the ODE (5.2) uniquely defines the *direction field* $\dot{\overline{x}}$ in terms of the λ-parametrization.

In order to avoid the specification of the parametrization, one may introduce the augmented variable

$$y := (x, \lambda) \in \mathbb{R}^{n+1}$$

and rewrite the above mapping (5.1) as

$$F(y) = 0$$

and the direction field (5.2) as

$$F'(y)\,t(y) = 0\,.$$

Whenever the condition

$$\operatorname{rank} F'(y) = n$$

or, equivalently,

$$\dim \ker F'(y) = 1\,,$$

then $t(y)$ is uniquely defined up to some *normalization*, which might be fixed as

$$\|t\|_2 = 1\,.$$

In general, whenever the local condition

$$\operatorname{rank}\ F'(y) = n - k$$

or, equivalently

$$\dim \ker F'(y) = k + 1$$

holds, then a *singularity* of order k occurs. A special role is played by *turning points*, which, with respect to the selected parameter λ, can be characterized by $k = 0$ and

$$\operatorname{rank}\ F'(y^*) = n\,, \quad \operatorname{rank}\ F_x(x^*, \lambda^*) = n - 1\,,$$

so that they are, formally speaking, singularities of order $k = 0$. For $k > 0$ the local direction field is *not* unique, its actual structure depending on properties of higher derivatives up to order $k + 1$. For $k = 1$ *simple bifurcation points* may occur, which require, however, some second derivative discriminant D to be *positive*: in this case, two distinct branch directions are defined; if $D = 0$, a so-called *isola* occurs. For $k > 1$ there exists a hierarchy of critical points, which we cannot treat here in full beauty. The complete solution structure of parameter dependent mappings, usually represented within a *bifurcation diagram*, may turn out to be rather complicated. Here we will restrict our attention to turning points and simple bifurcation points.

Every now and then, the scientific literature contains the suggestion to just integrate the Davidenko differential equation (5.1) numerically, which is not recommended here for the following reasons:

- In most applications only *approximations* of the Jacobian (n, n)-matrix $F_x(x, \lambda)$ are available.
- The numerical integration of (5.1) requires some *implicit* or at least *linearly implicit* discretization, which, in turn, requires the solution of linear equations of the kind

$$\Big(F_x(y) - \beta \Delta\lambda F_{\lambda x}(y)\Big)\Delta x + \beta \Delta\lambda F_{xx}[\dot{\bar{x}}, \Delta x] = -\Delta\lambda F_\lambda(y)\,,$$

obviously requiring second order derivative information. Even though any stiff integrators will solve the ODE, they will not assure the basic condition $F = 0$ up to sufficient accuracy due to global error propagation—this is the well-known 'drift' of the global discretization error.

Rather so-called *discrete continuation methods* are the methods of choice: they concentrate on the solution of $F = 0$ directly and only require sufficiently accurate evaluation of the mapping F and approximations of F'. Such methods consist of two essential parts:

- a *prediction method* that, from given solution points $(\overline{x}_\nu, \overline{\lambda}_\nu)$ on the homotopy path, produces some 'new' point $(\widehat{x}_{\nu+1}, \widehat{\lambda}_{\nu+1})$ assumed to be 'sufficiently close' to the homotopy path,

- an iterative *correction method* that, from a given starting point $(\widehat{x}_\nu, \widehat{\lambda}_\nu)$, supplies some solution point $(\overline{x}_\nu, \overline{\lambda}_\nu)$ on the homotopy path.

For the prediction step, *classical* or *tangent continuation* are the canonical choices—see below. Needless to say that, for the iterative correction steps, we here concentrate on local Newton and quasi-Newton methods (see Sections 2.1.1 and 2.1.4 above) as well as (rank-deficient) Gauss-Newton methods (see Section 4.4.1 above).

BIBLIOGRAPHICAL NOTE. The principle of local continuation has been suggested in 1892 by H. Poincaré [168] in the context of *analytical continuation*. The idea of *discrete continuation* seems to date back to E. Lahaye [142] in 1934. As for the analysis of higher order singular points, the interested reader may want to look up, e.g., the textbooks [109, 110, 111] of M. Golubitsky and coauthors.

Since the underlying homotopy path is a mathematical object in the domain space of the nonlinear mapping F, we select the *affine covariant* framework. In Section 5.1 below, we derive an adaptive pathfollowing algorithm as a *Newton continuation* method, which terminates locally in the presence of *critical points including turning points*. In the next Section 5.2, based on the preceding Section 4.4, we treat an adaptive *quasi-Gauss-Newton continuation* method. This method is able to follow the path *beyond turning points*, but still terminates in the neighborhood of any other critical point. In order to overcome such points as well, we exemplify a scheme to construct *augmented systems*, whose solutions are just selected critical points of higher order— see Section 5.3. This scheme is an appropriate combination of Lyapunov-Schmidt reduction and topological universal unfolding. Details of numerical realization are only worked out for the computation of diagrams including simple bifurcation points.

Before we begin with a presentation of any algorithmic details, we want to point out that, quite often, there is a *choice of embedding* to be made in view of computational complexity.

Example 5.1   *Choice of embedding.* Consider the problem from [99]

$$G(x) := x - \phi(x) = 0,$$

with

$$\phi_i(x) := \exp(\cos(i \cdot \sum_{j=1}^{10} x_j)), \quad i = 1, \ldots, 10.$$

In [99], K. Georg treated the unspecific embedding

$$F(x, \lambda) = \lambda F(x) + (1 - \lambda)x = x - \lambda\phi(x). \tag{5.3}$$

Assume that we know the solution at $\lambda = 0$, which is $x_i^0 = 0, i = 1, \ldots, 10$, and want to find the solution at $\lambda = 1$. The solution structure $x(\lambda)$ is given in Figure 5.1, left.

Alternatively, we might choose the more problem-oriented embedding (compare also [77, Section 4.4])

$$\tilde{F}_i(x, \lambda) := x_i - \exp(\lambda \cdot \cos(i \cdot \sum_{j=1}^{10} x_j)), \quad i = 1, \ldots, 10. \tag{5.4}$$

The solution at $\lambda = 0$ is now given by $x_i^0 = 1, i = 1, \ldots, 10$. The corresponding solution structure as given in Figure 5.1, right, is obviously much simpler.

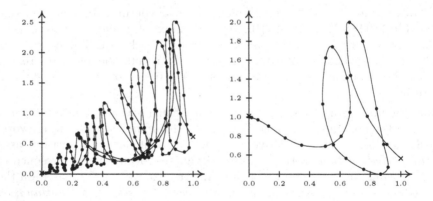

Fig. 5.1. Example 5.1. *Left:* unspecific embedding (5.3). *Right:* problem-oriented embedding (5.4).

All computations have been performed by the Gauss-Newton continuation code `ALCON1` to be described in Section 5.2 below. The dots in Figure 5.1 indicate the number of discrete continuation points as obtained from `ALCON1`: Observe that the computational complexity on the left is much higher than on the right. Look also at the quite different number of turning points. The cross-points arise from the projection $x_9(\lambda)$, not from bifurcations.

5.1 Newton Continuation Methods

This section deals with the situation that there exists a unique *homotopy path* $\bar{x}$ that can be explicitly parametrized with respect to λ over a finite interval of interest. A confirmation of this structure may often come directly from expert insight into the scientific or engineering problem to be solved. In this case the Jacobian (n, n)-matrix $F_x(x, \lambda)$ is known to be *nonsingular*, which excludes the occurrence of any type of critical points. As a consequence, *local Newton algorithms* may well serve as iterative correction methods within any discrete continuation method.

In order to treat the problem family (5.1), a *sequence of problems*

$$F(x, \lambda_\nu) = 0 , \ \nu = 0, 1 \ldots , \tag{5.5}$$

is solved instead, where the interval $[0.L]$ is replaced by the subdivision

$$0 = \lambda_0 < \lambda_1 < \cdots < \lambda_N = L .$$

In order to solve each of the problems (5.5) by a local Newton method, 'sufficiently good' starting points are required, which should be supplied by some suitable *prediction method*. Formally speaking, any starting points will lie on some *prediction path* $\hat{x}(\lambda)$ for $\lambda = \lambda_\nu$. The task therefore involves the choice of the prediction method (Section 5.1.1), the theoretical analysis of the coupling between prediction and Newton method (Section 5.1.2), which leads to a characterization of feasible stepsizes, and, on this theoretical basis, the adaptive choice of the stepsizes $\Delta\lambda_\nu = \lambda_{\nu+1} - \lambda_\nu$ in actual computation (Section 5.1.3). Since paths as mathematical objects live in the domain space of the mapping F, the *affine covariant* setting for both theory and algorithms is selected throughout Section 5.1.

5.1.1 Classification of continuation methods

As the first idea to choose a suitable *starting point* $\hat{x}(\lambda_{\nu+1})$ one will just take the previous *solution point* $\bar{x}(\lambda_\nu)$. This so-called *classical continuation method* is represented schematically in Figure 5.2. For this continuation method the prediction path is defined as

$$\hat{x}(\lambda) = \bar{x}(\lambda_\nu) , \ \lambda \geq \lambda_\nu .$$

A refinement of the above idea is to proceed along the tangent of the homotopy path in λ_ν. This is the so-called *tangent continuation method*, sometimes also called *method of incremental load* or *Euler continuation*, since it realizes the explicit Euler discretization of the ODE (5.2). The corresponding scheme is depicted in Figure 5.3. The associated prediction path is defined by

$$\hat{x}(\lambda) = \bar{x}(\lambda_\nu) + (\lambda - \lambda_\nu) \dot{\bar{x}}(\lambda_\nu) , \ \lambda \geq \lambda_\nu ,$$

wherein

$$\dot{\overline{x}}(\lambda_\nu) = -F_x\Big(\overline{x}(\lambda_\nu),\, \lambda_\nu\Big)^{-1} F_\lambda\Big(\overline{x}(\lambda_\nu),\, \lambda_\nu\Big).$$

Note that both the classical and the tangent prediction paths are given in *affine covariant* terms, which suggests that a theoretical classification of prediction methods should also be formulated in such terms that match with the *error oriented* local convergence analysis of Newton's method from Section 2.1 above.

Definition. Let $\Delta\lambda := \lambda - \lambda_\nu$. A continuation method defined via the prediction path $\widehat{x}(\lambda)$ is said to be of order p, if a constant η_p exists such that

$$\|\overline{x}(\lambda) - \widehat{x}(\lambda)\| \le \eta_p \cdot \Delta\lambda^p. \tag{5.6}$$

In order to illustrate this definition, a few examples are given first. For simplicity, let $\lambda_\nu := 0$ and $\lambda = \Delta\lambda$.

Classical continuation method. For the method represented in Figure 5.2 one immediately derives

$$\|\overline{x}(\lambda) - \widehat{x}(\lambda)\| = \|\overline{x}(\lambda) - \overline{x}(0)\| \le \lambda \cdot \max_{s \in [0,L]} \|\dot{\overline{x}}(s)\|.$$

Hence, this method is of the order $p = 1$ with order coefficient

$$\eta_1 := \max_{s \in [0,L]} \|\dot{\overline{x}}(s)\|.$$

Actually, both H. Poincaré [168] and E. Lahaye [142] had just thought of this simplest type of continuation.

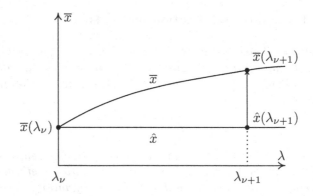

Fig. 5.2. Classical continuation method.

Tangent continuation method. For the method represented in Figure 5.3 one obtains

$$\left\|\overline{x}(\lambda) - \widehat{x}(\lambda)\right\| = \left\|\overline{x}(\lambda) - \overline{x}(0) - \lambda\dot{\overline{x}}(0)\right\| \leq \tfrac{1}{2}\lambda^2 \max_{s\in[0,L]} \left\|\ddot{\overline{x}}(s)\right\| . \tag{5.7}$$

So, this method is of the order $p = 2$ with order coefficient

$$\eta_2 := \tfrac{1}{2} \max_{s\in[0,L]} \left\|\ddot{\overline{x}}(s)\right\| . \tag{5.8}$$

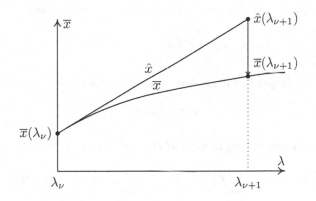

Fig. 5.3. Tangent continuation method.

Standard embedding. The simple embedding

$$F_0(x, \lambda) := F(x) - (1 - \lambda)F(x^0)$$

is rather popular. Note, however, that this homotopy 'freezes' the information of the starting point x^0. The least improvement, which can easily be realized, is to turn to a *damping* or *trust region strategy* (see Chapter 3), which may be understood as being formally based on the *homotopy chain* $(k = 0, 1, \ldots)$

$$F_k(x, \lambda) := F(x) - (1 - \lambda)F(x^k) ,$$

which brings in the information about the 'newest' iterate x^k. Observe, however, that this generally applicable homotopy does not exploit any special structure of the mapping.

Partial standard embedding. In the experience of the author the only sometimes successful variant has been the selection of only *one* component of the mapping, which leads to the so-called partial standard embedding

$$\overline{F}_0(x, \lambda) := PF(x) + P^{\perp}\Big(F(x) - (1 - \lambda)F(x^0)\Big) , \tag{5.9}$$

where P is an orthogonal projector and $P^\perp$ its complement. In what follows we give some comparative results on the classical versus the tangent continuation method for this special kind of embedding.

Lemma 5.1 *Consider the partial standard embedding* (5.9). *Notations as introduced in this section. Then, with the Lipschitz condition*

$$\left\| F'\left(\widehat{x}(\lambda)\right)^{-1}\left(F'(x) - F'(\widehat{x}(\lambda))\right)\right\| \leq \widehat{\omega}_0 \|x - \widehat{x}(\lambda)\|,$$
$$x, \widehat{x}(\lambda) \in D, 0 \leq \lambda \leq L,$$

the order coefficient for the classical continuation method is

$$\eta_1 = \sup_{\lambda \in [0,\bar{\lambda}]} \left\| F'\left(\overline{x}(\lambda)\right)^{-1} P^\perp F(x^0)\right\|, \tag{5.10}$$

the one for the tangent continuation is closely related as

$$\eta_2 := \tfrac{1}{2}\widehat{\omega}_0 \eta_1^2. \tag{5.11}$$

Proof. The extremely simple form of the embedding (5.9) leads to

$$\frac{\partial}{\partial x}\overline{F}_0(x, \lambda) = \frac{\partial}{\partial x}F(x) = F'(x), \frac{\partial}{\partial \lambda}\overline{F}_0(x, \lambda) = P^\perp F(x^0),$$

which implies

$$\dot{\overline{x}}(\lambda) = -F'(\overline{x})^{-1}P^\perp F(x^0)$$

and therefore (5.10). For the estimation of η_2, we must start with a Lipschitz condition for the directions

$$\|\dot{\overline{x}}(\lambda) - \dot{\overline{x}}(0)\| = \left\|\left(F'(\overline{x}(\lambda))\right)^{-1} - F'\left(\overline{x}(0)\right)^{-1})P^\perp F(x^0)\right\|$$

$$\leq \left\| F'\left(\widehat{x}(0)\right)^{-1}\left(F'(\overline{x}(\lambda)) - F'(\overline{x}(0))\right)\right\| \|\dot{\overline{x}}(\lambda)\|$$

$$\leq \widehat{\omega}_0 \|\overline{x}(\lambda) - \overline{x}(0)\| \eta_1 \leq \widehat{\omega}_0 \eta_1^2 \lambda.$$

Hence, with (5.8), one has the second result (5.11), which completes the proof. $\qquad \square$

As for the consequence of this Lemma for the feasible stepsizes see Lemma 5.6 below.

It is an easy exercise to construct further refinements of prediction methods beyond the tangent continuation method, just based on higher derivative information of F. In view of complex real life problems, however, this is not very promising, since this would also require accurate higher order derivative information, which may be rarely available.

Polynomial continuation In order to classify such methods, the monomials $\Delta\lambda^p$ in (5.6) must be replaced by some *strictly monotone increasing function* $\varphi(\Delta\lambda)$ with $\varphi(0) = 0$ such that

$$\|\overline{x}(\lambda) - \widehat{x}(\lambda)\| \leq \eta \cdot \varphi(\Delta\lambda). \tag{5.12}$$

In order to illustrate this definition, we consider two extrapolation methods (once more $\lambda_\nu := 0$ and $\lambda := \Delta\lambda$).

Standard polynomial extrapolation. Based on the data

$$\overline{x}(\lambda_{-q}), \ldots, \overline{x}(\lambda_{-1}), \ \overline{x}(0)$$

a prediction path can be defined as (with $\lambda_0 := 0$)

$$\widehat{x}_q(\lambda) := \sum_{m=-q}^{0} \overline{x}(\lambda_m) L_q^m(\lambda)$$

in terms of Lagrange polynomials $L(\cdot)$. Application of standard approximation error estimates then leads to

$$\|\overline{x}(\lambda) - \widehat{x}(\lambda)\| \leq C_{q+1} \cdot \lambda(\lambda - \lambda_{-1}) \cdot \ldots \cdot (\lambda - \lambda_{-q}),$$

which naturally defines

$$\varphi(\lambda) := \lambda(\lambda - \lambda_{-1}) \cdot \cdots \cdot (\lambda - \lambda_{-q}). \tag{5.13}$$

The numerical evaluation of the prediction path is, of course, done by the Aitken-Neville algorithm.

Hermite extrapolation. This type of polynomial extrapolation is based on the data

$$\overline{x}(\lambda_{-q}), \ \dot{\overline{x}}(\lambda_{-q}), \ldots, \overline{x}(0), \ \dot{\overline{x}}(0).$$

Evaluation of the prediction path $\widehat{x}_q(\lambda)$ is done here by a variant of the Aitken-Neville algorithm for pairwise confluent nodes. Proceeding as above leads to

$$\|\overline{x}(\lambda) - \widehat{x}(\lambda)\| \leq \overline{C}_{q+1} \cdot \lambda^2(\lambda - \lambda_{-1})^2 \cdot \cdots \cdot (\lambda - \lambda_{-q})^2,$$

which defines the monotone function

$$\varphi(\lambda) := \lambda^2(\lambda - \lambda_{-1})^2 \cdot \cdots \cdot (\lambda - \lambda_{-q})^2.$$

Note, however, that this prediction method requires quite accurate derivative information, which restricts its applicability.

BIBLIOGRAPHICAL NOTE. An affine contravariant definition of the order of a continuation method based on the residual F has been given in 1976 by R. Menzel and H. Schwetlick [150]. The here presented affine covariant alternative is due to the author [61] from 1979. Its extension to polynomial extrapolation is due to H.G. Bock [32]. For the standard embedding, convergence proofs were given by H.B. Keller [131] or M.W. Hirsch and S. Smale [119]; a general code has been implemented by A.P. Morgan et al. [154].

5.1.2 Affine covariant feasible stepsizes

Any of the discrete continuation methods are efficient only for 'sufficiently' small *local stepsizes*

$$\Delta\lambda_\nu := \lambda_{\nu+1} - \lambda_\nu ,$$

which must be chosen such that the local Newton method starting at the predicted value $x^0 = \widehat{x}(\lambda_{\nu+1})$ can be expected to converge to the value $x^* = \overline{x}(\lambda_{\nu+1})$ on the homotopy path. In what follows, a theoretical analysis of *feasible stepsizes* is worked out, which will serve as the basis for an adaptive stepsize control to be presented in Section 5.1.3.

Among the *local Newton methods* to be discussed as correction methods are

- the *ordinary* Newton method with a new Jacobian at each iterate (cf. Section 2.1.1),
- the *simplified* Newton method with the initial Jacobian throughout (cf. Section 2.1.2), and
- the quasi-Newton method starting with an initial Jacobian based on 'good' Broyden Jacobian updates (cf. Section 2.1.4).

Ordinary Newton method. We begin with the ordinary Newton method as correction method within any discrete continuation method. The simplest theoretical framework is certainly given by Theorem 2.3.

Theorem 5.2 *Notation as introduced in this Section. Let $F_x(x,\lambda)$ be non-singular for all $(x,\lambda) \in D \times [0,L]$. Let a unique homotopy path $\overline{x}(\lambda)$ exist in a sufficiently large local domain. Assume the affine covariant Lipschitz condition*

$$\|F_x(x,\lambda)^{-1}\left(F_x(y,\lambda) - F_x(x,\lambda)\right)(y-x)\| \le \omega\|y-x\|^2 . \tag{5.14}$$

Let $\widehat{x}(\lambda)$ denote a prediction method of order p as defined in (5.6) based on the previous solution point $\overline{x}(\lambda_\nu)$. Then the ordinary Newton method with starting point $\widehat{x}(\lambda_{\nu+1})$ converges towards the solution point $\overline{x}(\lambda_{\nu+1})$ for all stepsizes

$$\Delta\lambda_\nu \le \Delta\lambda_{\max} := \left(\frac{2}{\omega\eta_p}\right)^{1/p} \tag{5.15}$$

within the interval $[0,L]$.

Proof. Upon skipping any fine structure of the local domains assumed to be sufficiently large, we must merely check the hypothesis of Theorem 2.3 for the ordinary Newton method with starting point $x^0 = \widehat{x}(\lambda)$ and solution point $x^* = \overline{x}(\lambda)$. In view of the local contractivity condition (2.9), we will estimate

$$\|x^* - x^0\| = \|\overline{x}(\lambda) - \widehat{x}(\lambda)\| \le \eta_p \Delta\lambda^p .$$

Upon inserting the upper bound into (2.9), we arrive at the sufficient condition

$$\eta_p \Delta\lambda^p < 2/\omega ,$$

which is essentially (5.15). □

The above theorem requires some knowledge about the 'rather global' Lipschitz constant ω. In order to permit a finer tuning within the automatic stepsize control to be derived in Section 5.1.3, we next apply the affine covariant Newton-Kantorovich theorem (Theorem 2.1), which requires 'more local' Lipschitz information.

Theorem 5.3 *Notation and assumptions essentially as just introduced. Compared to the preceding theorem replace the Lipschitz condition (5.14) by the condition*

$$\left\| F_x\left(\widehat{x}(\lambda), \lambda\right)^{-1} \left(F_x(y, \lambda) - F_x(x, \lambda) \right) \right\| \le \widehat{\omega}_0 \|y - x\| , x, y \in D . \qquad (5.16)$$

Then the ordinary Newton method with starting point $\widehat{x}(\lambda_{\nu+1})$ converges towards the solution point $\overline{x}(\lambda_{\nu+1})$ for all stepsizes

$$\Delta\lambda_\nu \le \Delta\lambda_{\max} := \left(\frac{\sqrt{2} - 1}{\widehat{\omega}_0 \eta_p}\right)^{1/p} . \qquad (5.17)$$

Proof. The above Lipschitz condition (5.16) permits the application of Theorem 2.1, which (with $\lambda = \Delta\lambda$) requires that

$$\alpha(\lambda)\widehat{\omega}_0 \le \tfrac{1}{2} . \qquad (5.18)$$

So an upper bound $\|\Delta x^0(\lambda)\| \le \alpha(\lambda)$ for the first Newton correction needs to be derived. To obtain this, we estimate

$$\|\Delta x^0(\lambda)\| = \left\| F_x\left(\widehat{x}(\lambda), \lambda\right)^{-1} F\left(\widehat{x}(\lambda), \lambda\right) \right\| = \left\| F_x(\widehat{x}, \lambda)^{-1} \left(F(\widehat{x}, \lambda) - F(\overline{x}, \lambda) \right) \right\|$$

$$= \left\| F_x(\widehat{x}, \lambda)^{-1} \int_{s=0}^{1} F_x(\overline{x} + s(\widehat{x} - \overline{x}), \lambda)(\widehat{x} - \overline{x})ds \right\|$$

$$\le \|\widehat{x} - \overline{x}\| \left(1 + \tfrac{1}{2}\widehat{\omega}_0 \|\widehat{x} - \overline{x}\|\right) .$$

The application of the triangle inequality in the last step appears to be unavoidable.

With the definition of the order of a prediction method we are now able to derive the upper bound

$$\|\Delta x^0(\lambda)\| \leq \eta_p \cdot \lambda^p \left(1 + \tfrac{1}{2}\widehat{\omega}_0 \eta_p \lambda^p\right) =: \alpha(\lambda)\,. \tag{5.19}$$

Finally, combination of (5.18) and (5.19) yields

$$\widehat{\omega}_0 \eta_p \lambda^p \left(1 + \tfrac{1}{2}\widehat{\omega}_0 \eta_p \lambda^p\right) \leq \tfrac{1}{2}$$

or, equivalently

$$\widehat{\omega}_0 \eta_p \lambda^p \leq \sqrt{2} - 1\,,$$

which confirms the result (5.17). □

Corollary 5.4 *If the characterization* (5.6) *of a prediction method is replaced by the generalization* (5.12) *in terms of a strictly monotone increasing function* $\varphi(\Delta\lambda)$, *then the maximum feasible stepsize* (5.17) *is replaced by*

$$\Delta\lambda_{\max} := \varphi^{-1}\left(\frac{\sqrt{2}-1}{\widehat{\omega}_0 \eta}\right) \tag{5.20}$$

with φ^{-1} *the mapping inverse to* φ.

Proof. Instead of (5.19), we now come up with (once again $\lambda := \Delta\lambda$)

$$\alpha(\lambda) := \eta\varphi(\lambda)\left(1 + \tfrac{1}{2}\widehat{\omega}_0 \eta\varphi(\lambda)\right)\,,$$

which directly leads to

$$\widehat{\omega}_0 \eta\varphi(\lambda) \leq \sqrt{2} - 1$$

thus confirming (5.20). □

Simplified Newton method. In most applications, the *simplified* Newton method rather than the ordinary Newton method will be realized. For this specification, the following slight modifications hold.

Corollary 5.5 *Notation and assumptions as in Theorem* 5.3 *or Corollary* 5.4, *respectively. Let the ordinary Newton method therein be replaced by the simplified Newton method with the same starting point. Then, with the mere replacement of the Lipschitz constant* $\widehat{\omega}_0$ *via the slightly modified condition*

$$\left\|F_x\left(\widehat{x}(\lambda),\lambda\right)^{-1}\left(F_x(x,\lambda) - F_x(\widehat{x}(\lambda),\lambda)\right)\right\| \leq \widehat{\omega}_0 \|x - \widehat{x}(\lambda)\|\,,$$

the maximum feasible stepsizes (5.17) *and* (5.20) *still hold.*

Proof. The above two proofs can be essentially copied: the contraction condition (2.3) of Theorem 2.1 needs to be formally replaced by condition (2.17) of Theorem 2.5, which means the mere replacement of the Lipschitz condition (2.2) by (2.16). □

Partial standard embedding. In order to illustrate what has been said at the end of Section 5.1.1 about the embedding (5.9), we now study the consequences of Lemma 5.1 for the associated feasible stepsizes.

Lemma 5.6 *Notation and assumptions as in Lemma* 5.1. *Let* $\Delta\lambda_{\max}^{(1,2)}$ *denote the maximum feasible stepsizes for the classical continuation method* $(p = 1)$ *and for the tangent continuation method* $(p = 2)$. *Then the following results hold:*

$$\Delta\lambda_{\max}^{(2)} = \sqrt{2(\sqrt{2}+1)}\,\Delta\lambda_{\max}^{(1)} \approx 2.2\,\Delta\lambda_{\max}^{(1)}\,.$$

Proof. We merely insert the results (5.10) for $p = 1$ and (5.11) for $p = 2$ into the maximum stepsize formula (5.17) to verify the above results. □

Obviously, in this rather unspecific embedding, tangent continuation seems to require roughly double the number of continuation steps compared to classical continuation. This theoretically backed expectation has actually been observed in large scale problems. At the same time, however, an efficient implementation of the tangent continuation method will roughly require double the amount of work per step (see Section 5.1.3). Hence, there is no clear advantage on either side. In sensitive examples, however, smaller stepsizes increase robustness and reliability of the numerical pathfollowing procedure. That is why for this type of embedding *classical continuation* is generally recommended.

5.1.3 Adaptive pathfollowing algorithms

In the preceding section we analyzed discrete continuation methods with the simplified Newton method as correction method. In its actual realization in the code CONTI1, this is extended to some quasi-Newton correction method.

Simplified Newton method. This method keeps the Jacobian matrix $F'(x^0)$ fixed for all iterative steps, which implies that a single matrix decomposition at the beginning is sufficient throughout the iteration. As a convergence monitor, the *contraction factors* Θ_k in terms of the simplified Newton corrections are used. From (2.21) we require the local convergence criterion

$$\Theta_0 \leq \overline{\Theta} = \tfrac{1}{4}\,,$$

which is easily tested after the first Newton step.

Quasi-Newton method. After the first iterative step, we substitute the simplified Newton iteration by the quasi-Newton method based on 'good' Broyden updates, as documented by algorithm QNERR in Section 2.1.4. Let Θ_k denote the corresponding contraction factors in terms of the quasi-Newton corrections. In agreement with the convergence analysis in Theorem 2.7 we require that $\Theta_k \leq 1/2$. Hence, whenever the condition

$$\Theta_k > \tfrac{1}{2} \tag{5.21}$$

occurs, then the continuation step $\lambda_\nu \rightarrow \lambda_\nu + \Delta\lambda_\nu$ is repeated with reduced stepsize $\Delta\lambda_\nu$.

Discrete continuation method. The *classical continuation method* is certainly most simple to realize and needs no further elaboration. The *tangent continuation method* additionally requires the numerical solution of a linear system of the kind

$$F_x(\overline{x}(\lambda_\nu), \lambda_\nu)\dot{\overline{x}}(\lambda_\nu) = -F_\lambda(\overline{x}(\lambda_\nu), \lambda_\nu),$$

which has the same structure as the systems arising in Newton's method. Note, however, that in order to realize a continuation method of *actual* order $p = 2$, *sufficiently accurate* Jacobian approximations (F_x, F_λ) need to be evaluated not only at the starting points $\hat{x}$, but also at the solution points $\overline{x}$. In large scale problems, this requirement may roughly *double* the amount of work per continuation step compared to the classical method $(p = 1)$.

Adaptive stepsize control. In the globalization of local Newton methods by continuation, adaptive trust region strategies come up as adaptive stepsize strategies. Colloquially, the construction principle is as follows: choose stepsizes such that the initial guess $\hat{x}(\lambda_\nu)$ stays within the 'Newton contraction tube' around the homotopy path—see the theoretical stepsize results like (5.17) containing the unavailable theoretical quantities $\hat{\omega}_0$ and η_p.

Following once more our paradigm of Section 1.2.3, we replace the unavailable theoretical quantities by computationally available estimates $[\hat{\omega}_0] \leq \hat{\omega}_0$ and $[\eta_p] \leq \eta_p$—thus arriving at stepsize estimates

$$[\Delta\lambda_{\max}] := \left(\frac{\sqrt{2} - 1}{[\hat{\omega}_0][\eta_p]} \right)^{1/p} \geq \Delta\lambda_{\max}. \tag{5.22}$$

Again both a *prediction strategy* and a *correction strategy* will be needed. Of course, all formulas will be invariant under *rescaling or shifting* of the continuation parameter.

Suppose that, for given $\lambda_{\nu+1}$, the value Θ_0 has already been computed. From the convergence analysis of the simplified Newton method we know that

$$\Theta_0(\lambda) \leq \tfrac{1}{2}\hat{\omega}_0 \|\overline{\Delta x}^0(\lambda)\| \leq \tfrac{1}{2}\hat{\omega}_0\alpha(\lambda). \tag{5.23}$$

Insertion of $\alpha(\lambda)$ from (5.19) yields

$$\Theta_0 \leq \tfrac{1}{2}\hat{\omega}_0\eta_p\Delta\lambda^p \left(1 + \tfrac{1}{2}\hat{\omega}_0\eta_p\Delta\lambda^p \right)$$

or, equivalently,

$$\widehat{\omega}_0 \eta_p \Delta\lambda^p \geq g(\Theta_0)$$

in terms of the monotone increasing function

$$g(\Theta) := \sqrt{1 + 4\Theta} - 1.$$

From this, we may obtain the *a-posteriori estimate*

$$[\widehat{\omega}_0 \eta_p] := \frac{g\big(\Theta_0(\lambda)\big)}{\Delta\lambda^p} \leq \widehat{\omega}_0 \eta_p$$

and the associated stepsize estimate

$$[\Delta\lambda_{\max}] := \left(\frac{g\left(\overline{\Theta}\right)}{[\widehat{\omega}_0 \eta_p]}\right)^{1/p}.$$

Note that $g\left(\overline{\Theta}\right) = \sqrt{2} - 1$ as in formula (5.17)—a mere consequence of the fact that both formulas are based on the Kantorovich condition. Let $\Delta\lambda'_\nu$ denote some desirable stepsize corresponding to $\Theta_0 = \overline{\Theta}$, whereas the actual stepsize $\Delta\lambda_\nu$ corresponds to the actually computed value of Θ_0. Then the above derivation leads to the *stepsize correction* formula

$$\Delta\lambda'_\nu := \left(\frac{g\left(\overline{\Theta}\right)}{g(\Theta_0)}\right)^{1/p} \Delta\lambda_\nu . \tag{5.24}$$

For $\Theta_0 < \overline{\Theta}$ the actual stepsize $\Delta\lambda_\nu$ is acceptable, since $\Delta\lambda'_\nu > \Delta\lambda_\nu$. If, however, the termination criterion (5.21) is activated by some $\Theta_k > 1/2$, then the last continuation step should be repeated with stepsize

$$\Delta\lambda'_\nu := \left(\frac{g\left(\overline{\Theta}\right)}{g(\Theta_k)}\right)^{1/p} \Delta\lambda_\nu ,$$

which is a clear reduction, since

$$\Delta\lambda'_\nu < \left(\frac{\sqrt{2} - 1}{\sqrt{3} - 1}\right)^{1/p} \Delta\lambda_\nu \approx 0.57^{1/p} \Delta\lambda_\nu .$$

In order to derive *a-priori estimates*, we may exploit (5.23) again to obtain

$$[\widehat{\omega}_0] := \frac{2\Theta_0(\lambda_\nu)}{\|\overline{\Delta x}^0(\lambda_\nu)\|} \leq \widehat{\omega}_0$$

and just use the definition of the order of a prediction method in the form

$$[\eta_p] := \frac{\|\widehat{x}(\lambda_\nu) - \overline{x}(\lambda_\nu)\|}{|\Delta\lambda_{\nu-1}|^p} \le \eta_p \, .$$

Upon inserting these quantities into (5.22), a *stepsize prediction strategy* is defined via

$$\Delta\lambda_\nu^0 := \left(\frac{\|\overline{\Delta x}^0(\lambda_\nu)\|}{\|\widehat{x}(\lambda_\nu) - \overline{x}(\lambda_\nu)\|} \cdot \frac{g\left(\Theta\right)}{2\Theta_0} \right)^{1/p} \Delta\lambda_{\nu-1} \, . \qquad (5.25)$$

Note that this estimate is not sensitive to the computational accuracy of $\overline{x}$. Even if only a single Newton step is performed, which means that

$$\overline{x} \to \widetilde{x} := \widehat{x}(\lambda_\nu) + \overline{\Delta x}^0(\lambda_\nu) \, ,$$

then the predicted value degenerates to

$$\Delta\lambda_\nu^0 := \left(\frac{g\left(\Theta\right)}{2\Theta_0} \right)^{1/p} \Delta\lambda_{\nu-1} \, ,$$

which, compared with (5.24), is seen to be still a reasonable estimate. Precaution must be taken in the *nearly linear* case, characterized by

$$\Theta_0 \le \Theta_{\min} \ll 1 \, .$$

In this case, the stepsize estimate (5.25) should be replaced by

$$\Delta\lambda_\nu^0 := \left(\frac{g\left(\Theta\right)}{2\Theta_{\min}} \right)^{1/p} \Delta\lambda_{\nu-1} \, .$$

to avoid exponential overflow.

Polynomial continuation. Both the correction and the prediction strategy carry over to the more general case of continuation by polynomial extrapolation. Upon recalling definition (5.12), formula (5.24) must be modified such that

$$\Delta\lambda_\nu' = \varphi^{-1}\left(\frac{g\left(\Theta\right)}{g\left(\Theta_0\right)} \varphi(\Delta\lambda_\nu) \right) \, .$$

A comparable formula holds instead of (5.25). The above required evaluation of φ^{-1} is easily performed: since the typically arising functions φ are convex (see, e.g., (5.13)), both the ordinary and the simplified Newton method in $\mathbb{R}^1$ converge certainly monotonically; good starting guesses are available from the dominant monomial so that the methods converge even fast.

Graphical output. The automatically selected points typically give a really good representation on the basis of comparably few data—which is a nice side effect of any efficient stepsize control.

Classical continuation. In this case the data $\bar{x}(\lambda_\nu)$ are available, so that in the interactive mode only linear interpolation is possible, whereas in the batch mode cubic spline interpolation will be preferable. For details the reader may check, e.g., Section 7.4 in the textbook [77].

Tangent continuation. In this case both the nodal values $\bar{x}(\lambda_\nu)$ and the associated tangents $\dot{\bar{x}}(\lambda_\nu)$ are available data so that Hermite interpolation will be the method of choice—compare, e.g., Section 7.1.2 in [77].

Detection of critical points. Any critical point (x^*, λ^*) is characterized by the fact that $F_x(x^*, \lambda^*)$ is singular, which will show up as a convergence failure of the local Newton method for iterates sufficiently close to (x^*, λ^*). As a consequence, *turning points* with respect to λ are safely detected: beyond these points, the local Newton iteration will repeatedly activate the termination criterion (5.21). Generally speaking, the feasible stepsizes $\Delta\lambda_{\max}$ will shrink as $(x, \lambda) \to (x^*, \lambda^*)$. On the other hand, due to $[\Delta\lambda_{\max}] \geq \Delta\lambda_{\max}$, an equivalent behavior for the computational estimates cannot be guaranteed: since these estimates are only based on *pointwise sampling* of F and F' possible 'jumps beyond critical points without notice' cannot be excluded. Fortunately, extensive computational experience has demonstrated that the stepsize control derived herein is quite sensitive, typically exhibiting *marked stepsize reductions* in the neighborhood of critical points. Summarizing, critical points of order $k > 0$ are rather often, but not safely detected in continuation methods with explicit parametrization.

Jacobian ill-conditioning. When approaching a critical point on the homotopy path, the Jacobian condition number is known to increase, which might support the idea of estimating it along the continuation process. In connection with QR-decomposition, the *subcondition number* $\mathrm{sc}(F_x)$ is cheaply at hand [83]. Further condition number estimates may be found within `Matlab`. In the experience of the author, the most reliable technique for general linear equations has been found to be based on *iterative refinement with the same mantissa length* (cf. I. Jankowsky/H. Woźniakowski [122]): Let δx denote the correction computed from the linear residual equation and let ε denote the relative machine precision. Then a rough estimate of the condition number is

$$\mathrm{cond}\,(F_x) \doteq \mathrm{cd}\,(F_x) = \frac{\|\delta x\|_\infty}{\varepsilon\|x\|_\infty}.$$

BIBLIOGRAPHICAL NOTE. The stepsize control presented here is based on the author's habilitation thesis [61], there in the context of optimal control problems attacked by multiple shooting techniques—compare [71, Section 8.6.2]. In Section 7.1.3 below, the success of these methods is documented at a space shuttle problem (Example 7.1).

Explicit reparametrization beyond turning points. The mathematical reason, why turning points with respect to the parameter λ cannot be computed via Newton continuation is that in the neighborhood of these points the parametrization with respect to λ breaks down. In order to be able to pass beyond turning points, W.C. Rheinboldt and J.V. Burkardt [178] suggested a technique that selects any of the components $x_1, \ldots, x_{n+1}$ (identifying $\lambda = x_{n+1}$) for local parametrization, if only the curve can be locally parametrized by that component. This approach is rather popular in computational science and engineering, whenever the basic structure of a bifurcation diagram is known from insight into the problem at hand.

In the absence of such insight, explicit parametrization can also be automated: the selection may be based on the occurrence of 'small' pivots within Jacobian LU-decompositions during the Newton continuation process. If QR-decomposition with column permutation is realized, then the last column will be selected. Apart from the choice of a corresponding single component of x, any norm of x can be chosen likewise. The selected artificial parameter will then be used for discrete continuation, while the local Newton method runs over the remaining n components (code PITCON, which realizes the strategy of [178]).

Note, however, that this approach requires some switching between components of x, which introduces an element of *nondifferentiability* into the algorithm and, as a consequence, may cause some lack of robustness.

5.2 Gauss-Newton Continuation Method

In this section we again study the numerical solution of parameter dependent systems of nonlinear equations

$$F(y) = 0$$

in terms of the extended variable $y = (x, \lambda), x \in \mathbb{R}^n, \lambda \in \mathbb{R}^1$. In contrast to Section 5.1, however, local Newton methods are now replaced by local Gauss-Newton methods (see Section 4.4.1), which open the possibility to some smooth *pathfollowing beyond turning points*—as will be shown next.

5.2.1 Discrete tangent continuation beyond turning points

Throughout the present section we assume that a numerical solution $y^* = (x^*, \lambda^*)$ is at hand—either gained directly from insight into the problem or computed via a local or global Gauss-Newton method for underdetermined equations (see Section 4.4). As already discussed earlier, Newton methods with explicit λ-parameterization are bound to fail in the neighborhood of turning points, since there this parameterization breaks down.

Pseudo-arclength continuation. As a first idea to overcome this difficulty we may resort to *differential geometry*, where the smooth parametrization with respect to the *arclength* s is usually recommended: in addition to the n equations

$$F(y(s)) = 0$$

this parameterization includes the normalizing condition

$$\left\|\frac{dy}{ds}\right\|_2^2 = \left\|\frac{dx}{ds}\right\|_2^2 + \left(\frac{d\lambda}{ds}\right)^2 = 1\,,$$

which after discretization eventually leads to the *pseudo-arclength parametrization*

$$\|\Delta y\|_2^2 - \Delta s^2 = 0$$

as normalizing condition.

BIBLIOGRAPHICAL NOTE. The idea of pseudo-arclength continuation has been suggested and worked out in first details by H.B. Keller in [130]. Its most mature and popular implementation is in the code AUTO due to E. Doedel [89], a code known to be rather robust and reliable. Since a sound theoretically backed control of the above stepsize parameter Δs is hard to design, the code realizes some empirical stepsize control. From the invariance point of view, the concept of arclength is not even invariant under *rescaling* of the parameter λ—see Exercise 5.1, where an interesting limiting case is discussed.

Gauss-Newton continuation idea. Here we will follow an alternative idea: at any point $y \in \mathbb{R}^{n+1}$ *including turning points*, the *local tangent* is well-defined via the underdetermined system of equations

$$F'(y)t(y) = 0\,.$$

As long as rank $F'(y) = n$, the mapping $t(y)$ is known to vary smoothly along the parameter curve also beyond turning points. Suppose now we parameterize the curve $\overline{y}$ locally with respect to some coordinate $s > 0$ along the unique tangent direction t starting at the previous solution point $\overline{y}_\nu = \overline{y}(0)$. Continuation along t then defines some *prediction path* (for $\nu = 0, 1, \ldots$)

$$\widehat{y}_{\nu+1}(s) = \overline{y}_\nu + s_\nu t(\overline{y}_\nu), \quad s_\nu > 0\,. \tag{5.26}$$

The prediction path supplies possible starting points $y^0 = \widehat{y}_{\nu+1}$ for the *local quasi-Gauss-Newton iteration* towards the next solution point $\overline{y}_{\nu+1}$.

As derived in Section 4.4.1, we may construct a quasi-Gauss-Newton method, equivalent to a *local quasi-Newton method* in the n-dimensional hyperplane

$$H := \widehat{y}(s) \oplus \mathcal{N}^\perp\Big(F'(\widehat{y}(s))\Big)\,.$$

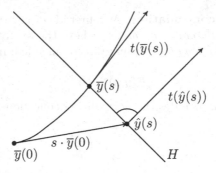

Fig. 5.4. Discrete tangent continuation in $y = (x, \lambda)$.

The geometric situation is represented in Figure 5.4. Observe that $\bar{y}_{\nu+1}$ is just defined as the intersection of H with the solution curve. A natural coordinate frame for this setting is $y = \hat{y}(s) + (u, \sigma), u \in H(s) = \mathbb{R}^n, \sigma \in \mathbb{R}^1$.

In view of the straightforward estimate

$$\|\bar{y}_\nu(s) - \hat{y}_\nu(s)\| \leq \tfrac{1}{2} \max_{\delta \in [0,s]} \|\ddot{\bar{y}}_\nu(\delta)\| s^2 \qquad (5.27)$$

the discrete tangent continuation method is seen to be of order 2—compare (5.7) and definition (5.6). In order to implement the *actual order* $p = 2$, tangent continuation requires a *sufficiently accurate* approximation of the Jacobian *both* at each solution point $\bar{y}_\nu$ and at each starting point $\hat{y}_\nu$.

Tangent computation via QR-decomposition. Assume that we realize the rank-deficient pseudoinverse J^+ of the Jacobian $(n, n+1)$-matrix J through the QR-Cholesky algorithm (4.77) as given in Section 4.4.1. Then, using the vector $w = R^{-1}S$ and the permutation Π as defined therein, we can compute the *normalized* kernel vector t as

$$t := \pm \Pi \begin{pmatrix} -w \\ 1 \end{pmatrix} \Big/ \sqrt{1 + w^T w} \, .$$

One method of fixing the arbitrary sign is to require the last component of $t(\bar{y}_\nu)$, say t_ξ, to have the same sign as $\hat{\xi}_\nu - \bar{\xi}_{\nu-1}$—thus defining a natural orientation also around turning points.

Tangent computation via LU-decomposition. In large sparse problems a *direct sparse solver* based on LU-decomposition will be applied. During the actual decomposition of the $(n, n+1)$-matrix J, the pivoting strategy with possible column permutations Π will give rise to a zero pivot. Upon dropping the associated column of J and setting the corresponding component to zero, say $\tilde{z}_\xi = 0$, a *particular* solution $\tilde{z}$ satisfying

$$J\tilde{z} = -F \qquad (5.28)$$

can be computed. In order to solve $Jt = 0$ for some (unnormalized) kernel vector t, we may set the component of t associated with the zero pivot column to some nonzero value, say $t_\xi = 1$. Upon using the relations

$$z := -J^+F = J^+J\tilde{z} = \left(I - \frac{t\,t^T}{t^T t}\right)\tilde{z}$$

we arrive at the computationally attractive representation

$$z = \tilde{z} - \frac{(t,\tilde{z})}{(t,t)}t \qquad (5.29)$$

in terms of the Euclidean inner product $(\cdot,\cdot)$.

Computation of quasi-Gauss-Newton corrections. The actual computation of the quasi-Gauss-Newton corrections

$$\Delta y^k = -J_k^+ F(y^k)$$

requires the computation of the Moore-Penrose pseudo-inverse of the Jacobian updates J_k. Since this quasi-Gauss-Newton update preserves the nullspace component, we can use a simple variant of the recursive quasi-Newton method, given as algorithm QNERR in Section 2.1.4: with $J_0 = F'(y^0)$, we only need to formally replace J_0^{-1} by J_0^+, wherever this term arises.

5.2.2 Affine covariant feasible stepsizes

In order to develop an adaptive stepsize control (see Section 5.2.3 below), theoretical feasible stepsizes are studied first. As in the case of the Newton continuation method, an *affine covariant* setting appears to be natural, since the path concept is the dominating one in continuation. Since the local *quasi-Gauss-Newton* iteration has been shown to be equivalent to a *quasi-Newton* iteration in H (see Figure 5.4), we may proceed as in the simpler Newton continuation method (Section 5.1.2) and model the situation just by the *simplified Newton* iteration in H.

Theorem 5.7 *Consider the discrete tangent continuation method* (5.26) *in combination with the simplified local Gauss-Newton iteration starting at* $y^0 := \hat{y}(s) = \hat{y}_\nu$ *and with Jacobian approximations* $J_k := F'(y^0)$ *for* $k \geq 0$. *Let* $F : D \subset \mathbb{R}^{n+1} \to \mathbb{R}^n$ *denote some* C^1*-mapping with* D *open, convex, and sufficiently large. Then, under the assumption of the affine covariant Lipschitz conditions*

$$\left\| F'(y^0)^+ \left(F'(y) - F'(y^0) \right) \right\| \leq \omega_H \|y - y^0\|, \; y, y^0 \in H$$

and

$$\left\| F'(y)^+ \Big(F'(u + \delta_2 t(u), -F'(u) \Big) t(u) \right\|_2 \leq \delta_2 \omega_t \| t(u), \|_2^2$$

with the (normalized) kernel vector $t(u) = \ker F'(u)$ *and*

$$y, u + \delta_2 t(u) \in D\,, 0 \leq \delta_2 \leq 1\,,$$

the simplified Gauss-Newton iteration converges for all

$$s \leq s_{\max} := 1/\sqrt{\omega_H \omega_t}\,. \tag{5.30}$$

Proof. For the simplified Newton iteration in H we may apply Theorem 2.5, which here requires the verification of the *sufficient* condition

$$\|\Delta y^0(s)\| \omega_H \leq \alpha_0(s) \omega_H \leq \tfrac{1}{2}\,. \tag{5.31}$$

For simplification, we introduce the notation

$$J(s) := F'(\widehat{y}(s))\,, \ F(s) := F(\widehat{y}(s))\,, \ t(s) := t(\widehat{y}(s))$$

so that

$$F(0) = 0\,, \ J(0)t(0) = 0\,.$$

Then the derivation of an appropriate $\alpha_0(s)$ may proceed as follows:

$$\left\| \Delta y^0(s) \right\| = \| J(s)^+ F(s) \| = \left\| J(s)^+ \Big(F(s) - F(0) \Big) \right\|$$

$$= \left\| J(s)^+ \int_{\delta=0}^{s} J(\delta)t(0)d\delta \right\| \leq \int_{\delta=0}^{s} \left\| J(s)^+ J(\delta)t(0) \right\| d\delta$$

$$= \int_{\delta=0}^{s} \left\| J(s)^+ \Big(J(\delta) - J(0) \Big) t(0) \right\| d\delta \leq \tfrac{1}{2}\omega_t s^2\,.$$

Hence

$$\alpha_0(s) := \tfrac{1}{2}\omega_t s^2\,, \tag{5.32}$$

which inserted above directly leads to the maximum feasible stepsize $s_{\max}$.
□

As an extension of the simpler Newton continuation case treated in Section 5.1.2 we next study the 'movement' of H along the parameter curve.

Theorem 5.8 *Assumptions and notation as in the preceding Theorem 5.7. Let*

$$\widehat{y}(s) := \overline{y}(0) + st\Big(\overline{y}(0)\Big)$$

denote a short-hand notation for the discrete tangent continuation (5.26). Let $t := t(\bar{s}) = t\left(\widehat{y}(\bar{s})\right)$, $\bar{s}$ fixed. Define $\bar{y}(s)$ as the intersection of $H(s)$ with the solution curve and let

$$c_s := t(\bar{s})^T t\left(\bar{y}(s)\right).$$

Then, for $c_s \neq 0$, one has

$$\left\|\ddot{\bar{y}}(s)\right\|_2 \leq \omega_t \frac{c_0^2}{|c_s|^3}. \tag{5.33}$$

Proof. For simplification we introduce

$$z(s) := \bar{y}(s) - \widehat{y}(s)$$

so that

$$
\begin{aligned}
\dot{z}(s) &= \dot{\bar{y}}(s) - t(0), & (5.34)\\
\ddot{z}(s) &= \ddot{\bar{y}}(s). & (5.35)
\end{aligned}
$$

For fixed $\bar{s}$ and $t(\bar{s}) = t(\widehat{y})$, we obtain

$$
\begin{aligned}
t(\bar{s})^T z(s) &\equiv 0,\\
t(\bar{s})^T \dot{z}(s) &\equiv 0, & (5.36)\\
t(\bar{s})^T \ddot{z}(s) &= 0 & (5.37)
\end{aligned}
$$

for $0 \leq s \leq \bar{s}$. Variation of s defines $\bar{y}(s)$ by virtue of

$$F\left(\bar{y}(s)\right) \equiv 0,$$

which implies

$$F'\left(\bar{y}(s)\right)\dot{\bar{y}}(s) \equiv 0, \tag{5.38}$$

$$F''\left(\bar{y}(s)\right)\left[\dot{\bar{y}}(s)\right]^2 + F'(\bar{y}(s))\ddot{\bar{y}}(s) \equiv 0.$$

Next, from rank $F'(y) = n$ for all $y \in D$ and (5.38), we may write

$$\dot{\bar{y}}(s) = \gamma(s)t\left(\bar{y}(s)\right)$$

in terms of some coefficient γ to be determined. From this we obtain

$$\ddot{z}(s) = \beta(s)t\left(\bar{y}(s)\right) - \eta(s) \tag{5.39}$$

in terms of some coefficient β and some vector

$$\eta \perp t\left(\bar{y}(s)\right).$$

Upon collecting these relations, we arrive at the expression

$$\eta(s) = F'\Big(\overline{y}(s)\Big)^{+} F''\Big(\overline{y}(s)\Big) \Big[\gamma(s)t\Big(\overline{y}(s)\Big)\Big]^{2}. \tag{5.40}$$

The determination of $\beta(s)$, $\gamma(s)$ starts from

$$\|\ddot{z}(s)\|_{2}^{2} = \|\eta(s)\|_{2}^{2} + \beta^{2}(s),$$

since t is normalized. Upon combining (5.34) and (5.36) we obtain

$$0 = t^{T}\dot{z}(s) = t^{T}\Big(\gamma(s)t(\overline{y}(s)) - t(0)\Big),$$

from which

$$\gamma(s) = \frac{c_{0}}{c_{s}} \quad \text{for } c_{s} \neq 0.$$

Application of the same procedure with (5.35), (5.39) and (5.37) yields

$$0 = t^{T}\ddot{z}(s) = t^{T}\Big(\beta(s)t(\overline{y}(s)) - \eta(s)\Big)$$

or, equivalently

$$\beta(s) = \frac{t^{T}\eta(s)}{c_{s}} \quad \text{for } c_{s} \neq 0. \tag{5.41}$$

As $t = t\Big(\widehat{y}(\overline{s})\Big)$, we may continue

$$t\Big(\widehat{y}(s)\Big)^{T}\eta(s) = 0.$$

Hence

$$
\begin{aligned}
|t^{T}\eta(s)| &= \left| t^{T}\eta(s) - c_{s}t\Big(\overline{y}(s)\Big)^{T}\eta(s) \right| \\
&= \left| t^{T}\eta(s) - t^{T}t\Big(\overline{y}(s)\Big)t\Big(\overline{y}(s)\Big)^{T}\eta(s) \right| \\
&= \left| t^{T}\Big(I - t(\overline{y}(s))t^{T}(\overline{y}(s))\Big)\eta(s) \right| \\
&\leq \left\| \Big(I - t(\overline{y}(s))t^{T}(\overline{y}(s))\Big)t \right\|_{2} \cdot \|\eta(s)\|_{2} \\
&= \sqrt{1 - c_{s}^{2}}\, \|\eta(s)\|_{2}.
\end{aligned}
$$

Insertion into (5.41) yields

$$\Big|\beta(s)\Big| \leq \frac{\sqrt{1 - c_{s}^{2}}}{|c_{s}|}\, \Big\|\eta(s)\Big\|_{2},$$

which leads to

$$\|\ddot{z}(s)\|_2^2 \le \|\eta(s)\|_2^2 + \frac{1 - c_s^2}{c_s^2}\|\eta(s)\|_2^2 = \frac{\|\eta(s)\|_2^2}{c_s^2}.$$

Finally, estimation of $\|\eta(s)\|$ in (5.40) supplies

$$\|\eta(s)\|_2 \le \omega_t(\gamma(s))^2$$

and, therefore

$$\|\ddot{z}(s)\|_2 \le \omega_t \cdot \left(\frac{c_0}{c_s}\right)^2 \cdot \frac{1}{|c_s|},$$

which with (5.35) completes the proof.   □

The result (5.33) shows that the Lipschitz constant ω_t just measures the *local curvature* $\ddot{\bar{y}}(s)$ of the parameter curve. Insertion into (5.27) specifies the second order coefficient. In view of actual computation we may eliminate the variation of c_s via the additional *turning angle restriction*

$$c_0 = \min_{\delta \in [0,\bar{s}]} c_\delta > 0,$$

which, in turn, yields a *stepsize restriction*, of course. Then the bound (5.33) can be replaced by the corresponding expression

$$\|\ddot{\bar{y}}(s)\|_2 \le \omega_t/c_0.\tag{5.42}$$

For an *affine contravariant* derivation of feasible stepsizes see Exercise 5.2.

5.2.3 Adaptive stepsize control

On the basis of the theoretical results above, we are now ready to derive computational estimates for feasible stepsizes s_ν according to (5.26). Recall again that the first quasi-Gauss-Newton step may be interpreted as an ordinary Newton step in H, whereas further quasi-Gauss-Newton steps are just quasi-Newton steps in H. Let $\overline{\Delta y}^1$ denote the first simplified Gauss-Newton correction, which is cheaply available in the course of the computation of the first quasi-Gauss-Newton correction. Then a first contraction factor Θ_0 must satisfy

$$\Theta_0 := \frac{\|\overline{\Delta y}^1\|}{\|\Delta y^0\|} \le \tfrac{1}{2}\omega_H \alpha_0,\tag{5.43}$$

which with (5.31) implies that

$$\Theta_0 \le \overline{\Theta} := \tfrac{1}{4}.$$

In the spirit of our paradigm in Section 1.2.3, we construct computational stepsize estimates $[\cdot]$ on the basis of the theoretical stepsizes (5.30) as

$$[s_{\max}] := 1/\sqrt{[\omega_H]\,[\omega_t]}\,.\tag{5.44}$$

As usual, since

$$[s_{\max}] \geq s_{\max}$$

both a prediction and a correction strategy need to be designed.

Correction strategy. From (5.43) we may obtain

$$[\omega_H] := \frac{2\Theta_0}{\|\Delta y^0\|} \leq \omega_H \tag{5.45}$$

and similarly from (5.32)

$$[\omega_t] := \frac{2\,\|\Delta y^0\|}{s^2} \leq \omega_t\,.$$

Upon inserting these estimates into (5.44), we are led to the stepsize suggestion

$$s'_\nu := \sqrt{\frac{\overline{\Theta}}{\Theta_0}}\,s_\nu\,.$$

This estimate requires the knowledge of an actual stepsize s_ν, which means that it may only serve within a correction strategy. Whenever the termination criterion (for $k \geq 0$)

$$\Theta_k > \tfrac{1}{2}$$

holds, then the quasi-Gauss-Newton iteration is terminated and the previous continuation step (5.26) is repeated supplying the new starting point

$$\widehat{y}'_{\nu+1} = \overline{y}_\nu + s'_\nu t(\overline{y}_\nu)\,,$$

wherein roughly

$$s'_\nu < 0.7 s_\nu$$

is guaranteed.

Prediction strategy. For the construction of a prediction strategy, we may combine the relations (5.42) and (5.27) to obtain

$$\|\overline{y}_\nu - \widehat{y}_\nu\| \leq \tfrac{1}{2}\omega_t s^2_{\nu-1}\frac{1}{c_0}\,.$$

This naturally defines the computational estimate

$$[\omega_t] := \frac{2c_0\|\overline{y}_\nu - \widehat{y}_\nu\|_2}{s^2_{\nu-1}} \leq \omega_t\,.$$

Together with $[\omega_H]$ from (5.45), the general formula (5.44) leads to the stepsize suggestion

$$s_\nu^0 := \left(\frac{\|\Delta y^0\|}{\|\overline{y}_\nu - \widehat{y}_\nu\|} \frac{\overline{\Theta}}{\Theta_0} \frac{1}{c_0} \right)^{1/2} s_{\nu-1}. \tag{5.46}$$

Clearly, this is a direct extension of the prediction formula (5.25) for the Newton continuation method. Recalling that $g(\overline{\Theta}) \approx 2\overline{\Theta}$, the main new item appears to be the trigonometric factor

$$c_0 = t^T(\widehat{y}_\nu) t\left(\overline{y}_{\nu-1}\right),$$

which roughly measures the 'turning angle' of the hyperplane $H_{\nu-1} \supset \overline{y}_{\nu-1}$ to $H_\nu \supset \overline{y}_\nu$—see again Figure 5.4.

Finally, as in the Newton continuation scheme, precaution must be taken for the *nearly linear* case

$$\Theta_0 \le \Theta_{\min} \ll 1,$$

in which case $\Theta_0 \to \Theta_{\min}$ in (5.46).

Detection of critical points. In the described tangent continuation *turning points* (as critical points of order $k = 0$) do not play any exceptional role. The occurrence of critical points of order $k \ge 1$, however, needs to be carefully monitored. For this purpose, we define by

$$d_\lambda := \det(F_x)$$

the determinant of the (n, n)-submatrix of the Jacobian $F'(y)$ that is obtained by dropping the λ-column, which is F_x, of course. Similarly, let d_ξ denote the determinant of the submatrix, where the last column has been dropped—which corresponds to the *internal parameter* ξ, in general different from the *external parameter* λ, when column permutations based on pivoting are involved. For the safe detection of critical points, we compute the *determinant pair* (d_ξ, d_λ) along the solution curve. The computation of d_ξ is easily done via $\det(R)$—with a possible sign correction, by $(-1)^n$ for n Householder reflections or for the actually performed permutations in the LU-decomposition. If $\lambda \ne \xi$, then the computation of d_λ requires the evaluation of the determinant of a Hessenberg matrix, which means $\mathcal{O}(n^2)$ operations only. Sign changes of this pair clearly indicate the occurrence of both turning points and simple (possibly unfolded) bifurcation points—see Figure 5.5 for an illustration of the typical situations. This device has a high degree of reliability in detecting turning and simple bifurcation points and a good chance of detecting higher order critical points. Safety,of course, cannot be guaranteed as long as only pointwise sampling is used.

Computation of turning points. Assume that the discrete continuation method has supplied some *internal embedding parameter* ξ (associated with the last column in the matrix decomposition) and some *interval* $[\underline{\xi}, \overline{\xi}]$ supposed to contain a turning point, say ξ^*. Then the implicit mapping $\lambda(\xi)$ will have a minimum or maximum value within that interval so that

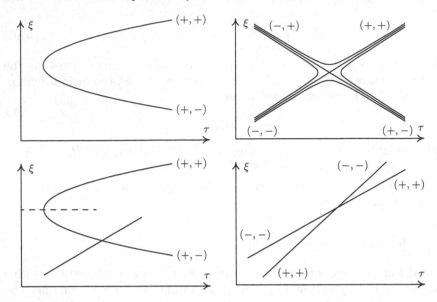

Fig. 5.5. Sign structure of determinant pair (d_ξ, d_λ). *Upper left*: turning point, *upper right*: detected simple bifurcation and unfoldings, *lower left*: pair of turning and bifurcation point or pitchfork bifurcation, *lower right*: possibly undetected bifurcation point due to too small branch angle.

$$\dot\lambda(\xi^*) = 0 \,. \tag{5.47}$$

On this basis, the following algorithm for the determination of turning points is recommended:

(I) Construct the *cubic Hermite polynomial* $p(\xi)$, $\xi \in \left[\underline\xi, \overline\xi\right]$ such that

$$p\left(\underline\xi\right) = \lambda\left(\underline\xi\right), \quad p\left(\overline\xi\right) = \lambda\left(\overline\xi\right),$$
$$\dot p\left(\underline\xi\right) = \dot\lambda\left(\underline\xi\right), \quad \dot p\left(\overline\xi\right) = \dot\lambda\left(\overline\xi\right).$$

As an approximation of the unknown implicit equation (5.47), we solve the quadratic equation

$$\dot p(\xi) = 0 \,. \tag{5.48}$$

The usual bisection assumption

$$\dot\lambda\left(\underline\xi\right)\dot\lambda\left(\overline\xi\right) \le 0 \,, \tag{5.49}$$

then assures that equation (5.48) has a real root $\widehat\xi \in \left[\underline\xi, \overline\xi\right]$.

(II) Perform a Gauss-Newton iteration of standard type with starting point $y^0 = \widehat y\left(\widehat\xi\right) = \left(\overline x\left(\widehat\xi\right), \widehat\xi\right)$. Let $\widehat y^*$ denote the point obtained on the solution curve.

(III) As soon as

$$\|\widehat{y} - \widehat{y}^*\| \le \varepsilon$$

holds for some prescribed (relative) accuracy ε, then $\widehat{y}^*$ is accepted as turning point approximation. Otherwise, $\widehat{\xi}^*$ replaces either $\underline{\xi}$ or $\overline{\xi}$ such that (5.49) holds and step (I) is repeated.

The above algorithm fits into the frame of a class of algorithms, for which *superlinear convergence* has been proved by H. Schwetlick [183].

Graphical output. Here both nodal data $\{\overline{y}_\nu\}$ and their local tangents $t(\{\overline{y}_\nu\})$ are usually given in different parametrizations corresponding to the different internal parameters ξ_ν. As a consequence, Bezier-Hermite splines appear to be the method of choice for the graphical output—compare, e.g., Section 7.3 in the textbook [77] or any book on computer aided design.

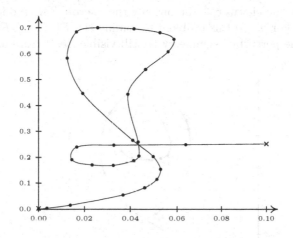

Fig. 5.6. Chemical reaction problem: $x_2(\lambda)$. Crosspoint just by projection.

Example 5.2 *Chemical reaction problem.* This model due to M. Kubiček [137] reads (see Figure 5.6):

$$\lambda(1 - x_3)\exp(10x_1/(1 + 0.01x_1)) - x_3 = 0$$

$$22\lambda(1 - x_3)\exp(10x_1/(1 + 0.01x_1)) - 30x_1 = 0$$

$$x_3 - x_4 + \lambda(1 - x_4)\exp(10x_2/(1 + 0.01x_2)) = 0$$

$$10x_1 - 30x_2 + 22\lambda(1 - x_4)\exp(10x_2/(1 + 0.01x_2)) = 0.$$

Example 5.3 *Aircraft stability problem.* In [149] R.G. Melhem and W.C. Rheinboldt presented the following problem:

$$-3.933x_1 + 0.107x_2 + 0.126x_3 - 9.99x_5 - 45.83\lambda$$
$$-0.727x_2x_3 + 8.39x_3x_4 - 684.4x_4x_5 + 63.5x_4\lambda \;=\; 0$$
$$-0.987x_2 - 22.95x_4 - 28.37u + 0.949x_1x_3 + 0.173x_1x_5 \;=\; 0$$
$$0.002x_1 - 0.235x_3 + 5.67x_5 - 0.921\lambda - 0.713x_1x_2$$
$$-1.578x_1x_4 + 1.132x_4\lambda \;=\; 0$$
$$x_2 - x_4 - 0.168u - x_1x_5 \;=\; 0$$
$$-x_3 - 0.196x_5 - 0.0071\lambda + x_1x_4 \;=\; 0.$$

Herein x_1, x_2, x_3 are the roll rate, pitch rate, and yaw rate, respectively, x_4 is the incremental angle of attack, and x_5 the sideslip angle. The variable u is the control for the elevator, λ the one for the aileron. The rudder deflection is set to zero. For $u = 0$ this problem is symmetric: $F(x, \lambda) = F(x, -\lambda)$. For $u = -0.008$ the perturbed symmetry is still visible, see Figure 5.7.

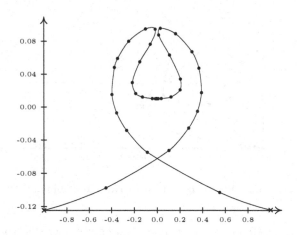

Fig. 5.7. Aircraft stability problem: $x_4(\lambda)$, perturbed symmetry.

BIBLIOGRAPHICAL NOTE. The adaptive pathfollowing algorithm, as worked out here, has been implemented in the code ALCON1 due to P. Deuflhard, B. Fiedler, and P. Kunkel [72].

5.3 Computation of Simple Bifurcations

Suppose that the numerical pathfollowing procedure described in Section 5.2 has produced some guess y^0 of an expected close-by simple bifurcation point y^*—see, e.g., the double determinant detection device presented in Section 5.2.3. Then the task is to either compute a bifurcation point y^* iteratively from the starting point y^0 or to decide that there is none in the neighborhood of y^0. In Section 5.3.1, we will first study the basic construction of *augmented systems* that have certain critical points of order $k > 0$ as locally unique solutions—excluding turning points $(k = 0)$, which can be computed easier as shown in the preceding section. The general construction scheme for augmented systems will be based on the theory of *universal unfolding of singularities*, which in the case of simple bifurcations specifies to the system of G. Moore. In Section 5.3.2, certain *Newton-like algorithms* for an efficient solution of that augmented system will be worked out in some detail. On the basis of *structure preserving* block elimination techniques for each Newton step, details of the *branching-off algorithm* are elaborated in Section 5.3.3—involving the computation of entering and emanating semi-branches as well as the restart of discrete tangent continuation on the new semi-branches.

5.3.1 Augmented systems for critical points

Let y^* denote a *perfect* or *unperturbed* singularity of order $k \geq 1$ with

$$F(y^*) = 0$$

and

$$\operatorname{rank} F'(y^*) = n - k. \tag{5.50}$$

Even though we will later only work out an algorithm for simple bifurcations $(k = 1)$, we include the more general case $k > 1$ here as well—to make the general construction of augmented systems transparent.

Lyapunov-Schmidt reduction. In the notation from above, let $A := F'(y^*)$, $\mathcal{N}(A) = \ker(A)$ its $(k + 1)$-dimensional nullspace, and $\mathcal{R}^{\perp}(A)$ its k-dimensional corange. If we again introduce the orthogonal projectors $P := A^+A$, $\overline{P} := AA^+$, we have that $P^{\perp}$ projects onto $\mathcal{N}(A)$ and $\overline{P}^{\perp}$ onto $\mathcal{R}^{\perp}(A)$. With this notation we may define the natural splitting

$$y = y^* + v + w, \quad w := P(y - y^*), \quad v := P^{\perp}(y - y^*)$$

in the space of the unknowns. From assumption (5.50) and the implicit function theorem, we know that there exists a function w^* such that

$$\overline{P}F(y^* + v + w) = 0 \iff w = w^*(v).$$

Replacement of the variable w by the function w^* then leads to a *reduced system* of k equations in $k + 1$ unknowns

$$f(v) := \overline{P}^\perp F(y^* + v + w^*(v)) = 0. \tag{5.51}$$

This is the well-known Lyapunov-Schmidt *reduction*, which stands at the beginning of every mathematical treatment of singularities—see, e.g., the classical book edited by P.H. Rabinowitz [173]. For actual computation we need to define *orthogonal bases* for both $\mathcal{N}$ and $\mathcal{R}^\perp$ as

$$\mathcal{N}(A) =: \langle t_1, \ldots, t_{k+1} \rangle \quad \mathcal{R}^\perp(A) =: \langle z_1, \ldots, z_k \rangle$$

or, in equivalent matrix notation, as

$$t := [t_1, \ldots, t_{k+1}], \quad z := [z_1, \ldots, z_k],$$

so that

$$\begin{aligned} At = 0, \quad t^T t = I_{k+1}, \quad P^\perp = tt^T, \\ A^T z = 0, \quad z^T z = I_k, \quad \overline{P}^\perp := zz^T. \end{aligned} \tag{5.52}$$

Note that t and z are only specified up to orthogonal transformations—which leaves $\dim O(k+1) = \frac{1}{2}k(k+1)$ degrees of freedom for t and $\dim O(k) = \frac{1}{2}k(k-1)$ degrees of freedom for z.

Upon introducing *local coordinates* $\xi \in \mathbb{R}^{k+1}$, $\gamma \in \mathbb{R}^k$ by virtue of

$$v = \sum_{i=1}^{k+1} \xi_i t_i = t\xi, \qquad f(v) = \sum_{j=1}^{k} \gamma_j z_j = z\gamma,$$

the reduced equations (5.51) can be rewritten in the form

$$\gamma(\xi) := z^T f(t\xi) = z^T F(y^* + t\xi + w^*(t\xi)) = 0. \tag{5.53}$$

For the actual determination of singularities, higher order derivatives of both sides will play an important role.

Lemma 5.9 *Assumptions as just introduced. Let* $y^* := 0$ *for convenience and* $a_i \in \mathbb{R}^{k+1}$. *Then the following relations hold:*

$$\dot\gamma(0)a_1 \;=\; 0 \tag{5.54}$$

$$\ddot\gamma(0)[a_1, a_2] \;=\; z^T F''[ta_1, ta_2] \tag{5.55}$$

$$\begin{aligned} \dddot\gamma(0)[a_1, a_2, a_3] \;=\; & z^T F'''[ta_1, ta_2, ta_3] \\ & - z^t F''[ta_1, A^+ F''[ta_2, ta_3]] \\ & - z^T F''[ta_2, A^+ F''[ta_3, ta_1]] \\ & - z^T F''[ta_3, A^+ F''[ta_1, ta_2]] \end{aligned} \tag{5.56}$$

Proof. As a consequence of $y^* = 0$ we have $v^* = 0$, $w^*(0) = 0$ and $\xi^* = 0$. We start from

$$\gamma(\xi) = z^T F(t\xi + w^*(t\xi)) \tag{5.57}$$

and

$$\overline{P}F(t\xi + w^*(t\xi)) \equiv 0. \tag{5.58}$$

Differentiation of (5.57) with respect to ξ yields

$$\dot{\gamma}(\xi)a_1 = z^T F'(t\xi + w^*(t\xi))(ta_1 + w_\xi^*(t\xi)a_1)$$

and, after insertion of $z^T A = 0$

$$\dot{\gamma}(0)a_1 = z^T A(ta_1 + w_\xi^*(0)a_1) = 0,$$

which confirms (5.54). Differentiation of (5.58) yields

$$\overline{P}A(ta_1 + w_\xi^*(0)a_1) = 0.$$

which, with $\overline{P}A = A$, $At = 0$ and $w_\xi^*(0)a_1 \in \mathcal{N}^\perp$, can be solved by

$$w_\xi^*(0)a_1 = 0. \tag{5.59}$$

Upon differentiating (5.57) once more and inserting the expression above we obtain

$$\ddot{\gamma}(0)[a_1, a_2[= z^T F''(0)[ta_1, ta_2],$$

which confirms (5.55). Differentiation of (5.58) once more leads to

$$\overline{P}F''(0)[ta_1, ta_2] + \overline{P}Aw_{\xi\xi}^*(0)[a_1, a_2] = 0.$$

With arguments as just used before, the latter equation can be solved to yield

$$w_{\xi\xi}^*(0)[a_1, a_2] = -A^+ F''(0)[ta_1, ta_2]. \tag{5.60}$$

Upon differentiating (5.57) for a third time, we eventually arrive at

$$\begin{aligned}
\dddot{\gamma}(0)[a_1, a_2, a_3] &= z^T F'''(0)[ta_1, ta_2, ta_3] \\
&\quad + z^T F''(0)[ta_1, w_{\xi\xi}^*(0)[a_2, a_3]] \\
&\quad + z^T F''(0)[ta_2, w_{\xi\xi}^*(0)[a_3, a_1]] \\
&\quad + z^T F''(0)[ta_3, w_{\xi\xi}^*(0)[a_1, a_2]].
\end{aligned}$$

Insertion of $w_{\xi\xi}^*(0)$ then finally confirms (5.56). $\qquad\square$

Universal unfolding. It is clear from the above derivation that the function $\gamma : \mathbb{R}^{k+1} \to \mathbb{R}^k$ contains all essential information needed to classify the local structure of a singularity. Since only derivatives of γ up to a certain order are required, it suffices to study simple polynomial *germs* $g(\xi)$, which then may stand for the whole function class

$$\Gamma(g) := \{\gamma(\xi) = \beta(\xi)g(h(\xi)) \mid \beta, h \ C^\infty\text{-diffeomorphism}, \ h(0) = 0\}$$

called the *contact equivalence class*. From a geometrical point of view all germs within one equivalence class show a similar solution structure around $\xi = 0$. We may say that the reduced mapping γ is contact equivalent to a representative germ $g \in \Gamma(\gamma)$ or, vice versa, $\gamma \in \Gamma(g)$. As examples, the germ

$$g_s(\xi) := \xi_1^2 - \xi_2^2 \quad (k = 1)$$

represents a *simple bifurcation*, whereas the germ

$$g_c(\xi) := \xi_1^2 - \xi_2^3 \quad (k = 1)$$

characterizes an *asymmetric cusp*. We then obtain

$$\beta(\xi)g(h(\xi)) = z^T F(y^* + t\xi + w^*(t\xi)) = 0$$

in terms of certain diffeomorphisms β, h.

Up to now, our analytical presentation has only covered *perfect* singularities. An efficient algorithm will have to deal with *imperfect* or *unfolded* singularities y^* as well—even without knowing in advance about the structure of the perturbations. As an immediate consequence, we will encounter $\gamma(0) \neq 0$ and the associated Jacobian matrix $F'(y^*)$ may no longer be exactly rank-deficient, but still 'close to' a rank-deficient matrix. In this situation, the structure of *topological perturbations* is important, which are known to lead to an *unfolding of nongeneric singularities*. In the just introduced framework, such perturbations may be written as polynomial perturbations $p(\xi, \alpha)$ of the germs g replacing them by the *perturbed germs*

$$G(\xi, \alpha) := g(\xi) + p(\xi, \alpha), \tag{5.61}$$

wherein $p(\xi, 0) \equiv 0$ and the parameters α denote the *unfolding parameters*. The *minimal number of unfolding parameters* is a characteristic of each type of singularity and is called its *codimension* q. In case of this minimal parameterization, the representation (5.61) is called *universal unfolding*. A special feature of any universal unfolding is that monomials arising in p are *at least 2 orders less* than corresponding monomials arising in g. As examples again, g_s for the *simple bifurcation* is replaced by

$$G_s(\xi, \alpha) := \xi_1^2 - \xi_2^2 + \alpha, \quad (k = 1, q = 1),$$

whereas g_c for the *asymmetric cusp* is replaced by

$$G_c(\xi, \alpha_1, \alpha_2) := \xi_1^2 - \xi_2^3 + \alpha_1 + \alpha_2\xi_2, \quad (k = 1, q = 2). \tag{5.62}$$

BIBLIOGRAPHICAL NOTE. For a general thorough treatment of unfolded singularities, the reader may refer to M. Golubitsky and D. Schaeffer [109] and their textbooks [110, 111]. Since these authors treat dynamical systems $\dot{x} = F(x, \lambda)$ with state variables x, they give the *explicit parameter* λ an extra role—in contrast to the present section here, which treats all $n + 1$ components of $y = (x, \lambda)$ the same.

Construction of augmented systems. Summarizing, we may assume that specific diffeomorphisms β, h exist such that

$$\beta(\xi)G(h(\xi), \alpha) = z^T F(y^* + t\xi + w^*(t\xi)) \tag{5.63}$$

with, in general,

$$G(h(0), \alpha) = G(0, \alpha) = p(0, \alpha) \neq 0.$$

Therefore, for perturbed singularities, the reduced system (5.53) must be replaced by

$$z^T F(y^*) = p(0, \alpha).$$

These k equations, together with the $n - k$ equations $\overline{P}F = 0$, then lead to the n equations

$$F(y^*) = z\, p(0, \alpha)$$

in terms of the $(k + 1)n + q + 1$ unknowns (y, z, α).

Simple bifurcation. Let us now return to the special case $k = 1, q = 1$. Here we arrive at the *augmented system* of G. Moore [103]:

$$F'(y)^T z = 0, \tag{5.64}$$

$$F(y) + \alpha z = 0, \tag{5.65}$$

$$\tfrac{1}{2}(z^T z - 1) = 0. \tag{5.66}$$

It comprises $(2n + 2)$ nonlinear equations for the $(2n + 2)$ unknowns (y, z, α). The Jacobian $J(y, z, \alpha)$ of this mapping is *nonsingular* for sufficiently small perturbation parameter α. The proof of this fact is postponed to Section 5.3.2 below, since it stimulates an algorithmic idea for the iterative solution of the above augmented system.

In order to make sure that a geometrical bifurcation really exists locally, we must impose the additional *second derivative* condition

$$z^T F''(y^*)[t, t] \qquad \text{nondegenerate, indefinite}. \tag{5.67}$$

This condition assures the existence of two local *branch directions* as will be shown next. In most of the established analysis treatise—compare, e.g., M.G. Crandall and P.H. Rabinowitz [46]—one of the intersecting branches is assumed to be the trivial one—which is acceptable for a merely analytical treatment, but unsatisfactory for the construction of efficient numerical algorithms. Therefore we here include a theorem that treats both branches as indistinguishable.

Theorem 5.10 *Assumptions and notation as just introduced. Let $F \in C^k$, $k \geq 3$. Then, in a neighborhood of y^*, the solution set $F + \alpha z = 0$ consists of two one-dimensional C^{k-2}-branches $\gamma_1(s)$, $\gamma_2(s)$ such that*

$$\gamma_i(0) = y^*, \quad i = 1, 2,$$

$$\mathcal{N} = \langle \dot\gamma_1(0), \dot\gamma_2(0) \rangle,$$

$$z^T F''(y^*)\left[\dot\gamma_i(0), \dot\gamma_i(0)\right] = 0.$$

Proof. (*Sketch*) For convenience, assume again that $y^* = 0$. Let a standard Lyapunov-Schmidt reduction have been performed in terms of a parametrization of the two-dimensional nullspace $\mathcal{N}$. It is then sufficient to study the mapping $\overline{F} = \overline{P}^{\perp} F : \mathcal{N} \to \mathbb{R}$. Introducing polar coordinates, define a blow-up version of $\overline{F}$ by

$$\Phi(r, \Theta) := \begin{cases} 2\overline{F}\left(r\, e(\Theta)\right) \big/ r^2 & r > 0 \\ \overline{F}''(0)\left[e(\Theta), e(\Theta)\right] & r = 0, \end{cases}$$

where $e(\Theta) := (\cos \Theta, \sin \Theta)$ denotes the unit vector in the direction $\Theta \in [0, \pi]$. Obviously, $\Phi = 0$ holds if $\overline{F} = 0$. Moreover, $\overline{F} \in C^k$ implies $\Phi \in C^{k-2}$ and Φ can also be formally continued to $r < 0$. By assumption (5.67) there exist directions $\Theta_i (i = 1, 2)$ such that

$$\Phi(0, \pm\Theta_i) = 0,$$

$$\Phi_\Theta(0, \pm\Theta_i) = F''(0)\left[e_\Theta(\Theta_i), e_\Theta(\Theta_i)\right] \neq 0.$$

Hence, by the implicit function theorem, there are four C^{k-2}–semi-branches

$$\gamma_i^{\pm}(r) = r\, e(\Theta_i^{\pm}(r)), \quad i = 1, 2 \quad r \geq 0$$

for sufficiently small r. At $r = 0$, the functions $\gamma_i^+(r)$ and $\gamma_i^-(-r)$ have the same derivatives up to order $k - 2$. Therefore, combining $\gamma_i^+(r)$ and $\gamma_i^-(-r)$, $i = 1, 2$, yields two C^{k-2}-branches, which completes the proof.   □

From this result, the desired local tangent directions

$$t_i^* = \dot\gamma_i(y^*), \quad i = 1, 2,$$

are seen to be defined from the *quadratic equation*

$$z^T F''(y^*)\, [t_i^*, t_i^*] = 0,$$

which, under the assumption (5.67) has the two distinct real roots t_1^*, t_2^*.

Asymmetric cusp. This possibly unfolded singularity has also rank-deficiency $k = 1$, but codimension $q = 2$. Recall the perturbed germ $G_c = G$ from (5.62) to derive the characterization (with $h(0) = 0$):

$$G_c(0, \alpha) = \alpha_1, \tag{5.68}$$

$$G_{\xi_1}(0, \alpha) = 0,$$
$$G_{\xi_2}(0, \alpha) = \alpha_2, \tag{5.69}$$

$$G_{\xi_2\xi_2}(0, \alpha) = 0,$$
$$G_{\xi_1\xi_2}(0, \alpha) = 0. \tag{5.70}$$

Note that all nonvanishing derivatives beyond (5.69) and (5.70) such as

$$G_{\xi_1\xi_1}(0, \alpha) = 2$$

do not show up, since they are arbitrary due to the arbitrary C^∞-diffeomorphic transformation $\beta(\xi)$ in (5.63). Upon differentiating the right-hand side of (5.63) with $h(\xi) = \xi$ and $\beta(\xi) = 1$, we may obtain (as in the simple bifurcation)

$$z^T F(y^*) = \alpha_1$$

from (5.68), which leads to

$$F(y^*) = \alpha_1 z. \tag{5.71}$$

From (5.69) and (5.59) we may verify that

$$F'(y^*)t_1 = 0, \ F'(y^*)t_2 = \alpha_2 z. \tag{5.72}$$

Finally, from (5.70), (5.60) and $z^T \overline{P}^\perp = z^T$ we arrive at

$$z^T F''(y^*)[t_2, t_2] = 0,$$
$$z^T F''(y^*)[t_1, t_2] = 0.$$

Of course, we will add the perturbed corange condition

$$F'(y^*)^T z = \alpha_2 t_2 \tag{5.73}$$

to be compatible with (5.72). For normalization we will choose the four equations

$$z^T z = 1, \ t_1^T t_1 = t_2^T t_2 = 1, \ t_1^T t_2 = 0. \tag{5.74}$$

Upon combining (5.71) up to (5.74) we would arrive at an overdetermined system, $4n + 8$ equations in $4n + 5$ unknowns. Careful examination for general $h(\xi)$ (with still $\beta(\xi) = 1$ w.l.o.g) leads to a replacement of (5.72) and (5.73) by the *perturbed Lyapunov-Schmidt reduction* (originally suggested by R. Menzel)

$$F'(y^*)^T z = \gamma_{11} t_1 + \gamma_{21} t_2,$$
$$F'(y^*) t_1 = \beta_{11} z, \qquad\qquad (5.75)$$
$$F'(y^*) t_2 = \beta_{21} z.$$

Thus we end up with the following *augmented system* for the asymmetric cusp

$$F(y) = \alpha z,$$
$$F'(y)^T z = \gamma_{11} t_1 + \gamma_{21} t_2,$$
$$F'(y) t_1 = \beta_{11} z_1,$$
$$F'(y) t_2 = \beta_{21} z_2,$$
$$z^T F''(y)[t_2, t_2] = 0,$$
$$z^T F''(y)[t_1, t_2] = 0,$$
$$z^T z = 1,$$
$$t_1^T t_1 = t_2^T t_2 = 1, \quad t_1^T t_2 = 0.$$

This system comprises $4n + 7$ equations in the $4n + 8$ unknowns y, z, t_1, t_2, α_1, γ_{11}, γ_{21}, β_{11}, β_{21}—which means that the system is *underdetermined*.

The associated augmented Jacobian can be shown to have full row rank for sufficiently small perturbation parameters α_1, γ_{11}, γ_{21}, β_{11}, β_{21}.

Higher order critical points. The perturbed system (5.75) is a special case of the general perturbed Lyapunov-Schmidt reduction, wherein $A = F'(y^*)$ has rank-deficiency $k > 0$:

$$
\begin{aligned}
A^T z_j &= \sum_{i=1}^{k+1} \gamma_{ji} t_i, \\
A t_i &= \sum_{j=1}^{k} \beta_{ij} z_j, \qquad\qquad (5.76) \\
z_j z_l &= \delta_{j,l}, \qquad l \le j = 1, \ldots, k, \\
t_i t_m &= \delta_{i,m}, \qquad m \le i = 1, \ldots, k+1.
\end{aligned}
$$

This *underdetermined* system comprises $(2k+1)(n+k+1) - k^2$ equations in $(2k+1)(n+k+1)$ unknowns. It has been suggested by P. Kunkel in his thesis [138, 139, 140] and worked out using tree structures in [141]. The extended Jacobian has full row rank at a *perfect* singularity, where

$$\beta_{ij}^* = \gamma_{ij}^* = 0.$$

Note that, also for an *imperfect* singularity y^*, we can verify that

$$\gamma_{ij} = \beta_{ij} = z_j^T A t_i, \qquad i = 1, \ldots k, \quad j = 1, \ldots, k+1.$$

However, had we identified $\gamma_{ij} = \beta_{ij}$ from the start, then the system would no longer be uniquely solvable.

The missing k^2 degrees of freedom come from $\frac{1}{2}k(k-1)$ arbitrary degrees in z and $\frac{1}{2}k(k+1)$ arbitrary degrees in t due to orthogonal transformation— compare (5.76) and (5.52) and the discussion thereafter.

Generally speaking, for *higher order* singularities the number of equations and of dependent variables does not agree as nicely as in the simple bifurcation case. Consequently, quite complicated augmented systems may arise, usually *underdetermined* as in the *cusp* case. Part of such systems are still under investigation including the problem of their automatic generation by means of computer algebra systems—see, e.g., D. Armbruster [6]. In principle, a general bifurcation algorithm would need to represent a whole *hierarchy of augmented systems*—which, however, will be limited for obvious reasons.

5.3.2 Newton-like algorithm for simple bifurcations

We return to the augmented system (5.64) of G. Moore. The associated *extended Jacobian* has the block structure

$$J(y,z,\alpha) := \begin{bmatrix} C & A^T & 0 \\ A & \alpha I_n & z \\ 0 & z^T & 0 \end{bmatrix}$$

in terms of the submatrices

$$C := \left(F'(y)^T z\right)' = \sum_{i=1}^{n} f_i''(y)z_i \, , \quad A := F'(y) \, .$$

Note that C and therefore J are symmetric matrices.

Theorem 5.11 *At a simple (possibly perturbed) bifurcation point y^* with sufficiently small perturbation parameter α^* the extended Jacobian $J(y^*, z^*, \alpha^*)$ is nonsingular.*

Proof. First $J(y,z,0)$ is shown to be nonsingular. We start with applying the singular value decomposition

$$A = U \sum V^T, \; \sum = \begin{bmatrix} \sum' & 0 \\ 0 & 0 \end{bmatrix} ,$$

$$\begin{aligned} U \quad &: \quad \text{orthogonal } (n,n)\text{-matrix,} \\ V^T \quad &: \quad \text{orthogonal } (n+1, n+1)\text{-matrix,} \\ \textstyle\sum' \quad &= \quad \text{diag}\,(\sigma_1, \ldots, \sigma_{n-1}), \; \sigma_i > 0\,. \end{aligned}$$

Inserting this decomposition into $J(y,z,0)$ yields after proper transformation

$$J = \begin{bmatrix} C & A^T & 0 \\ A & 0 & z \\ 0 & z^T & 0 \end{bmatrix} \rightarrow \begin{bmatrix} \overline{C} & \Sigma^T & 0 \\ \Sigma & 0 & \overline{z} \\ 0 & \overline{z}^T & 0 \end{bmatrix} =: \overline{J}$$

with $\overline{z} := U^T z$, $\overline{C} := VCV^T = \overline{C}^T$. In the above notation, the equations $F'(y)^T z = 0$ now read

$$\Sigma^T \overline{z} = 0,$$

which implies that $\overline{z}^T = (0, \ldots, 0, 1)$. If, in addition, we introduce the corresponding partitioning

$$\overline{C} = \begin{bmatrix} \overline{C}_{11} & \overline{C}_{12} \\ \overline{C}_{12}^T & \overline{C}_{22} \end{bmatrix}, \quad \overline{C}_{22} = \overline{C}_{22}^T \quad (2,2)\text{-matrix},$$

then

$$\overline{J} = \begin{bmatrix} \overline{C}_{11} & \overline{C}_{12} & \Sigma' & 0 & 0 \\ \overline{C}_{12} & \overline{C}_{22} & 0 & 0 & 0 \\ \Sigma' & 0 & 0 & 0 & 0 \\ 0 & 0 & 0 & 0 & 1 \\ 0 & 0 & 0 & 1 & 0 \end{bmatrix}.$$

Hence

$$\text{rank}(J) = \text{rank}\left(\overline{J}\right) = 2n + \text{rank}\left(\overline{C}_{22}\right).$$

Upon recalling assumption (5.67), we obtain here

$$\left(z^T F'\right)'[t, t] = \overline{C}_{22},$$

which assures that $\overline{C}_{22}$ is certainly nonsingular and

$$\text{rank}\left(J(y^*, z^*, 0)\right) = 2n + 2.$$

Finally, by the usual perturbation lemma for symmetric matrices with symmetric perturbation, $J(y^*, z^*, \alpha^*)$ is nonsingular for α^* 'sufficiently small', which completes the proof.   □

The above Theorem 5.11 assures that the *ordinary* Newton method (dropping the iteration index)

$$\begin{bmatrix} C & A^T & 0 \\ A & \alpha I & z \\ 0 & z^T & 0 \end{bmatrix} \begin{bmatrix} \Delta y \\ \Delta z \\ \Delta \alpha \end{bmatrix} = - \begin{bmatrix} F'^T z \\ F + \alpha z \\ \frac{1}{2}(z^T z - 1) \end{bmatrix}$$

is well-defined in a neighborhood of the simple bifurcation point. In the special situation a *Newton-like* method characterized by the replacement

$$J(y, z, \alpha) \quad \rightarrow \quad J(y, z, 0),$$
$$A = F'(y) \quad \rightarrow \quad \widetilde{A} \approx F'(y^*)$$

seems to be preferable, since the associated linear block system (again dropping the iteration index)

$$\begin{bmatrix} C & \widetilde{A}^T & 0 \\ \widetilde{A} & 0 & z \\ 0 & z^T & 0 \end{bmatrix} \begin{bmatrix} \Delta y \\ \Delta z \\ \Delta \alpha \end{bmatrix} = - \begin{bmatrix} A^T z \\ F + \alpha z \\ \frac{1}{2}(z^T z - 1) \end{bmatrix} \qquad (5.77)$$

is easier to solve. On the basis of the theory of Section 2.1, the thus defined iteration will converge *superlinearly* for *perfect* bifurcations ($\alpha^* = 0$), but only *linearly* for *imperfect* bifurcations. Since excellent starting points are available in the present setting, one may even keep the *initial* Jacobian approximation—thus implementing a variant of the *simplified* Newton method. This method permits even further computational savings per iteration step.

Distinction between perfect and imperfect bifurcations. Whenever a *perfect bifurcation* arises, then the Newton-like iterates $\{\alpha^k\}$ will approach *zero* superlinearly so that the criterion

$$|\alpha^{k+1}| \le \tfrac{1}{4} |\alpha^k|$$

will be passed. Otherwise, leading digits of α^* will show up.

In order to compute the two branch directions t_1^*, t_2^* easily, the above extended Jacobian matrix needs to be decomposed in some *structure preserving* way. We work out two possibilities.

Implementation based on QR-decomposition. Let $A = F'(y)$ denote the Jacobian $(n, n + 1)$-matrix to be decomposed according to

$$A = Q \begin{bmatrix} R & S \\ 0 & \varepsilon^T \end{bmatrix} \Pi^T, \text{ where } \begin{array}{ll} Q : & \text{orthogonal } (n, n)\text{-matrix,} \\ \Pi : & \text{permutation } (n + 1, n + 1)\text{-matrix,} \\ R : & \text{upper triangular } (n - 1, n - 1)\text{-matrix,} \\ S : & (n - 1, 2)\text{-matrix,} \\ \varepsilon : & 2\text{-vector.} \end{array}$$

For y 'close to' y^*, R will be nonsingular and ε 'small'. Hence, the approximation

$$\widetilde{A} := Q \begin{bmatrix} R & S \\ 0 & 0 \end{bmatrix} \Pi^T$$

will be appropriate within a Newton-like iteration.

Starting points for Newton-like iteration. A starting guess y^0 is available from the path-following procedure, typically from linear interpolation between the two points $\bar{y}_1$, $\bar{y}_2$ on the solution curve (see Figure 5.8) that had activated the device for the detection of critical points (see Figure 5.5). A starting guess z^0 can be obtained from solving

$$\tilde{A}^T z = 0, \qquad \|z\|_2^2 = 1,$$

which, upon inserting the QR-decomposition directly leads to

$$z^0 = Q e_n, \ e_n = (0,\dots,0,1).$$

From this, a natural choice of α^0 can be seen to be

$$\alpha^0 = -\left(Q^T F(y^0)\right)_n.$$

Block elimination. Insertion of the above QR-decomposition into the block system (5.77) suggests the following partitioning

$$\hat{C} := \Pi^T C \Pi = \begin{pmatrix} C_{11} & C_{12} \\ C_{12} & C_{22} \end{pmatrix}, \quad C_{22} : (2,2)\text{-matrix},$$

$$\bar{z} := Q^T z = \begin{pmatrix} w \\ \zeta \end{pmatrix}, \qquad\qquad w \in \mathbb{R}^{n-1}, \ \zeta \in \mathbb{R},$$

$$\Pi^T \Delta y = \begin{pmatrix} \Delta u \\ \Delta v \end{pmatrix}, \qquad\qquad \Delta u \in \mathbb{R}^{n-1}, \ \Delta v \in \mathbb{R}^2,$$

$$\Delta\bar{z} = Q^T \Delta z = \begin{pmatrix} \Delta w \\ \Delta\zeta \end{pmatrix}, \qquad \Delta w \in \mathbb{R}^{n+1}, \ \Delta\zeta \in \mathbb{R},$$

$$\Pi^T A^T z = \begin{pmatrix} f_1 \\ f_2 \end{pmatrix}, \qquad\qquad f_1 \in \mathbb{R}^{n-1}, \ f_2 \in \mathbb{R}^2,$$

$$Q^T(F + \alpha z) = \begin{pmatrix} g_1 \\ g_2 \end{pmatrix}, \qquad\qquad g_1 \in \mathbb{R}^{n-1}, \ g_2 \in \mathbb{R}$$

$$h := \tfrac{1}{2}(z^T z - 1).$$

In this notation, (5.77) now reads

$$\begin{bmatrix} C_{11} & C_{12} & R^T & 0 & 0 \\ C_{12}^T & C_{22} & S^T & 0 & 0 \\ R & S & 0 & 0 & w \\ 0 & 0 & 0 & 0 & \zeta \\ 0 & 0 & w^T & \zeta & 0 \end{bmatrix} \begin{bmatrix} \Delta u \\ \Delta v \\ \Delta w \\ \Delta\zeta \\ \Delta\alpha \end{bmatrix} = - \begin{bmatrix} f_1 \\ f_2 \\ g_1 \\ g_2 \\ h \end{bmatrix}.$$

Simplified Newton iteration. The initial guess z^0 is equivalent to

$$w^0 = 0, \ \zeta^0 = 1,$$

which may be used to decouple the last two equations. In this simplified Newton method the last two equations then yield (dropping the index k):

$$\Delta\alpha = -g_2, \qquad \Delta\zeta = -h.$$

In order to solve the remaining three equations, we compute once

$$R\overline{S} \ \ = \ \ S, \tag{5.78}$$
$$\overline{C}_{22} \ := \ C_{22} - (C_{12}^T\overline{S} + \overline{S}^T C_{12}) + \overline{S}^T C_{11}\overline{S} \tag{5.79}$$

and repeatedly for each new right-hand side

$$
\begin{aligned}
R\overline{g}_1 \ \ &= \ \ g_1, \\
\overline{C}_{22}\Delta v \ \ &= \ \ -f_2 + \left(C_{12}^T - \overline{S}^T C_{11}\right)\overline{g}_1 + \overline{S}^T f_1, \\
\Delta u \ \ &= \ \ -\overline{g}_1 - \overline{S}\Delta v, \\
R^T\Delta w \ \ &= \ \ -f_1 - C_{11}\Delta u - C_{12}\Delta v.
\end{aligned}
$$

Note that the symmetric $(2,2)$-matrix $\overline{C}_{22}$ is just the one used in the proof of Theorem 5.11—which means that $\overline{C}_{22}$ may be assumed to be nonsingular in a neighborhood of a simple bifurcation point. If it appears to be singular when decomposed, then there will be *no bifurcation point* locally. Finally, back substitution yields

$$\Delta y = \Pi \left(\begin{array}{c} \Delta u \\ \Delta v \end{array} \right), \qquad \Delta z = Q \left(\begin{array}{c} \Delta w \\ \Delta \zeta \end{array} \right).$$

Implementation based on LU-decomposition. Assume that we start with the decomposition

$$A(\varepsilon) := L \left[\begin{array}{cc} R & S \\ 0 & \varepsilon^T \end{array} \right] \Pi^T,$$

where L is a lower triangular matrix obtained from some sparse elimination technique. Due to conditional pivoting within a sparse solver, say, the entries ε may be 'small'. Then A may be replaced according to

$$\widetilde{A} := A(0).$$

For reasons to be understood below, it is advisable to modify the *normalizing condition* $z^T z = 1$ such that

$$h := \tfrac{1}{2}(\|L^T z\|_2^2 - 1) = 0.$$

Starting points for Newton-like iteration. With y^0 given, initial guesses z^0 are easily derived from

$$\widetilde{A}^T z \;=\; 0\,,$$
$$\|L^T z\|_2^2 \;=\; 1\,.$$

which, with $L = (l_{ij})$, leads to

$$L^T z^0 \;:=\; e_n\,,$$
$$\alpha^0 \;:=\; (F(y^0))_n / l_{nn}\,.$$

Block elimination. The above LU-decomposition is now inserted into the block system (5.77) with the slight change of normalization as just described. In comparison with the QR-variant we can keep the notation for $\widehat{C}$, Δu, Δv, f_1, f_2 unchanged and introduce the following modified quantities:

$$\widehat{z} := L^{-1} z = \begin{pmatrix} \widehat{w} \\ \widehat{\zeta} \end{pmatrix}\,, \qquad \widehat{w} \in \mathbb{R}^{n-1}\,,\ \widehat{\zeta} \in \mathbb{R}\,,$$

$$\overline{z} := L^T z = \begin{pmatrix} w \\ \zeta \end{pmatrix}\,, \qquad w \in \mathbb{R}^{n-1}\,,\ \zeta \in \mathbb{R}\,,$$

$$L^T \Delta z = \begin{pmatrix} \Delta w \\ \Delta \zeta \end{pmatrix}\,, \qquad \Delta w \in \mathbb{R}^{n+1}\,,\ \Delta \zeta \in \mathbb{R}\,,$$

$$L^{-1}(F + \alpha z) = \begin{pmatrix} g_1 \\ g_2 \end{pmatrix}\,, \qquad g_1 \in \mathbb{R}^{n-1}\,,\ g_2 \in \mathbb{R}\,.$$

In this notation, (5.77) now reads

$$\begin{bmatrix} C_{11} & C_{12} & R^T & 0 & 0 \\ C_{12}^T & C_{22} & S^T & 0 & 0 \\ R & S & 0 & 0 & \widehat{w} \\ 0 & 0 & 0 & 0 & \widehat{\zeta} \\ 0 & 0 & w^T & \zeta & 0 \end{bmatrix} \begin{bmatrix} \Delta u \\ \Delta v \\ \Delta w \\ \Delta \zeta \\ \Delta \alpha \end{bmatrix} = - \begin{bmatrix} f_1 \\ f_2 \\ g_1 \\ g_2 \\ h \end{bmatrix}\,.$$

Simplified Newton iteration. In order to implement this variant, just verify that the choice of z^0 is again equivalent to

$$w^0 = 0\,,\ \zeta^0 = 1\,,$$

which once more implies that

$$\Delta \alpha = -g_2\,,\quad \Delta \zeta = -h\,.$$

With $\overline{S}$ and $\overline{C}_{11}$ from equations (5.78) and (5.79) the remaining system can be solved as follows

$$R\overline{g}_1 \;=\; g_1 + \widehat{w}\Delta\alpha\,,$$
$$\overline{C}_{22}\Delta v \;=\; -f_2 + \left(C_{12}^T - \overline{S}^T C_{11} \right)\overline{g}_1 + \overline{S}^T f_1\,,$$
$$\Delta u \;=\; -\overline{g}_1 - \overline{S}\Delta v\,,$$
$$R^T \Delta w \;=\; -f_1 - C_{11}\Delta u - C_{12}\Delta v\,.$$

Finally, back substitution yields

$$\Delta y = \Pi \left(\begin{array}{c} \Delta u \\ \Delta v \end{array} \right) L^T \Delta z = \left(\begin{array}{c} \Delta w \\ \Delta \zeta \end{array} \right).$$

5.3.3 Branching-off algorithm

Suppose that a simple bifurcation point y^* has been computed. Then y^* is the intersection of exactly two solution branches associated with the mapping $F + \alpha z$. In order to continue the numerical pathfollowing beyond bifurcations, one will need to first compute the directions of these branches and second to design an efficient restart strategy along each new semi-branch.

Computation of branch directions. As described above, the local tangent directions t_i^*, $i = 1, 2$ are computed from the quadratic equation (5.67). Starting from any of the two presented decompositions, the following parametrization of $\mathcal{N}$ is natural:

$$t_i^* := \Pi \left(\begin{array}{c} -\overline{S} \, e_i \\ e_i \end{array} \right), \qquad e_i := (\cos \Theta_i, \ \sin \Theta_i)$$

with $\overline{S}$ as defined by (5.78). In the present notation this equation can be rewritten as

$$t_i^{*T} C \, t_i^* = 0,$$

which reduces to

$$e_i^T \overline{C}_{22} e_i = 0$$

in terms of the symmetric $(2, 2)$-matrix $\overline{C}_{22}$ known to be nonsingular when a simple bifurcation point exists locally. This is again a quadratic equation in either $\tan \Theta_i$ or $\cot \Theta_i$ with two different real roots under the assumption (5.67). In case $\overline{C}_{22}$ turns out to be semi-definite or degenerate, then a *non-simple* bifurcation point is seen to occur. Complex conjugate roots indicate an *isola* (which, however, would be hard to detect by just pathfollowing with respect to one parameter!).

Stepsize control restart. Suppose we have the situation of one *entering* semi-branch (already computed) and three *emanating* semi-branches (to be computed next) as depicted in Figure 5.8.

Let, formally, $t_3^* := -t_2^*$. In order to start the path-following procedure along each emanating semi-branch, one is required to define starting points $\widehat{y}$ for the quasi-Gauss-Newton iteration ($i = 1, 2, 3$):

$$\widehat{y}_i(s) := y^* + s \cdot t_i^*.$$

Herein an efficient control of the stepsize s requires some care. Two antagonistic conditions are to be matched:

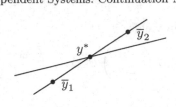

Fig. 5.8. Simple bifurcation point: branch situation.

- The Jacobian $F'(\widehat{y}(s))$ must have full (numerical) rank, which leads to a lower bound $s > s_{\min}$.
- The local quasi-Gauss-Newton iteration starting at $\widehat{y}(s)$ should converge sufficiently fast, which leads to an upper bound $s \leq s_{\max}$.

Prediction strategy. If we use the fact that the Gauss-Newton iteration had converged on the entering semi-branch towards some point $\overline{y}_{\mathrm{old}}$, we are led to the choice

$$s_0 := \rho \frac{\|y^* - \overline{y}_{\mathrm{old}}\|}{\|t_1^*\|} , \quad \rho < 1$$

with some safety factor ρ. With $\widehat{y}(s_0)$, the Jacobian $F'(\widehat{y}(s_0))$ and the contraction factors $\Theta(s_0)$ are available in the course of the Gauss-Newton iteration. Therefore, numerical estimates $[s_{\min}]$, $[s_{\max}]$ can be computed.

In view of (I) above, we require that

$$\overline{\varepsilon}\,\mathrm{cond}\left(F'(\widehat{y}(s))\right) < 1$$

with some prescribed $\overline{\varepsilon} > eps$, the relative machine precision. Near y^* all determinants (such as d_ξ, d_λ) are $O(s)$. So we have

$$\mathrm{cond}\left(F'(\widehat{y}(s))\right) > \frac{\gamma}{s} ,$$

which leads to

$$s > \overline{\varepsilon}\gamma =: s_{\min} .$$

As γ is unknown, a numerical condition number estimate $[\mathrm{cond}\,(\cdot)]$ is required to derive the estimate

$$[s_{\min}] := \overline{\varepsilon}\left[\mathrm{cond}\left(F'(\widehat{y}(s_0))\right)\right] \cdot s_0 .$$

In view of (II) above, a careful analysis shows that the contraction factor

$$\Theta_0(s) = O(s)$$

close to y^* instead of $O(s^2)$ in the neighborhood of a regular point $\overline{y}$. As a consequence of the Newton-Kantorovich theorem, we require

$$\Theta_0 \leq \overline{\Theta} = \tfrac{1}{4}\,,$$

which leads to the stepsize estimate

$$[s_{\max}] = \frac{\overline{\Theta}}{\Theta_0(s_0)} \cdot s_0\,.$$

In case we had obtained

$$[s_{\min}] > [s_{\max}]$$

the computation would have to be terminated suggesting higher precision arithmetic—thus lowering $[s_{\min}]$.

In the author's experience, such a situation has never occurred up to now. In the standard situation

$$[s_{\min}] \ll [s_{\max}]$$

some initial steplength $\overline{s}_0$ can be selected. A typical choice will be

$$\overline{s}_0 := \rho \cdot [s_{\max}]$$

for some sufficiently large $\rho < 1$.

Construction of complete bifurcation diagrams. The implementation of the whole algorithm—as described in the previous Section 5.2 and the present Section 5.3—requires careful book-keeping of critical points and of entering and emanating semi-branches to avoid endless cycling. As an example, *before* actually iterating towards some conjectured bifurcation point y^*, the corresponding starting point y^0 should be tested: if it is within the Kantorovich neighborhood of some formerly computed bifurcation point, then identity of old and new bifurcation point can be assumed; as usual, the test is based on the local contraction factor criterion $\Theta_0 \leq 1/4$ in agreement with the sufficient Kantorovich condition ($h_0 \leq 1/2$). Whenever $\Theta_0 > 1/4$, i.e., when the Kantorovich condition is locally violated, then a possible local nonuniqueness of a solution is indicated.

BIBLIOGRAPHICAL NOTE. The computation of simple bifurcations via the QR-implementation of Moore's extended system has been worked out in the paper [72] by P. Deuflhard, B. Fiedler, and P. Kunkel. The here presented algorithm with quasi-Gauss-Newton method, adaptive stepsize control, computation of turning points and simple bifurcation points has been implemented in the code ALCON2. An advanced descendant of ALCON2 is the code SYMCON due to K. Gatermann and A. Hohmann [95] for equivariant parameter dependent nonlinear systems. In this algorithm, symmetries are exploited such that along each branch symmetry transformations are performed based on Schur's lemma. As a consequence, symmetry breaking or symmetry preserving *higher order* bifurcations often just show up as *simple* generic or non-generic bifurcations and can be treated as such. Due to this property the algorithm is considerably more robust than its predecessor ALCON2. In passing, dynamical stability of the solutions along each branch can be identified. In order to illustrate the kind of additional results available from SYMCON, we give a rather challenging illustrative example.

Example 5.4 *Hexagonal lattice dome.* This well-known challenging equivariant bifurcation problem from continuum mechanics is due to T.J. Healey [117]. It is known to contain a large number of all kinds of higher order singularities connected with symmetries of the mechanical construction, the most dominant of which is the symmetry D_6. The problem has been tackled by SYMCON [95] and is documented in detail in [94]. Figure 5.9 gives parts of the rather complex total bifurcation diagram associated with two different subsymmetries. As it turned out, the bifurcation diagram computed by SYMCON revealed hitherto unknown parts.

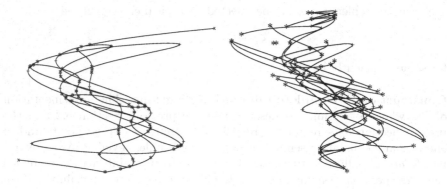

Fig. 5.9. Hexagonal lattice dome: Bifurcation subdiagrams associated with partial symmetries. *Left:* Kleinian group and D_6. *Right:* D_3 and Z_2^4.

Exercises

Exercise 5.1   Consider the *pseudo-arclength* continuation method as discussed at the beginning of Section 5.2.1. Study the effect of rescaling of the parameter

$$\lambda \longrightarrow \sigma = \lambda\kappa.$$

What kind of continuation method is obtained in the limiting case $\kappa \to 0$?

Exercise 5.2   Derive feasible stepsize bounds for the classical and the tangent Newton-continuation method using

a) the affine contravariant Newton-Mysovskikh theorem for the ordinary Newton method (Theorem 2.12),

b) the affine contravariant Newton-Kantorovich theorem for the simplified Newton method (Theorem 2.13).

c) On this theoretical basis, design computational estimates for use within an adaptive stepsize control strategy.

Exercise 5.3   *Classical continuation method for nonlinear least squares problems.* For given real parameter λ, let the prediction path be $\hat{x}(\lambda) = \overline{x}(0)$. In the residual based formulation of the Gauss–Newton method as given in Section 4.2, we write the homotopy for the path $\overline{x}(\lambda)$ as

$$\overline{P}(\overline{x}(\lambda), \lambda) F(\overline{x}(\lambda), \lambda) \equiv 0 \,,$$

where

$$\overline{P}(x, \lambda) = F'(x, \lambda) F'(x, \lambda)^- \,, \quad \overline{P}^{\perp}(x, \lambda) = I_m - \overline{P}(x, \lambda)$$

are the corresponding projectors, assumed to be orthogonal.

a) Show that the classical continuation method is of order $p = 1$.

b) Derive an affine contravariant formula for the feasible stepsize.

c) Design an affine contravariant computational estimate for the order coefficient and consider details for the corresponding *adaptive* continuation algorithm.

Exercise 5.4   *Tangent continuation method for nonlinear least squares problems.* The notation is the same as in Exercise 5.3. The only new aspect is that the prediction path now reads

$$\hat{x}(\lambda) = \overline{x}(0) + \lambda \dot{\overline{x}}(0) \,.$$

a) We need an expression for the local path direction $\dot{\overline{x}}(0)$. Verify the result

$$F'(\overline{x}(0), 0) \dot{\overline{x}}(0) + F'(\overline{x}(0), 0)^{-T} F''(\overline{x}(0), 0) [\overline{P}^{\perp}(\overline{x}(0), 0) F(\overline{x}(0), 0), \dot{\overline{x}}(0)] =$$
$$- \left(\overline{P}(\overline{x}(0), 0) F_\lambda(\overline{x}(0), 0) + F'(\overline{x}(0), 0)^{-T} F'_\lambda(\overline{x}(0), 0) [\overline{P}^{\perp}(\overline{x}(0), 0) F(\overline{x}(0), 0)] \right).$$

Hint: For the symmetric projector $\overline{P} = A A^-$ in terms of the generalized (inner) inverse A^- apply the formula

$$D\overline{P} = \overline{P}^{\perp}(DA) A^- + \left(\overline{P}^{\perp}(DA) A^- \right)^T$$

and, in addition, use special properties at $(\overline{x}(0), 0)$.

b) Show that under the assumption

$$\overline{P}^{\perp}(\overline{x}(\lambda), \lambda)F(\overline{x}(\lambda), \lambda) \equiv 0$$

the above equation shrinks to

$$\|F'(\overline{x}(0), 0)\dot{\overline{x}}(0) + \overline{P}(\overline{x}(0), 0)F_{\lambda}(\overline{x}(0), 0)\| = \min,$$

which can be satisfied by

$$\dot{\overline{x}}(0) = -F'(\overline{x}(0), 0)^{-}F_{\lambda}(\overline{x}(0), 0).$$

c) Discuss the necessary steps to be taken toward an *adaptive* tangent continuation algorithm.

Part II

DIFFERENTIAL EQUATIONS

6 Stiff ODE Initial Value Problems

This chapter deals with *stiff* initial value problems for ODEs

$$\dot{x} = F(x), \ x(0) = x_0 .$$

The discretization of such problems is known to involve the solution of non-linear systems per each discretization step—in one way or the other.

In Section 6.1, the contractivity theory for linear ODEs is revisited in terms of *affine similarity*. Based on an affine similar convergence analysis for a simplified Newton method in *function space*, a *nonlinear contractivity* theory for stiff ODE problems is derived in Section 6.2, which is quite different from the theory given in usual textbooks on the topic. The key idea is to replace the Picard iteration in function space, known as a tool to show uniqueness in nonstiff initial value problems, by a simplified Newton iteration in function space to characterize stiff initial value problems. From this point of view, *linearly implicit* one-step methods appear as direct realizations of the simplified Newton iteration in function space. In Section 6.3, exactly the same theoretical characterization is shown to apply also to *implicit* one-step methods, which require the solution of a nonlinear system by some finite dimensional Newton-type method at each discretization step.

Finally, in a deliberately longer Section 6.4, we discuss a class of algorithms called *pseudo-transient continuation* algorithms, whereby steady state problems are solved via stiff integration. The latter type of algorithm is particularly useful, when the Jacobian matrix is singular due to hidden dynamical invariants (such as mass conservation). The affine similar theoretical characterization permits the derivation of an *adaptive (pseudo-)time step strategy* and an accuracy matching strategy for a residual based inexact Newton algorithm.

6.1 Affine Similar Linear Contractivity

For the time being, consider a *linear* ODE system of the kind

$$\dot{x} = Ax, \ x(0) = x_0 . \tag{6.1}$$

Formally, system (6.1) can be solved in terms of the matrix exponential

$$x(t) = \exp(At)x_0 \,.$$

In view of *affine similarity* as discussed in Section 1.2.2, we start from the (possibly complex) Jordan decomposition

$$A = TJT^{-1} \,,$$

wherein J is the *Jordan canonical form* consisting of elementary Jordan blocks for each separate eigenvalue $\lambda(A)$. Then the (possibly complex) transformation

$$z := T^{-1}(x - \hat{x})$$

has been shown in Section 1.2.2 to generate an affine similar coordinate frame. In what follows we will have to work with norms $\| \cdot \|$ induced by certain inner products $(\cdot, \cdot)$. For simplicity, we may think of the Euclidean norm $\| \cdot \|$ induced by the (possibly complex) Euclidean inner product $(u, v) = u^*v$ with u^* the adjoint. If we phrase our subsequent theoretical statements in terms of the *canonical norm*

$$|u| := \|T^{-1}u\| \,, \tag{6.2}$$

induced by the *canonical inner product*

$$\langle u, v \rangle = (T^{-1}u, T^{-1}v) \,,$$

then such statements will automatically meet the requirement of affine similarity. In this setting, we may define some constant $\mu = \mu(A)$, allowed to be *positive, zero,* or *negative,* such that

$$\langle u, Au \rangle \le \mu(A)|u|^2 \,. \tag{6.3}$$

This definition is obviously equivalent to

$$(\bar{u}, J\bar{u}) \le \mu(A)\|\bar{u}\|^2 \,, \tag{6.4}$$

wherein $\bar{u} = T^{-1}u$. Assuming that the quantity μ is chosen best possible, it can be shown to satisfy

$$\mu(A) = \max_{u \neq 0} \frac{\langle u, Au \rangle}{|u|^2} \ge \max_i \Re\lambda_i(A) + \epsilon \,, \quad \epsilon \ge 0 \,. \tag{6.5}$$

Herein $\epsilon = 0$ and equality holds, if the eigenvalue defining $\mu(A)$ is simple. It is an easy task to show that

$$\mu(BAB^{-1}) = \mu(A) \tag{6.6}$$

for any nonsingular matrix B, which confirms that this quantity is indeed *affine similar.* In the canonical norm we may obtain the estimate

$$|x(t)| \leq \exp(\mu t)|x_0|.$$

Whenever

$$\mu \leq 0 \qquad (6.7)$$

holds, then

$$|x(t)| \leq |x(0)|,$$

which means that the linear dynamical system (6.1) is *contractive*.

For computational reasons, the Euclidean product (possibly in a scaled variant) is preferred to the canonical product. Suppose that we therefore replace the above definition (6.3) in terms of the canonical inner product by the analogous definition

$$\nu(A) = \max_{u \neq 0} \frac{(u, Au)}{\|u\|^2} \qquad (6.8)$$

in terms of the Euclidean product. The thus defined quantity can be expressed as

$$\nu(A) = \lambda_{\max}\left(\tfrac{1}{2}(A + A^T)\right),$$

where $\lambda_{\max}$ is the maximum (real) eigenvalue of the symmetric part of the matrix A. Upon comparison with (6.4) we immediately observe that

$$\mu(A) = \nu(J) = \lambda_{\max}\left(\tfrac{1}{2}(J + J^T)\right),$$

which directly leads to the above result (6.5) —see, e.g., [71, Section 3.2]. From this we see that the quantities $\nu(A)$ and $\mu(A)$ may be rather different —in fact, unless A is symmetric, not even the signs may be the same. Moreover, in contrast to (6.6), we now have the undesirable property

$$\nu(BAB^{-1}) \neq \nu(A),$$

i.e., this quantity is *not affine similar*. Consequently, contractivity in the canonical norm $|\cdot|$ does *not* imply contractivity in the original norm $\|\cdot\|$. Whenever a relation of the kind

$$|u| \leq |v|$$

is transformed back to the original norm, we can only prove that

$$\|u\| \leq \mathrm{cond}(T)\|v\|$$

in terms of the condition number $\mathrm{cond}(T) = \|T^{-1}\| \cdot \|T\| \geq 1$, which here arises as an unavoidable *geometric distortion* factor. This distortion factor also indicates possible *ill-conditioning* of the Jordan decomposition as a whole—which may affect the theoretical presentation in terms of canonical inner products and norms. Nevertheless, we will stick to a formal notation in terms of the canonical norm $|\cdot|$ below to make the underlying structure transparent.

6.2 Nonstiff versus Stiff Initial Value Problems

Reliable numerical algorithms are, in one way or the other, appropriate implementations of uniqueness theorems of the underlying analytic problem.
The most popular uniqueness theorem for ODEs is the well-known Picard-
Lindelöf theorem: it is based on the *Picard fixed point iteration* in function
space (Section 6.2.1) and characterizes the growth of the solution by means
of the Lipschitz constant of the right-hand side—a characterization known
to be appropriate for *nonstiff* ODEs, but inappropriate for *stiff* ODEs. As
will be shown in this section, an associated uniqueness theorem for stiff ODEs
can be derived on the basis of a *simplified Newton iteration* in function space,
wherein the above Lipschitz constant is circumvented by virtue of a one-sided
linear contractivity constant. As a natural spin-off, this theory produces some
common *nonlinear contractivity* concept both for ODEs (Section 6.2.2) and
for implicit one-step discretizations (see Section 6.3 below).

6.2.1 Picard iteration versus Newton iteration

Consider again the *nonlinear* initial value problem

$$\dot{x} = F(x), \ x(0) = x_0.$$

For the subsequent presentation, its equivalent formulation as a Volterra operator equation (of the second kind) is preferable:

$$G(x, \tau) := x(\tau) - x_0 - \int_{t=0}^{\tau} F(x(t))dt = 0. \tag{6.9}$$

This equation defines a *homotopy* in terms of the interval length $\tau \geq 0$. Let Γ
denote some neighborhood of the graph of a solution of (6.9). Then Peano's
existence theorem requires that

$$L_0 := \sup_{\Gamma} \|F(x)\| < \infty$$

in terms of some pointwise norm $\|\cdot\|$ in $\mathbb{R}^n$.

In order to prove *uniqueness*, the standard approach is to construct the so-
called Picard iteration

$$x^{i+1}(\tau) = x_0 + \int_{t=0}^{\tau} F(x^i(t))dt \tag{6.10}$$

to be started with $x^0(t) \equiv x_0$. From this fixed point iteration, one immediately derives

$$\|x^{i+1}(\tau) - x^i(\tau)\| \leq \int_{t=0}^{\tau} \|F(x^i(t)) - F(x^{i-1}(t))\| dt.$$

Hence, in order to study contraction, the most natural theoretical characterization is in terms of the Lipschitz constant L_1 defined by

$$\|F(u) - F(v)\| \leq L_1 \|u - v\|.$$

With this definition, the sequence $\{x^i\}$ can be shown to converge to some solution x^* such that

$$\|x^*(\tau) - x_0\| \leq L_0 \tau \varphi(L_1 \tau)$$

with

$$\varphi(s) := \left\{ \begin{array}{cc} (\exp(s) - 1)/s & s \neq 0 \\ 1 & s = 0. \end{array} \right. \tag{6.11}$$

Moreover, x^* is unique in Γ. This is the main result of the well-known Picard-Lindelöf theorem—ignoring for simplicity any distinction between local and global Lipschitz constants.

A similar term arises in the analysis of *one-step discretization* methods for ODE initial value problems. Let $p \geq 1$ denote the consistency order of such a method and τ a selected uniform stepsize assumed to be sufficiently small. Then the discretization error between the continuous solution x and the discrete solution x_τ at some final point $T = n\tau$ can be represented in the form (see, e.g., the ODE textbook by E. Hairer and G. Wanner [114]):

$$\|x_\tau(T) - x(T)\| \leq C_p \cdot \tau^p \cdot T \cdot \varphi(\bar{L}_1 T).$$

For *explicit* one-step methods, the coefficient C_p just depends on some bound in terms of higher derivatives of F. The above discrete Lipschitz constant $\bar{L}_1 \geq L_1$ is an analog of the continuous Lipschitz constant L_1, this time for the increment function of the one-step method. In order to assure that the notion of a consistency order p is meaningful at all, a restriction of the kind

$$L_1 \tau \leq C, \quad C = \mathcal{O}(1) \tag{6.12}$$

will be needed. Consequently, this characterization is appropriate only for *nonstiff* discretization methods.

HISTORICAL NOTE. Originally, it had first been thought that the use of *implicit* discretization methods would be the essential item to overcome the observed difficulties in the numerical integration of what have been called stiff ODEs—see, for instance, the early fundamental paper by G. Dahlquist [47]. For implicit one-step methods, the above coefficient C_p is bounded only, if the discrete solution can be locally continued over each discretization step of length τ. This aspect will be studied in detail in the subsequent Section 6.3. In the next stage of the development of stiff integration, however, it was recognized that the solution method for the thus arising algebraic equations is equally important: the early paper of W. Liniger and R.A. Willoughby [145] pointed out that any fixed point iteration based only on F-evaluations for the algebraic equations would again bring in restriction (6.12), whereas a *Newton-like iteration* could, in principle, avoid the restriction. Much later, so-called semi-implicit or linearly-implicit discretization methods (such as Rosenbrock methods, W-methods, or extrapolation methods) were constructed that only apply one single Newton-like iteration per discretization step. Therefore the present essence of insight seems to be that nonstiff integration is character-ized by sampling of F only, whereas stiff integration requires the additional sampling of $F'(x)$ or an approximation.

With these preparations, a natural approach towards a uniqueness theorem covering stiff ODEs as well will be to replace the Picard iteration (6.10) by a Newton iteration. For the *ordinary* Newton method we would obtain

$$G'(x^i)\Delta x^i = -G(x^i), \quad x^{i+1} = x^i + \Delta x^i$$

or, in more explicit notation

$$\Delta x^i(\tau) - \int_{t=0}^{\tau} F'(x^i(t))\Delta x^i(t)dt = -\left[x^i(\tau) - x_0 - \int_{t=0}^{\tau} F(x^i(t))dt\right]. \quad (6.13)$$

However, the above iteration requires global sampling of the Jacobian $F'(x)$ rather than just pointwise sampling as in numerical stiff integration. There-fore, the *simplified* Newton method will be chosen instead: we just have to replace

$$G'(x^i) \to G'(x^0), \quad x^0(t) \equiv x_0$$

or, equivalently,

$$F'(x^i(t)) \to F'(x_0) =: A.$$

The corresponding replacement in (6.13) then leads to

$$x^{i+1}(\tau) - A\int_0^{\tau} x^{i+1}(t)dt = x_0 + \int_0^{\tau} [F(x^i(t)) - Ax^i(t)]dt. \quad (6.14)$$

Note that this may be interpreted as a Picard iteration associated with the formally modified ODE

$$\dot{x} - Ax = F(x) - Ax, \quad x(0) = x_0,$$

which is the basis for linearly-implicit one-step methods.

6.2.2 Newton-type uniqueness theorems

The above simplified Newton-iteration (6.14) is now exploited with the aim of proving uniqueness theorems for ODE IVP's that cover stiff ODEs as well. In order to guarantee *affine similarity*, we will define coordinates $x \in \mathbb{R}^n$ in such a way that any required (possibly approximate) Jacobian A is already in its Jordan canonical form J. Consequently, the selected vector norm $\| \cdot \|$ is identical to canonical norm as defined in (6.2)—see also the discussion in Section 6.1. For the time being, we assume that we have an exact initial Jacobian

$$F'(x_0) = A = J.$$

A discussion of the case of an approximate Jacobian will follow subsequently.

Theorem 6.1 *In the above notation let $F \in C^1(D)$, $D \subseteq \mathbb{R}^n$. For the Jacobian $A := F'(x_0)$ assume a one-sided Lipschitz condition of the form*

$$(u, Ju) \le \mu \|u\|^2,$$

where $(\cdot, \cdot)$ denotes the inner product that induces the norm $\|\cdot\|$. In this norm, assume that

$$\|F(x_0)\| \le L_0 \qquad for \quad x_0 \in D$$

$$\|(F'(x) - F'(x_0))u\| \le L_2 \|x - x_0\| \|u\| \qquad for \quad x, x_0, u \in D. \qquad (6.15)$$

Then, for D sufficiently large, existence and uniqueness of the solution of the ODE IVP is guaranteed in $[0, \tau]$ such that

$$\tau \text{ unbounded }, \text{ if } \mu\bar{\tau} \le -1,$$

$$\tau \le \bar{\tau}\Psi(\mu\bar{\tau}) , \text{ if } \mu\bar{\tau} > -1$$

with $\bar{\tau} := (2L_0 L_2)^{-1/2}$ and

$$\Psi(s) := \begin{cases} \ln(1+s)/s & s \ne 0 \\ 1 & s = 0. \end{cases}$$

Proof. Upon performing the variation of constants, we rewrite (6.14) as

$$\Delta x^i(\tau) = \int_{t=0}^{\tau} \exp(A(\tau - t))\left(F(x^i(t)) - \frac{d}{dt}x^i(t)\right) dt, \qquad (6.16)$$

where $\exp(At)$ denotes the matrix exponential. Within this proof let $|\cdot|$ denote the standard C^0-norm:

$$|u| := \max_{t \in [0,\tau]} \|u(t)\|.$$

In order to study convergence, we set the initial guess $x^0(t) \equiv x_0$ and apply Theorem 2.5 from Section 2.1.2, which essentially requires that

$$|\Delta x^0| \le \alpha \,, \tag{6.17}$$

$$|G'(x^0)^{-1}(G'(x) - G'(x^0))| \le \omega|x - x^0| \,, \tag{6.18}$$

$$\alpha\omega \le \tfrac{1}{2} \,. \tag{6.19}$$

The rest of the assumptions holds for D sufficiently large. The task is now to derive upper bounds α, ω and to assure (6.19). With x^0 as set above, the first Newton correction satisfies—compare (6.16)

$$\Delta x^0(\tau) = \int_{t=0}^{\tau} \exp(A(\tau - t))F(x_0)dt \,.$$

Hence

$$\|\Delta x^0(\tau)\| \le \int_{s=0}^{\tau} \| \exp(As)F(x_0)\|ds \le$$

$$\le L_0 \int_{s=0}^{\tau} \exp(\mu s)ds = L_0\tau\varphi(\mu\tau) =: \alpha(\tau)$$

with φ as introduced in (6.11). In order to derive $\omega(\tau)$, we introduce the operator norm in (6.18) by

$$z := G'(y^0)^{-1}(G'(x^0 + w) - G'(x^0))u \,,$$
$$|z| \le \omega \cdot |u| \cdot |w| \,.$$

Once more by variation of constants, we obtain

$$\|z(\tau)\| \le \int_{t=0}^{\tau} \| \exp(A(\tau - t))[F'(x_0 + w) - F'(x_0)]u\|dt \,,$$

which, similar as above, yields

$$|z| \le L_2 \cdot \tau \cdot \varphi(\mu\tau) \cdot |u| \cdot |w| \,.$$

Hence, a natural definition is

$$\omega(\tau) := L_2\tau\varphi(\mu\tau) \,.$$

Insertion into the Kantorovich condition (6.19) produces

$$(\tau\varphi(\mu\tau))^2 \le (2L_0L_2)^{-1} =: \bar{\tau}^2$$

or, equivalently,

$$\tau\varphi(\mu\tau) \le \bar{\tau} \,.$$

Since μ may have either sign, the main statements of the theorem are an immediate consequence.   □

A graphical representation of the above monotone decreasing function Ψ is given below in Figure 6.3.

Remark 6.1   Upon using the same characterizing quantities $\mu, \bar{\tau}$ as above, an improved theorem has been shown by W. Walter [195] using differential inequalities—compare his book [194]. Since this theorem is nonconstructive and does not make a difference for the discretizations to be treated in Section 6.3, it is omitted here (details can be found in the paper [66]).

Nonlinear contractivity.   As can be seen in Fig. 6.1 below, the above theorem (as well as the one in [66]) comes up with a pole at $s = -1$, which reflects the condition

$$\mu\bar{\tau} \leq -1$$

for *global boundedness* of the solution. This condition may be rewritten as

$$\mu + \sqrt{2L_0L_2} \leq 0. \qquad (6.20)$$

As $L_2 = 0$ in the linear case, the above condition is immediately recognized as a direct generalization of the linear contractivity condition (6.7). In other words: the above pole represents *global nonlinear contractivity*, involving local contractivity via μ and the part from the nonlinearity in well-separated form.

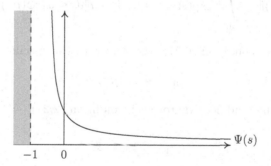

Fig. 6.1. Nonlinear contractivity: function Ψ as defined in Theorem 6.1 .

If, instead of the *exact* Jacobian $A = F'(x_0)$, an approximation error

$$\delta A = A - F'(x_0)$$

must be taken into account, then a modification of the above theorem will be necessary. In view of an affine similar presentation we again assume that

$$A = J,$$

which means that the approximate Jacobian is already in Jordan canonical form and the norm $\| \cdot \|$ is identical to the canonical norm.

Theorem 6.2 *Notation and assumptions as in Theorem 6.1. In addition, let the Jacobian approximation error be bounded as*

$$\|\delta A\| \le \delta_0 \ , \quad \delta_0 \ge 0 \,.$$

Then the results of Theorem 6.1 hold with $\bar{\tau}$ replaced by

$$\hat{\tau} := \bar{\tau}/(1 + \delta_o \bar{\tau}) \,.$$

Proof. For the proof we apply Theorem 2.6, the convergence theorem for Newton-like iterations, with $G'(x^0)$ now replaced by $M(x^0)$, which means replacing $F'(x_0)$ by $A \ne F'(x_0)$. With μ now associated with the Jacobian *approximation A* , the estimates $\alpha(\tau), \omega(\tau)$ carry over. In addition, the assumptions (6.17) up to (6.19) must be extended by

$$|M(x^0)^{-1}(G'(x^0) - M(x^0))| \le \bar{\delta}_0 < 1 \,. \tag{6.21}$$

Upon defining

$$z := M(x^0)^{-1}(G'(x^0) - M(x^0))u$$

a similar estimate as in the proof of Theorem 6.1 leads to

$$\|z(\tau)\| \le \int_{t=0}^{\tau} \| \exp(A(\tau - t)) \cdot \delta A \cdot u\| dt \le \delta_0 \tau \varphi(\mu\tau)|u| \,.$$

Hence, the above condition (6.21) shows up with the specification

$$\bar{\delta}_0 := \delta_0 \tau \varphi(\mu\tau) \,.$$

Insertion into the modified Kantorovich condition (2.23)

$$\frac{\alpha\omega}{(1 - \bar{\delta}_0)^2} \le \tfrac{1}{2}$$

then yields

$$\tau\varphi(\mu\tau) \le \bar{\tau}/(1 + \delta_0\bar{\tau}) =: \hat{\tau} \,.$$

Note that condition (6.21) is automatically satisfied, since

$$\bar{\delta}_0 = \delta_0\tau\varphi(\mu\tau) \le \delta_0\bar{\tau}/(1 + \delta_0\bar{\tau}) < 1 \,,$$

which completes the proof.   □

Finally, we want to emphasize that all above results also hold, if the norm $\|\cdot\|$ is not identified with the canonical norm $|\cdot|$, but allowed to be a general vector norm. However, as already worked out at the end of Section 6.1, this would include a tacit deterioration of all results, since then the one-sided Lipschitz constant μ may be rather off scale, if not even nonnegative—compare again the definitions (6.3) and (6.8).

Remark 6.2   The experienced reader will be interested to know whether these results carry over to differential-algebraic equations (DAEs) as well. Unfortunately, this causes some difficulty, which can already be seen in the simple separable DAE of the form

$$y' = f(y, z) , \quad 0 = g(y, z) .$$

In this case, the differential part f and the variable y suggest affine similarity, whereas the equality constrained part g would require some affine covariance or contravariance. For this reason, a common affine invariant theoretical framework is hard to get, if at all possible. Up to now, more subtle estimation techniques use a characterization in terms of perturbation parameters ϵ, which by construction do not allow for any affine invariance concept.

6.3 Uniqueness Theorems for Implicit One-step Methods

A natural requirement for any discretization π of the above ODE initial value problem will be that it passes basic symmetries of the underlying continuous problem on to the generated discrete problem. In particular, we will require that the diagram in Figure 6.2 commutes, which ensures that $y_\pi = Bx_\pi$ holds whenever $y = Bx$. Among the discretization methods satisfying this requirement, we will restrict our attention to implicit and linearly implicit one-step methods, also called Runge-Kutta methods.

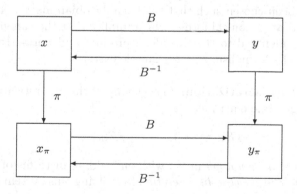

Fig. 6.2. Affine invariance under discretization π

As an extension of Section 6.1, any affine similar discretization of linear ODEs can also be treated in affine similar terms. This idea directly leads to G. Dahlquist's linear scalar model equation [47]

$$\dot{x} = \lambda x, \quad x(0) = 1.$$

Therefore, *linear contractivity* of implicit discretizations as well as of linearly implicit discretizations can be treated just as described in usual numerical ODE textbooks—see, e.g., [114, 115].

Things are different with respect to *nonlinear contractivity*. First, recall from Section 6.2.1 that the simplified Newton iteration (6.14) for the continuous ODE problem may also be regarded as a Picard iteration (6.10) for the ODE

$$\dot{x} - Ax = F(x) - Ax.$$

This ODE is the starting point for linearly implicit one-step discretizations (such as Rosenbrock methods, W-methods, or extrapolation methods), which just discretize the above left hand side implicitly and the above right hand side explicitly. Therefore, *linearly implicit discretizations* may be interpreted as *direct realizations of the simplified Newton iteration in function space*. Of course, they should also observe the local timestep restrictions as worked out for the continuous problem in Section 6.2.2. For the special case of the linearly implicit Euler discretization we refer to the residual analysis given in the subsequent Section 6.4.

Here we concentrate on *implicit* one-step discretizations. In such discretizations the discrete system comprises a nonlinear algebraic system, which again brings up the question of local continuation. We will be interested to see, in which way some kind of nonlinear contractivity is inherited from the continuous initial value problem to various implicit one-step methods. In order to permit a comparison with Section 6.2.2, we will again assume that the coordinates have been chosen such that the local Jacobian matrix $A \approx F'(x_0)$ is already in Jordan canonical form—which implies that the canonical product and norm are identical to the Euclidean product and norm. To start with, we exemplify the formalism at a few simple cases.

Implicit Euler discretization. In each step of this discretization, we must solve the n algebraic equations

$$G(x, \tau) := x - x_0 - \tau F(x) = 0, \tag{6.22}$$

which represent a *homotopy* in $\mathbb{R}^n$ with embedding in terms of the stepsize τ—say $\tau \geq 0$. The *Newton-like iteration* for solving this system is

$$(I - \tau A)\Delta x_i = -(x_i - x_0 - \tau F(x_i)), \quad x_{i+1} = x_i + \Delta x_i, \tag{6.23}$$

where $\delta A := A - F'(x_0) \neq 0$ will be assumed.

Theorem 6.3 *Assumptions and notation as in Theorems 6.1 and 6.2 above. Then the Newton-like iteration (6.23) for the implicit Euler discretization converges to a unique solution for all stepsizes*

$$\tau \text{ unbounded}, \text{ if } \mu\hat{\tau} \leq -1,$$
$$\tau \leq \hat{\tau}\Psi_D(\mu\hat{\tau}), \text{ if } \mu\hat{\tau} > -1,$$

where

$$\Psi_D(s) := (1+s)^{-1}.$$

Proof. Once more, Theorem 2.6 is applied, here to the finite-dimensional homotopy (6.22). The Jacobian approximation $A \approx F'(x_0)$ leads to the approximation

$$I - \tau A =: M(x_0) \approx F'(x_0),$$

which is used in the definition of the affine covariant Lipschitz constant

$$\|M(x_0)^{-1}(F'(u) - F'(v))\| \leq \omega(\tau)\|u - v\|,$$

the first correction bound

$$\|\Delta x_0\| = \|M(x_0)^{-1}F(x_0)\| \leq \alpha(\tau)$$

and the approximation measure

$$\|M(x_0)^{-1}(M(x_0) - F'(x_0))\| \leq \bar{\delta}_0(\tau) < 1.$$

With these definitions, the modified Kantorovich condition here reads

$$\frac{\alpha\omega}{(1 - \bar{\delta}_o)^2} \leq \tfrac{1}{2}. \tag{6.24}$$

Upon using similar techniques as in the proof of Theorem 6.2 above, we come up with the estimates:

$$\alpha(\tau) := \tau L_0/(1 - \mu\tau), \quad \omega(\tau) := h L_2/(1 - \mu\tau), \quad \bar{\delta}_0(\tau) := \tau\delta_0/(1 - \mu\tau),$$

where the quantities L_0, L_2, δ_0 are the same as in Section 6.2.2. Insertion into condition (6.24) then yields, for $\mu\tau < 1$:

$$\frac{\tau}{1 - \mu\tau} \leq \hat{\tau}$$

or, equivalently,

$$\tau \leq \hat{\tau}/(1 + \mu\hat{\tau}).$$

This is the main statement of the theorem. Finally, note that for $\mu > 0$

$$\mu\tau \leq \mu\hat{\tau}/(1 + \mu\hat{\tau}) < 1,$$

which assures the above requirement. The case $\mu \leq 0$ is trivial.   □

The intriguing similarity of Theorems 6.2 and 6.3 is illustrated in Figure 6.3.

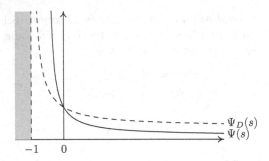

Fig. 6.3. Nonlinear contractivity inherited: function Ψ (continuous case) versus function Ψ_D (discrete case).

Implicit trapezoidal rule. This discretization requires the solution of the n in general nonlinear equations

$$G(x) := x - x_0 - \tfrac{1}{2}\tau \left(F(x) + F(x_0) \right) = 0 \,. \tag{6.25}$$

Standard Newton-like iteration leads to the steplength restriction

$$
\begin{aligned}
&\text{a)} \quad \tau \text{ unbounded, if } \mu\bar{\tau} \leq -\sqrt{2}\,, \\
&\text{b)} \quad \tau \leq \bar{\tau}\sqrt{2}\Psi_D \left(\frac{\mu\bar{\tau}}{\sqrt{2}} \right)\,.
\end{aligned} \tag{6.26}
$$

Observe that the pole of Ψ at $s = -1$ is *not preserved* here, so that nonlinear contractivity is not correctly inherited from the continuous case.

Implicit midpoint rule. This discretization leads to the n equations

$$G(x) := x - x_0 - \tau F \left(\frac{x + x_0}{2} \right) = 0 \,, \tag{6.27}$$

which, along similar lines of derivation, yields the stepsize bounds

$$
\begin{aligned}
&\text{a)} \quad \tau \text{ unbounded, if } \mu\bar{\tau} \leq -1\,, \\
&\text{b)} \quad \tau \leq 2\bar{\tau}\Psi_D(\mu\bar{\tau})\,.
\end{aligned} \tag{6.28}
$$

Here the pole is correctly preserved. Moreover, less restrictive bounds appear.

Summarizing, the implicit trapezoidal rule and the implicit midpoint rule have the same *linear* contractivity properties, but different *nonlinear* contractivity properties. From the nonlinear contractivity point of view, the implicit midpoint rule is clearly preferable. Both proofs are just along the lines of the proof for the implicit Euler method and therefore left as Exercise 6.1. Of course, one would really like to characterize the whole subclass of those one-step methods that preserve the pole exactly—a question left to future research.

6.4 Pseudo-transient Continuation for Steady State Problems

In this section we consider the case that the solution of a nonlinear system $F(x) = 0$ can be interpreted as *steady state* of a *dynamical system* of the kind

$$\dot{x} = F(x). \qquad (6.29)$$

Already from mere geometrical insight it is clear that such an approach will only work, if the fixed point of the dynamical system is *attractive* in a sufficiently large neighborhood. As an example, stiff integration towards a *hyperbolic* fixed point might come close to the fixed point for a while and run away afterwards. Exceptions will be possible only for a measure zero set of starting points x_0.

Dynamical invariants. This type of invariants occurs rather frequently in dynamical systems causing *singular* Jacobian matrices $F'(x)$ for all arguments x—which prohibits the application of standard Newton methods.

Example: mass conservation. Suppose the above ODE (6.29) describes some reaction kinetic model. Then mass conservation shows up as

$$e^T x(t) = e^T x_0,$$

where $e^T = (1, \ldots, 1)$. This implies

$$e^T \dot{x} = e^T F(x) \equiv 0, \quad x \in D \subset \mathbb{R}^n, \quad F(x) \neq 0.$$

By differentiation with respect to x we obtain

$$e^T F'(x) F(x) \equiv 0, \quad F(x) \neq 0$$

and hence every Jacobian has a zero eigenvalue with left eigenvector e. If we define the orthogonal projectors

$$P^\perp := \frac{1}{n} e e^T, \quad P = I - P^\perp,$$

then we can write equivalently

$$P^\perp F'(x) = 0. \qquad (6.30)$$

Of course, naive application of any standard Newton method would fail in this situation. In this *special* case, a modification is possible that makes the Newton methods nevertheless work—see, e.g., Exercise 6.3.

In the *general* case, however, more than one dynamical invariant exists, most of them unspecified or even unknown, so that (6.30) holds again, now for an *unknown* projector P such that

$$P^\perp \dot{x} = P^\perp F(x) = 0 \quad \Longrightarrow \quad P^\perp F'(x) = 0. \qquad (6.31)$$

Clearly, Newton methods cannot be modified to work without full knowledge about all dynamical invariants and are therefore bound to fail.

Fixed point iterations. In contrast to the other affine invariance classes of nonlinear problems, affine similarity also holds for *fixed point* iterations

$$\Delta x^k = x^{k+1} - x^k = \alpha F(x^k)$$

with a parameter α to be adapted. From equation (6.31) we see that such an iteration automatically realizes

$$P^\perp \Delta x = 0 \,.$$

Pseudo-transient continuation. The same property can be shown to hold for any linear combination of Newton and fixed point iteration. A popular technique is the so-called pseudo-transient continuation method

$$(I - \tau A)\, \Delta x = F(x_0), \quad x(\tau) = x_0 + \tau \Delta x \tag{6.32}$$

with timestep τ to be adapted and $A = F'(x_0)$ or a Jacobian approximation. The above iteration is just a special stiff discretization of the ODE (6.29), known as the *linearly implicit Euler discretization*.

Of course, in order to obtain the solution, we may directly solve the time dependent system (6.29) by any numerical stiff integrator up to the steady state. In what follows, however, we want to restrict our attention to the simple case of the linearly implicit Euler discretization.

6.4.1 Exact pseudo-transient continuation

We now want to study an iterative method for the numerical solution of the nonlinear System $F(x) = 0$ based on the linearly implicit Euler discretization (6.32). Throughout this section we will assume that we can evaluate an *exact Jacobian* $A = F'(x)$ and solve the linear system (6.32) by *direct* elimination techniques.

As worked out in detail in Section 6.1, the problem itself is invariant under affine similarity transformation, which would suggest some theoretical treatment in terms of canonical norms and inner products. Usual stiff integration focuses on the *accuracy of the solution* which naturally belongs to an affine covariant setting. For reasons of numerical realization, however, we need to study the convergence of the iteration in terms of its *residual* behavior—which leads to an *affine contravariant* setting. For that reason, we will need to replace the canonical norm $|\cdot|$ (see Section 6.1) by some Euclidean norm $\|\cdot\|$, possibly *scaled*. Accordingly $(\cdot, \cdot)$ will denote the Euclidean inner product, also possibly scaled.

Let $x(\tau)$ denote the *homotopy path* defined by (6.32) and starting at the point $x(0) = x_0$. Before we actually study the residual norm $\|F(x(\tau))\|$, the following auxiliary result will be helpful.

Lemma 6.4 *Notation as just introduced with $A \approx F'(x_0)$. Then the residual along the homotopy path $x(\tau)$ starting at x_0 satisfies*

$$F(x(\tau)) = (I - \tau A)^{-1} F(x_0) + \int_{\sigma=0}^{\tau} (F'(x(\sigma)) - A) (I - \sigma A)^{-2} F(x_0) d\sigma . \quad (6.33)$$

Proof. Taylor's expansion of the residual yields

$$
\begin{aligned}
F(x(\tau)) &= F(x_0) + \int_{\sigma=0}^{\tau} F'(x(\sigma)) \dot{x}(\sigma) d\sigma \\
&= F(x_0) + A(x(\tau) - x_0) + \int_{\sigma=0}^{\tau} (F'(x(\sigma)) - A) \dot{x}(\sigma) d\sigma .
\end{aligned}
$$

Upon differentiating the homotopy (6.32) with respect to τ, we obtain

$$(I - \tau A)\dot{x} = F(x_0) + A(x(\tau) - x_0) = F(x_0) + \tau A(I - \tau A)^{-1} F(x_0)$$

and therefore

$$\dot{x}(\tau) = (I - \tau A)^{-2} F(x_0),$$

which then readily leads to the result of the lemma.   □

Discussion of Lipschitz conditions. With the above representation at hand, the question is now how to formulate first and second order Lipschitz conditions in view of theoretical estimates. The switch from the canonical norms in Sections 6.2 and 6.3 to the Euclidean norm here implies changes in all our definitions of first and second order Lipschitz constants below. Needless to mention that we are bound to lose the nice property of affine similarity in all our characterizing quantities. Instead all of our estimates will now depend on the scaling of the residual (to be carefully handled).

First order Lipschitz condition: linear contractivity. We may employ (6.8) to define some one-sided Lipschitz constant ν. Recall, however, that due to dynamical invariants, zero eigenvalues will occur in the Jacobian, which implies that $\nu \geq 0$—just apply the definition (6.8) again. Therefore, in order to take care of dynamical invariants, we will restrict our attention to iterative corrections in the subspace

$$S_P = \{ u \in \mathbb{R}^n \mid P^{\perp} u = 0 \} .$$

Then the inequality

$$(u, Au) \leq \nu \|u\|^2, \quad u \in S_P$$

is equivalent to

$$(Pu, (PAP)Pu) \leq \nu \|Pu\|^2 .$$

With this modified definition, the case $\nu < 0$ may well happen even in the presence of dynamical invariants.

Since $\Delta x(\tau) \in S_P$, we may insert it into the above definition and obtain

$$\hat{\nu}(\tau) = \frac{(\Delta x, A \Delta x)}{\|\Delta x\|^2} \leq \nu. \qquad (6.34)$$

If we multiply equation (6.32) by Δx from the left, we obtain

$$
\begin{aligned}
\|\Delta x\|^2 &= \tau(\Delta x, A \Delta x) + (\Delta x, F(x_0)) \\
&= \tau \hat{\nu}(\tau)\|\Delta x\|^2 + (\Delta x, F(x_0)) \\
&\leq \tau \hat{\nu}(\tau)\|\Delta x\|^2 + \|\Delta x\|\|F(x_0)\| \\
&\leq \tau \nu \|\Delta x\|^2 + \|\Delta x\|\|F(x_0)\|,
\end{aligned}
$$

which then leads to the estimates

$$\|\Delta x\| \leq \frac{\|F(x_0)\|}{1 - \hat{\nu}\tau} \leq \frac{\|F(x_0)\|}{1 - \nu\tau}. \qquad (6.35)$$

Moreover, since

$$\Delta x(\tau) = F(x_0) + \mathcal{O}(\tau),$$

we also have

$$\hat{\nu}(0) = \frac{(F(x_0), AF(x_0))}{\|F(x_0)\|^2} \leq \nu. \qquad (6.36)$$

This quantity can be monitored even before the linear equation (6.32) is actually solved. It plays a key role in the residual reduction process, as shown in the following lemma.

Lemma 6.5 *Let $\hat{\nu}(0) < 0$ as defined in (6.36). Then there exists some $\tau^* > 0$ such that*

$$\|F(x(\tau))\| < \|F(x_0)\| \quad \text{and} \quad \hat{\nu}(\tau) < 0 \quad \text{for all} \quad \tau \in [0, \tau^*[.$$

Proof. By differentiating the residual norm with respect to τ we obtain

$$\frac{d}{d\tau}\|F(x(\tau))\|^2|_{\tau=0} = 2(F'(\cdot)^T F(\cdot), \dot{x}(\tau))|_{\tau=0}$$

$$= 2(F(x_0), AF(x_0)) = 2\hat{\nu}(0)\|F(x_0)\|^2 < 0.$$

Since both $F(x(\tau))$ and the norm $\| \cdot \|$ are continuously differentiable, there exists some nonvoid interval w.r.t. τ, wherein the residual norm decreases— compare the previous Lemma 3.2. The proof of the statement for $\hat{\nu}(\tau)$ uses the same kind of argument. $\qquad \square$

In other words: if at the given starting point x_0 the condition $\hat{\nu}(0) < 0$ is not satisfied, then the pseudo-transient continuation method based on residual reduction cannot be expected to work at all. Recall, however, the discussion at the end of Section 6.1 which pointed out that residual reduction is not coupled to canonical norm reduction.

Second order Lipschitz condition. Here we may start from the affine similar Lipschitz condition (6.15) and replace the canonical norm $|\cdot|$ therein by the Euclidean norm $\|\cdot\|$. Thus we take the fact into account that in the affine similar setting domain and image space transform in the same way.

Convergence analysis. With these preparations we are now ready to state our main result.

Theorem 6.6 *Notation as in the preceding Lemma 6.4, but with $A = F'(x_0)$ and partly $L_0 = \|F(x_0)\|$. Let dynamical invariants show up via the properties $F(x) \in S_P$. Assume the one-sided first order Lipschitz condition*

$$(u, Au) \leq \nu \|u\|^2 \quad for \quad u \in S_P, \quad \nu < 0$$

and the second order Lipschitz condition

$$\|(F'(x) - F'(x_0))u\| \leq L_2 \|x - x_0\| \|u\| . \tag{6.37}$$

Then the following estimate holds

$$\|F(x(\tau))\| \leq \left(1 + \frac{\frac{1}{2}L_0 L_2 \tau^2}{1 - \nu\tau}\right) \frac{\|F(x_0)\|}{1 - \nu\tau} . \tag{6.38}$$

From this, residual monotonicity

$$\|F(x(\tau))\| \leq \|F(x_0)\|$$

is guaranteed for all $\tau \geq 0$ satisfying the sufficient condition

$$\nu + (\tfrac{1}{2}L_0 L_2 - \nu^2)\tau \leq 0 . \tag{6.39}$$

Moreover, if

$$L_0 L_2 > \nu^2 , \tag{6.40}$$

then the theoretically optimal pseudo-timestep is

$$\tau_{opt} = \frac{|\nu|}{L_0 L_2 - \nu^2} \tag{6.41}$$

leading to a residual reduction

$$\|F(x(\tau))\| \leq \left(1 - \frac{\frac{1}{2}\nu^2}{L_0 L_2}\right) \|F(x_0)\| < \|F(x_0)\| .$$

Proof. We return to the preceding Lemma 6.4. Obviously, the first and the second right hand terms in equation (6.33) are independent. Upon recalling (6.35) for the first term, we immediately recognize that, in order to be able to prove residual reduction, we necessarily need the condition $\nu < 0$, which means $\nu = -|\nu|$ throughout the proof. For the second term we may estimate, again recalling (6.35),

$$
\int_{\sigma=0}^{\tau} \| \left(F'(x(\sigma)) - F'(x_0)\right)(I - \sigma A)^{-2} F(x_0)\| d\sigma
$$

$$
\leq L_2 \int_{\sigma=0}^{\tau} \|x(\sigma) - x_0\| \|(I - \sigma A)^{-2} F(x_0)\| d\sigma
$$

$$
\leq L_2 \int_{\sigma=0}^{\tau} \frac{\sigma \|F(x_0)\|^2}{(1 - \sigma\nu)^3} d\sigma
$$

$$
= \tfrac{1}{2} L_2 \|F(x_0)\|^2 \tau^2 (1 - \nu\tau)^{-2} .
$$

Combination of the two estimates then directly confirms (6.38), which we here write as

$$
\|F(x(\tau))\| \leq \alpha(\tau) \|F(x_0)\| ,
$$

in terms of

$$
\alpha(\tau) = \left(1 - \nu\tau + \tfrac{1}{2} L_0 L_2 \tau^2\right) / (1 - \nu\tau)^2 .
$$

Upon requiring $\alpha(\tau) \leq 1$, we obtain the equivalent sufficient condition (6.39). Finally, in order to find the optimal residual reduction, a short calculation shows that

$$
\dot{\alpha}(\tau) = \left(\nu + \frac{L_0 L_2 \tau}{1 - \nu\tau}\right) / (1 - \nu\tau)^2 .
$$

An interior minimum can arise only for $\dot{\alpha}(\tau) = 0$, which is equivalent to (6.41) under the condition (6.40). Insertion of τ_{opt} into the expression for $\alpha(\tau)$ then completes the proof. □

From the above condition (6.39) we may conclude: if

$$
\nu + \tfrac{1}{2}\sqrt{2 L_0 L_2} \leq 0 ,
$$

then τ is *unbounded* for local continuation. Obviously, this is the residual oriented *nonlinear contractivity* condition to be compared with the error oriented relation (6.20). (The difference in the prefactor just indicates that there we needed to show uniqueness in addition.) If

$$
\nu + \tfrac{1}{2}\sqrt{2 L_0 L_2} > 0 , \tag{6.42}
$$

then the pseudo-timestep is bounded according to

$$\tau \leq \frac{|\nu|}{\frac{1}{2}L_0 L_2 - |\nu|^2} \,.$$

Note that condition (6.40) is less restrictive than (6.42) so that either unbounded or bounded optimal timesteps may occur.

Adaptive (pseudo-)timestep strategy. In the spirit of the whole book we now want to derive an adaptive strategy based on the theoretical optimal pseudo-timestep (6.41), which we repeat for convenience

$$\tau_{opt} = \frac{|\nu|}{L_0 L_2 - \nu^2} \,.$$

The above expression can be rewritten in implicit form as

$$\tau_{opt} = \frac{|\nu|(1 - \nu\tau_{opt})}{L_0 L_2} \,. \tag{6.43}$$

In passing we note that from this representation we roughly obtain

$$\tau_{opt} \sim \frac{1}{L_0} = \frac{1}{\|F(x_0)\|} \,,$$

which gives some justification for a quite popular *heuristic strategy*: new timesteps are proposed on the basis of successful old ones via

$$\tau_{new} = \frac{\|F(x_{old})\|}{\|F(x_{new})\|} \, \tau_{old} \,. \tag{6.44}$$

For reference see, e.g., the recent paper [133] by C.T. Kelley and D.E. Keyes, where also a whole class of further heuristics is mentioned. A different approach is taken by S.B. Hazra, V. Schulz, J. Brezillon, and N. Gauger in [116] where in a fluid dynamical problem no overall timestep exists; this approach is not treated here.

Here, however, we want to exploit the structure of (6.43) in a different way by rewriting it in the form

$$\|\Delta x(\tau_{opt})\| L_2 \tau_{opt} \leq \frac{L_0 L_2}{1 - \nu\tau_{opt}} = |\nu| \,.$$

On this basis, we replace τ_{opt} by the upper bound

$$\bar{\tau}_{opt} = \frac{|\nu|}{L_2 \|\Delta x(\tau)\|} \geq \tau_{opt} \,.$$

So we are left with the task of identifying cheap computational estimates $[\nu] \leq \nu < 0, [L_2] \leq L_2$ to replace the unknown theoretical quantities ν, L_2. Once this is achieved, we can compute the corresponding pseudo-timestep

$$[\tau_{opt}] = \frac{\|[\nu]\|}{[L_2]\|\varDelta x(\tau)\|} \geq \bar{\tau}_{opt} \geq \tau_{opt}. \qquad (6.45)$$

As for the estimation of ν, we may exploit (6.34) in a double way: First, whenever

$$\|\varDelta x(\tau)\| \geq \|F(x_0)\|,$$

then we know that $\nu \geq [\nu] \geq 0$ is guaranteed, which means that we should terminate the iteration. Second, we may recognize that

$$[\nu]\tau = \hat{\nu}(\tau)\tau = \tau \frac{(\varDelta x, A\varDelta x)}{\|\varDelta x\|^2} = \frac{(\varDelta x, \varDelta x - F(x_0))}{\|\varDelta x\|^2} \leq \nu\tau \qquad (6.46)$$

gives us a quite cheap estimation formula for ν. As for the estimation of L_2, we may rearrange terms in the proof of Theorem 6.6 to obtain

$$\|F(x(\tau)) - \varDelta x(\tau)\| \leq \int_{\sigma=0}^{\tau} \| (F'(x(\sigma)) - F'(x_0)) (I - \sigma A)^{-2} F(x_0)\| d\sigma$$

$$\leq L_2 \int_{\sigma=0}^{\tau} \|x(\sigma) - x_0\| \|(I - \sigma A)^{-1} \varDelta x(\sigma)\| d\sigma.$$

If we approximate the integral by the trapezoidal rule, we arrive at

$$\|F(x(\tau)) - \varDelta x(\tau)\| \leq \tfrac{1}{2}L_2\|\varDelta x(\tau)\|^2\tau^2/(1 - \nu\tau) + \mathcal{O}(\tau^4)$$

$$\leq \tfrac{1}{2}L_2\|\varDelta x(\tau)\|^2\tau^2 + \mathcal{O}(\tau^4).$$

Note that already the approximation term, ignoring the $\mathcal{O}(\tau^4)$ term, gives rise to the upper bound

$$\|F(x(\tau)) - \varDelta x(\tau)\| \leq \tfrac{1}{2}L_2\|F(x_0)\|^2\tau^2/(1 - \nu\tau)^2,$$

which is the basis of the derivation of Theorem 6.6. Hence, we may well regard

$$[L_2] = \frac{2\|F(x(\tau)) - \varDelta x\|}{\tau^2\|\varDelta x\|^2} \leq L_2 + \mathcal{O}(\tau^2)$$

as a suitable computational estimate for L_2. Upon collecting all above estimates and inserting them into (6.45), we arrive at the following pseudo-timestep suggestion

$$[\tau_{opt}] = \frac{|(\varDelta x(\tau), F(x_0) - \varDelta x(\tau))|}{2\|\varDelta x(\tau)\| \|F(x(\tau)) - \varDelta x(\tau)\|}\, \tau.$$

On this basis, an *adaptive* τ-*strategy* can be realized in the usual two modes, a *correction* and a *prediction* strategy: If in the iterative step $x_0 \longrightarrow x(\tau)$ the residual norm does *not* decrease, then the actual step size τ is replaced by

$[\tau_{opt}] < \tau$; if the residual norm decreases successfully, then the next step is started with the trial value $[\tau_{opt}]$. Finally, note that the above strategy will just terminate, if the steady state to be computed is not attractive in the (residual) sense that $[\nu] \geq 0$. For $[\nu] \to 0^-$, the suggested stepsize behaves like $[\tau_{opt}] \to 0^+$ - as to be reasonably required.

6.4.2 Inexact pseudo-transient continuation

Suppose the linear system (6.32) is so large that we cannot but solve it *iteratively* ($i = 0, 1, \dots$):

$$(I - \tau A)\delta x_i = F(x_0) - r_i, \quad x_i(\tau) = x_0 + \tau \delta x_i . \qquad (6.47)$$

Herein r_i represents the iterative linear residual, δx_i the corresponding inexact correction, and $x_i(\tau)$ the approximate homotopy path instead of the exact $x(\tau)$. To start the iteration, let $x_0(\tau) = x_0$ so that $\delta x_0 = 0$ and $r_0 = F(x_0)$.

If we want to *minimize the residuals* within each iterative step, we are directly led to GMRES—see Section 1.4 and the notation therein. In terms of the Euclidean norm $\| \cdot \|$ we define the approximation quantities

$$\eta_i := \frac{\|r_i\|}{\|F(x_0)\|} < 1 \quad for \quad i = 1, 2, \dots .$$

Recall that GMRES assures $\eta_{i+1} \leq \eta_i$, in the generic case even $\eta_{i+1} < \eta_i$. Moreover, due to the residual minimization property and $r_0 = F(x_0)$, we have

$$\|F(x_0) - r_i\|^2 = (1 - \eta_i^2)\|F(x_0)\|^2 .$$

In the present context of pseudo-transient continuation, we may additionally observe that GMRES realizes the special structure

$$\delta x_i(\tau) = V_i z_i(\tau) \quad and \quad H_i(\tau) = (I_i, 0)^T + \tau \hat{H}_i .$$

Herein V_i is just the orthonormal basis of the Krylov space $\mathcal{K}_i(r_0, A)$ and $\hat{H}_i$ is a Hessenberg matrix like $H_i(\tau)$, but also independent of τ. On this basis, we see that dynamical invariants are correctly treated throughout the iteration. The proof of these properties is left as Exercise 6.5. The special structure permits computational savings when the same system is solved for different pseudo-timesteps τ.

Convergence analysis. As before, we first analyze the convergence behavior theoretically as a basis for the subsequent derivation of an adaptive algorithm, which here will have to include the matching of inner and outer iteration. For this purpose we need to modify Lemma 6.4.

Lemma 6.7 *Notation as in Lemma 6.4 with $A \approx F'(x_0)$. Then the residual along the approximate homotopy path $x_i(\tau)$ starting at x_0 satisfies*

$$F(x(\tau)) - r_i = (I - \tau A)^{-1} (F(x_0) - r_i)$$

$$+ \int_{\sigma=0}^{\tau} (F'(x_i(\sigma)) - A) (I - \sigma A)^{-2} (F(x_0) - r_i) \, d\sigma \,.$$

Proof. The proof is just an elementary modification of the proof of Lemma 6.4. For example, if we differentiate the homotopy (6.47) with respect to τ, we now obtain

$$\dot{x}_i(\tau) = (I - \tau A)^{-2} (F(x_0) - r_i) \,.$$

Further details can be omitted.   □

Theorem 6.8 *Notation as in the preceding Lemma 6.7. Let $A = F'(x_0)$ and partly $\tilde{L}_0 = \sqrt{1 - \eta_i^2} \|F(x_0)\|$. Assume that dynamical invariants show up via the properties $F(x) \in S_P$. Then, with the Lipschitz conditions*

$$(u, Au) \leq \nu \|u\|^2, \quad \nu < 0, \quad for \quad u \in S_P$$

and

$$\| (F'(x) - F'(x_0)) u \| \leq L_2 \|x - x_0\| \|u\| \,,$$

the estimates

$$\|F(x(\tau)) - r_i\| \leq \left(1 + \frac{\frac{1}{2}\tilde{L}_0 L_2 \tau^2}{1 - \nu\tau} \right) \frac{\|F(x_0) - r_i\|}{1 - \nu\tau}$$

and

$$\|F(x(\tau))\| \leq \left(\eta_i + \frac{\sqrt{1 - \eta_i^2}}{1 - \nu\tau} \left(1 + \frac{\frac{1}{2}\tilde{L}_0 L_2 \tau^2}{1 - \nu\tau} \right) \right) \|F(x_0)\|$$

hold. Let

$$s(\eta_i) := \sqrt{\frac{1 - \eta_i}{1 + \eta_i}} > \tfrac{1}{2} \quad or, \; equivalently, \quad \eta_i < \tfrac{3}{5} \,. \tag{6.48}$$

Then residual monotonicity

$$\|F(x(\tau))\| \leq \|F(x_0)\|$$

is guaranteed for all $\tau \geq 0$ satisfying the sufficient condition

$$1 - s(\eta_i) + (2s(\eta_i) - 1)\nu\tau + \left(\tfrac{1}{2}\tilde{L}_0 L_2 - s(\eta_i)\nu^2 \right) \tau^2 \leq 0 \,. \tag{6.49}$$

Assume further that

$$\tfrac{4}{5}L_0L_2 > \nu^2 \tag{6.50}$$

and that GMRES *has been continued until*

$$\eta_i + \sqrt{1 - \eta_i^2} < 1 + \frac{\tfrac{1}{2}\nu^2}{L_0L_2}. \tag{6.51}$$

Then the theoretically optimal pseudo-timestep is

$$\tau_{opt} = \frac{|\nu|}{\tilde{L}_0L_2 - \nu^2} \tag{6.52}$$

leading to the estimate

$$\|F(x(\tau)) - r_i\| \le \left(1 - \frac{\tfrac{1}{2}\nu^2}{\tilde{L}_0L_2}\right)\|F(x_0) - r_i\| < \|F(x_0) - r_i\|$$

and to the residual reduction

$$\|F(x(\tau))\| \le \left(\eta_i + \sqrt{1 - \eta_i^2} - \frac{\tfrac{1}{2}\nu^2}{L_0L_2}\right)\|F(x_0)\| < \|F(x_0)\|. \tag{6.53}$$

Proof. We return to the preceding Lemma 6.7 and modify the proof of Theorem 6.6 carefully step by step. For example, the second order term may be estimated as

$$\int_{\sigma=0}^{\tau} \|\left(F'(x(\sigma)) - F'(x_0)\right)(I - \sigma A)^{-2}(F(x_0) - r_i)\|d\sigma$$
$$\le \tfrac{1}{2}L_2\|F(x_0) - r_i\|^2\tau^2(1 - \nu\tau)^{-2}.$$

Combination of estimates then directly confirms

$$\|F(x_i(\tau)) - r_i\| \le \bar{\alpha}_i(\tau)\|F(x_0) - r_i\|$$

in terms of

$$\bar{\alpha}_i(\tau) = \left(1 - \nu\tau + \tfrac{1}{2}\tilde{L}_0L_2\tau^2\right)/(1 - \nu\tau)^2,$$

from which we obtain

$$\|F(x_i(\tau))\| \le \alpha_i(\tau)\|F(x_0)\|$$

with

$$\alpha_i(\tau) = \eta_i + \sqrt{1 - \eta_i^2}\,\bar{\alpha}_i(\tau).$$

Upon requiring $\alpha(\tau) \le 1$, we have

$$\eta_i + \sqrt{1 - \eta_i^2}\,\bar{\alpha}_i(\tau) \le 1,$$

which is equivalent to

$$\bar{\alpha}_i(\tau) \le s(\eta_i) \le 1. \tag{6.54}$$

From this, we immediately verify the sufficient condition (6.49). Note that $2s - 1 > 0$, which is just condition (6.48), is necessary to have at least one negative term in the left hand side of (6.49).

Finally, in order to find the *optimal* residual reduction, a short calculation shows that

$$\dot{\alpha}(\tau) = \sqrt{1 - \eta_i^2}\,\dot{\bar{\alpha}}(\tau) = \frac{\sqrt{1 - \eta_i^2}}{(1 - \nu\tau)^2}\left(\nu + \frac{\tilde{L}_0 L_2 \tau}{1 - \nu\tau}\right).$$

For the interior minimum we require $\dot{\bar{\alpha}}(\tau) = 0$, which is equivalent to (6.52) under the condition (6.50), where

$$\sqrt{1 - \eta_i^2} \ge \sqrt{1 - (\tfrac{3}{5})^2} = \tfrac{4}{5}$$

has been used. Insertion of τ_{opt} into the expression for $\alpha(\tau)$ then leads to

$$\|F(x_i(\tau)) - r_i\| \le \left(1 - \frac{\frac{1}{2}\nu^2}{\tilde{L}_0 L_2}\right)\|F(x_0) - r_i\|$$

and eventually to (6.53). In order to assure an actual *residual reduction*, condition (6.54) must also hold for τ_{opt}, which confirms the necessary condition (6.51). Note that the scalar function $\eta_i + \sqrt{1 - \eta_i^2}$ is monotonically increasing for $\eta_i < \frac{1}{2}\sqrt{2} \approx 0.7$, hence also for $\eta_i < \frac{3}{5} = 0.6$. Therefore GMRES may be just continued until the relation (6.51) is satisfied. This completes the proof.
□

Adaptive (pseudo-)timestep strategy. We follow the line of the derivation for the exact pseudo-transient continuation in Section 6.4.1. For convenience, we repeat the expression

$$\tau_{opt} = \frac{|\nu|}{\tilde{L}_0 L_2 - \nu^2},$$

which can be rewritten in implicit form as

$$\tau_{opt} = \frac{|\nu|(1 - \nu\tau_{opt})}{\tilde{L}_0 L_2}.$$

Recall now that

$$\|\delta x_i(\tau)\| = \|(I - \tau A)^{-1}(F(x_0) - r_i)\| \le \frac{\|F(x_0) - r_i\|}{1 - \nu\tau} = \frac{\tilde{L}_0}{1 - \nu\tau}, \tag{6.55}$$

which directly implies

$$\bar{\tau}_{opt} = \frac{|\nu|}{L_2 \|\delta x_i(\tau)\|} \geq \tau_{opt}.$$

So we need to compute the pseudo-timestep

$$[\tau_{opt}] = \frac{|[\nu]|}{[L_2] \|\delta x_i(\tau)\|} \geq \bar{\tau}_{opt} \geq \tau_{opt} \tag{6.56}$$

in terms of the appropriate estimates of the unknown theoretical quantities ν, L_2.

As for the estimation of ν, we exploit (6.55). Whenever

$$\|\delta x_i(\tau)\| \geq \|F(x_0) - r_i\|,$$

then we know that $\nu \geq 0$ is guaranteed and the iteration must be terminated. Moreover, the relation

$$[\nu]\tau = \tau \frac{(\delta x_i, A\delta x_i)}{\|\delta x_i\|^2} = \frac{(\delta x_i, \delta x_i - F(x_0) + r_i)}{\|\delta x_i\|^2} \leq \nu\tau$$

supplies an estimation formula for ν. As for the estimation of L_2, we revisit Lemma 6.7 to obtain

$$\|F(x_i(\tau)) - r_i - \delta x_i(\tau)\|$$

$$\leq \int_{\sigma=0}^{\tau} \| (F'(x_i(\sigma)) - F'(x_0)) (I - \sigma A)^{-2} (F(x_0) - r_i) \| d\sigma$$

$$\leq L_2 \int_{\sigma=0}^{\tau} \|x_i(\sigma) - x_0\| \|(I - \sigma A)^{-1} \delta x_i(\sigma)\| d\sigma$$

$$\leq \tfrac{1}{2} L_2 \tau^2 \frac{\tilde{L}_0^2}{(1 - \nu\tau)^2}.$$

If we approximate the above integral by the trapezoidal rule (*before* using the final estimate), we arrive at

$$\|F(x_i(\tau)) - r_i - \delta x_i(\tau)\| \leq \tfrac{1}{2} L_2 \|\delta x_i(\tau)\|^2 \tau^2 / (1 - \nu\tau) + \mathcal{O}(\tau^4)$$

$$\leq \tfrac{1}{2} L_2 \tau^2 \|\delta x_i(\tau)\|^2 + \mathcal{O}(\tau^4).$$

Already the first right hand term gives rise to the above upper bound—compare (6.55). Hence, as in Section 6.4.1, we will pick

$$[L_2] = \frac{2 \|F(x_i(\tau)) - r_i - \delta x_i(\tau)\|}{\tau^2 \|\delta x_i(\tau)\|^2} \leq L_2 + \mathcal{O}(\tau^2)$$

as computational estimate for L_2. Upon inserting the two derived estimates into (6.56), we arrive at the pseudo-timestep estimate

$$[\tau_{opt}] = \frac{|(\delta x_i(\tau), F(x_0) - r_i - \delta x_i(\tau))|}{2\|\delta x_i(\tau)\|\|F(x_i(\tau)) - r_i - \delta x_i(\tau)\|} \, \tau \, .$$

On this basis, an *adaptive τ-strategy* can again be realized as in the case of the *exact* pseudo-transient continuation method.

Finally, we want to mention that the iterative version of the pseudo-transient continuation method still works in the case of *unbounded* timestep. To see this, just rewrite (6.47) in the form

$$\left(\frac{1}{\tau}I - A\right)(x_i(\tau) - x_0) = F(x_0) - r_i \, .$$

Herein $\tau \to \infty$ is possible leaving $x_i(\tau) - x_0$ well-defined even in the presence of singular Jacobian A caused by *dynamical invariants*: This is due to the fact that GMRES (like any Krylov solver) keeps the nullspace components of the solution unchanged, so that $P^\perp(x_i(\infty) - x_0) = 0$ is guaranteed throughout the iteration.

Preconditioning. If we multiply the nonlinear system by means of some nonsingular matrix M from the left as

$$M\dot{x} = MF(x) = 0 \, ,$$

then GMRES will have to work on the preconditioned residuals Mr_i and adaptivity must be based on norms $\|M \cdot \|$. Note that it is totally unclear, whether such a transformation leads to the necessary linear contractivity result $\nu(MA) < 0$ for the preconditioned system with $A \approx F'(x^0)$.

Preconditioning from the right will just influence the convergence speed of GMRES without changing the above derived adaptivity devices.

Matrix-free realization. Sometimes the inexact pseudo-continuation method is realized in a matrix-free variant using the first order approximation

$$A\delta x \approx F(x + \delta x) - F(x) \, .$$

A numerically stable realization will use *automatic differentiation* as suggested by A. Griewank [112].

Exercises

Exercise 6.1 Prove the results (6.26) for the implicit trapezoidal rule (6.25) and (6.28) for the implicit midpoint rule (6.27).

Exercise 6.2 Consider the linearly implicit Euler (LIE) discretization for the ODE system $y' = f(y)$, which reads (for $k = 0, 1, \ldots$)

$$y_{k+1} = y_k + (I - \tau A)^{-1} f(y_k),$$

where $A = f_y(y_k)$. This scheme is usually monitored to run in some 'neighborhood' of the implicit Euler (IE) discretization

$$F(y) = y - y_k - \tau f(y) = 0.$$

For this purpose the LIE is interpreted as the first iterate of IE and local contraction within that IE scheme is required. Most LIE codes realize this requirement via an error oriented criterion introduced in [66]. Here we want to look at a residual based variant due to [120].

a) On the basis of the residual based Newton-Mysovskikh theorem derive a computational monitor that is cheap to evaluate.

b) Of which order $O(\tau^s)$ is this contraction factor? Derive an adaptive step-size procedure on that basis.

   Hint: Interpret the method as a continuation method with embedding parameter τ.

c) Compare the error oriented and the residual based variant in terms of computational amount per discretization step.

d) *Optional:* Implement the two variants within an adaptive integrator (like LIMEX) and compare them at several ODE examples.

Exercise 6.3   Consider the system of n nonlinear differential equations (with time variable t)

$$\dot{x} = F(x), \quad x(0) = x^0$$

modeling some process $x(t)$. Assume that there exists a *dynamical invariant* (such as mass conservation) of the form

$$e^T x(t) = e^T x^0, \quad e^T = (1, \ldots, 1) \in \mathbb{R}^n.$$

In many cases, one is only interested in a steady state solution $x^* = x(\infty)$ defined by

$$F(x^*) = 0.$$

Since, in general, x^* will depend on the initial value x^0, uniqueness of the solution is not guaranteed.

a) Show that the Jacobian $F'(x^*)$ is singular, which makes a naive application of Newton methods impossible.

b) As a remedy, consider the iterative method

$$\begin{bmatrix} F'(x^k) \\ e^T \end{bmatrix} \Delta x^k = - \begin{bmatrix} F(x^k) \\ 0 \end{bmatrix}, \quad x^{k+1} := x^k + \Delta x^k$$

started by some initial guess x^0. Show that the thus produced iterates satisfy

$$e^T x^k = e^T x^0 .$$

What kind of restriction is necessary for the choice of x^0?

c) Develop a program to treat the described problem type—test examples may come from chemical kinetics, where mass conservation is often tacitly assumed without explicitly stating it.

Exercise 6.4   Consider the pseudo-transient continuation method with *approximate* Jacobian $A \approx F'(x^0)$. Upon using the notation of Section 6.4.1 and, in addition,

$$\|(A - F'(x^0))u\| \le \delta |\nu| \|u\| , \quad \delta < 1 ,$$

prove a variant of Theorem 6.6, containing results on the residual descent and the optimal pseudo-timestep.

Check: For $\nu < 0$, the optimal timestep τ_{opt} comes out to be

$$\tau_{opt} = \frac{(1 - \delta)|\nu|}{L_0 L_2 - (1 - \delta)\nu^2}$$

in the terms defined—assuming, of course, that the denominator is positive.

Exercise 6.5   How can the iterative linear solver GMRES be optimally adapted to pseudo-transient continuation? Design a special version, which saves computing time and storage.

7 ODE Boundary Value Problems

In this chapter, we consider *two-point boundary value problems* (BVPs) for *ordinary differential equations* (ODEs)

$$y' = f(y) \,, \quad f \in C^2 \,, \quad r(y(a), y(b)) = 0 \,, \quad r \in C^2 \,,$$

wherein both the right side f (autonomous for ease of writing) and the boundary conditions r are of dimension n and may be nonlinear. Algorithms for the solution of such problems can be grouped in two classes: initial value methods and global discretization methods. The presentation and notation here closely relates to Chapter 8 in the textbook [71].

Initial value methods. This kind of methods transforms the BVP into a sequence of *initial value problems* (IVPs), which are solved by means of numerical integrators. The most prominent method of this type is the *multiple shooting method*, which is a good choice only for problems, wherein a well-conditioned IVP direction exists, i.e. for so-called *timelike* BVPs. The name comes from the fact that in this problem class the independent variable t typically represents a time (or time related) variable. As a rule, there exists no generalization to boundary value problems for partial differential equations (PDEs).

Global discretization methods. Conceptually, this kind of BVP methods does not depend on any preferable direction and is therefore also applicable to cases, where a well-conditioned IVP direction does not exist, i.e. to so-called *spacelike* boundary value problems. In this type of BVP the independent variable t typically represents a space (or space related) variable, which implies that a generalization to BVPs for PDEs is possible. Such methods include, e.g., *finite difference* and *collocation methods*.

In Section 7.1, the realization of Newton and discrete continuation methods within the standard *multiple shooting* approach is elaborated. Gauss-Newton methods for parameter identification in ODEs are discussed in Section 7.2. For periodic orbit computation, Section 7.3 presents Gauss-Newton methods, both in the shooting approach (Sections 7.3.1 and 7.3.2) and in a collocation approach based on Fourier series (Galerkin-Urabe method in Section 7.3.3).

In Section 7.4 we concentrate on polynomial *collocation methods*, which have reached a rather mature status including affine covariant Newton methods. In Section 7.4.1, the possible discrepancy between discrete and continuous solutions is studied including the possible occurrence of so-called 'ghost solutions' in the nonlinear case. On this basis, the realization of *quasilinearization* seems to be preferable in combination with collocation. The following Section 7.4.2 is then devoted to the key issue that quasilinearization can be interpreted as an *inexact Newton method in function space*: the approximation errors in the infinite dimensional setting just replace the inner iteration errors arising in the finite dimensional setting. With this insight, an adaptive multilevel control of the collocation errors can be realized to yield an adaptive inexact Newton method in function space—which is the bridge to adaptive Newton multilevel methods for PDEs (compare Section 8.3).

BIBLIOGRAPHICAL NOTE. Affine invariant global Newton methods—now called *affine covariant* Newton methods—have first been developed in the frame of multiple shooting techniques by P. Deuflhard [60, 62, 61]. Therein they have turned out to be of crucial importance for the overall performance, especially in challenging real life optimal control problems. These Newton techniques have then quickly been adopted by U.M. Ascher and R.D. Russell within their adaptive collocation methods [8]—with comparable success, see also their textbook [9]. They have also played an important role within parameter identification algorithms and their convergence analysis as worked out by H.G. Bock [29, 31, 32] since 1981.

7.1 Multiple Shooting for Timelike BVPs

In this approach the interval $[a, b]$ is subdivided into a partition

$$\Delta = \{a = t_1 < t_2 < \cdots < t_m = b\}, \quad m > 2 \,.$$

Let $x_j \in \mathbb{R}^n$, $j = 1, \ldots, m$ denote estimates of the unknown values at the nodes t_j. Then, in terms of the flow Φ, we may define those $m - 1$ subtrajectories

$$y_j(t) = \Phi^{t, t_j} x_j, \quad t \in [t_j, t_{j+1}], \quad j = 1, \ldots, m - 1$$

that solve $(m-1)$ independent IVPs. The situation is illustrated in Figure 7.1. For the solution of the problem the sub-trajectories have to be joined continuously and hence at the intermediate nodes the n *continuity conditions*

$$F_j(x_j, x_{j+1}) = \Phi^{t_{j+1}, t_j} x_j - x_{j+1} = 0, \quad j = 1, \ldots, m - 1$$

have to hold.

In addition, we have to satisfy the n *boundary conditions*

$$F_m(x_1, x_m) = r(x_1, x_m) = 0.$$

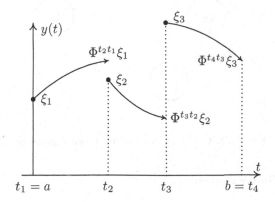

Fig. 7.1. Multiple shooting ($m = 4$).

The overall full nm-dimensional system is written in the form

$$x = \begin{pmatrix} x_1 \\ \vdots \\ x_m \end{pmatrix} \in \mathbb{R}^{n \cdot m}, \quad F(x) = \begin{pmatrix} F_1(x_1, x_2) \\ \vdots \\ F_m(x_1, x_m) \end{pmatrix} = 0. \tag{7.1}$$

This nonlinear system has a *cyclic* block structure as indicated in Figure 7.2. For the solution of the above cyclic nonlinear system (7.1) we compute the *ordinary Newton correction* as usual by solving the linear system

$$F'(x^k)\Delta x^k = -F(x^k), \quad x^{k+1} = x^k + \Delta x^k, \quad k = 0, 1, \ldots .$$

The corresponding Jacobian matrix has the cyclic block structure

$$J = F'(x) = \begin{bmatrix} G_1 & -I & & & \\ & \ddots & & \ddots & \\ & & G_{m-1} & & -I \\ A & & & & B \end{bmatrix}.$$

Herein the matrices A, B are the derivatives of the boundary conditions r with respect to the boundary values $(x(a), x(b)) = (x_1, x_m)$. The propagation matrices G_j on each of the sub-intervals, also called Wronskian matrices, read

$$G_j = \frac{\partial \Phi^{t_{j+1}, t_j} x_j}{\partial x_j}, \quad j = 1, \ldots, m - 1.$$

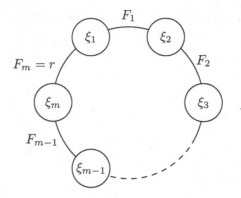

Fig. 7.2. Cyclic system of nonlinear equations

7.1.1 Cyclic linear systems

The block structure of the Jacobian matrix gives rise to a *block cyclic linear system* of the following kind:

$$G_1 \Delta x_1 \quad -\Delta x_2 \qquad\qquad\qquad\qquad = -F_1$$

$$\ddots \qquad\qquad \ddots$$

$$G_{m-1} \Delta x_{m-1} \quad -\Delta x_m \quad = -F_{m-1}$$

$$A \Delta x_1 \qquad\qquad\qquad\qquad +B \Delta x_m \quad = -F_m = -r \,.$$

If this linear system were just solved by some (sparse) direct elimination method, then global Newton methods as described in the preceding sections could be directly taken off the shelf.

For timelike BVPs, however, there exists an efficient alternative option, which opens the door to the construction of interesting specific Gauss-Newton methods. This option dates back to a suggestion of J. Stoer and R. Bulirsch [187]. It is often called *condensing* algorithm, since it requires only the decomposition of a 'condensed' (n, n)-matrix E instead of that of the total Jacobian (nm, nm)-matrix J. In order to convey the idea, we present the idea first for the case $m = 3$:

$$(1) \quad G_1 \Delta x_1 \quad -\Delta x_2 \qquad\qquad = \quad -F_1$$

$$(2) \qquad\qquad G_2 \Delta x_2 \quad -\Delta x_3 \quad = \quad -F_2$$

$$(3) \quad A \Delta x_1 \qquad\qquad\qquad +B \Delta x_3 \quad = \quad -r \,.$$

First we multiply (1) by G_2 from the left and add the result

$$G_2 G_1 \Delta x_1 - G_2 \Delta x_2 = -G_2 F_1$$

to equation (2). This gives

$$G_2 G_1 \Delta x_1 - \Delta x_3 = -(F_2 - G_2 F_1),$$

which after multiplication by B yields

$$B G_2 G_1 \Delta x_1 - B \Delta x_3 = -B(f_2 - G_2 F_1).$$

Finally, by addition of equation (3) it follows that

$$\underbrace{(A + B G_2 G_1)}_{=E} \Delta x_1 = \underbrace{-r - B(F_2 - G_2 F_1)}_{=-u}.$$

Hence, in the general case $m \geq 2$ we obtain the following algorithm:

a)   Evaluate by recursion over $j = 1, \ldots, m - 1$

$$E := A + B G_{m-1} \cdots G_1,$$

$$u := r + B \left[F_{m-1} + G_{m-1} F_{m-2} + \cdots + G_{m-1} \ldots G_2 F_1 \right].$$

b)   Solve the linear $(n, n) -$ system   (7.2)

$$E \Delta x_1 = -u.$$

c)   Execute the explicit recursion

$$\Delta x_{j+1} := G_j \Delta x_j + F_j, \quad j = 1, \ldots, m - 1.$$

The memory required by this algorithm is essentially $m \cdot n^2$. The computational cost is dominated by the accumulation of the matrix E as an $(m - 1)$-fold product of (n, n)-matrices. Together with the decomposition of E this results in a cost of $O(m \cdot n^3)$ operations, where terms of order $O(n^2)$ have been neglected as usual.

The large sparse Jacobian matrix J and the small matrix E are closely connected as can be seen by the following lemma.

Lemma 7.1 *Notation as just introduced. Define* $W_j = G_{m-1} \cdots G_j$ *and* $E := A + B W_1$. *Then*

$$\det(J) = \det(E).$$ (7.3)

Moreover, if E is nonsingular, one has the decomposition

$$LJR = S, \quad J^{-1} = R S^{-1} L$$ (7.4)

in terms of the block matrices

$$
L := \begin{bmatrix} BW_2 \dots & B, & I \\ -I & & \\ & \ddots & \\ & & -I, & 0 \end{bmatrix}, \qquad R^{-1} := \begin{bmatrix} I & & \\ & -G_1, I & \\ & & \ddots \ddots \\ & & -G_{m-1}, I \end{bmatrix},
$$

$$
S := \mathrm{diag}(E, I, \dots, I), \qquad S^{-1} = \mathrm{diag}(E^{-1}, I, \dots, I).
$$

Proof. The decomposition (7.4) is a direct formalization of the above block Gaussian elimination. The determinant relation (7.3) follows from

$$
\det(L) = \det(R) = \det(R^{-1}) = 1.
$$

With

$$
\det(J) = \det(S)
$$

the proof is completed. □

Interpretation. The matrix E is an approximation of the special *sensitivity matrix*

$$
E(a) = \frac{\partial r}{\partial y_a} = A + BW(b, a)
$$

corresponding to the BVP as a whole. Herein $W(\cdot, \cdot)$ denotes the propagation matrix of the variational equation. Generically this means that, whenever the underlying BVP has a locally uniqueness solution, a locally unique solution $x^* = (x_1^*, \dots, x_m^*)$ is guaranteed—independent of the partitioning Δ.

Separable linear boundary conditions. This case arises when part of the boundary conditions fix part of the components of x_1 at $t = a$ and part of the components of x_m at $t = b$. The situation can be conveniently described in terms of certain projection matrices P_a, $\bar{P}_a$, P_b, $\bar{P}_b$ such that

$$
\bar{P}_a A = P_a, \quad \bar{P}_a B = 0,
$$
$$
\mathrm{rank}(\bar{P}_a) = \mathrm{rank}(P_a) = n_a < n,
$$

$$
\bar{P}_b B = P_b, \quad \bar{P}_b A = 0,
$$
$$
\mathrm{rank}(\bar{P}_b) = \mathrm{rank}(P_b) = n_b < n
$$

with $n_a + n_b \leq n$. Of course, we will choose initial guesses x_1^0, x_m^0 for the Newton iteration so that the separable boundary conditions

$$
\bar{P}_a r = 0, \quad \bar{P}_b r = 0
$$

automatically hold. Then the linearization of these conditions

$$A\Delta x_1 + B\Delta x_m = -r$$

directly implies

$$P_a \Delta x_1 = 0 \,, \quad P_b \Delta x_m = 0 \,.$$

Consequently, the variables $P_a x_1$ and $P_b x_m$ can be seen to satisfy

$$P_a x_1 = P_a x_1^0 \,, \quad P_b x_m = P_b x_m^0$$

throughout the iteration. This part can be realized independent of any elimination method by carefully analyzing the sparsity pattern of the matrices A and B within the algorithm. As a consequence, the sensitivity matrix E also has the projection properties

$$\bar{P}_a E = 0, \quad \bar{P}_b E = 0, \qquad E P_a = 0, \quad E P_b = 0 \,.$$

Iterative refinement sweeps. The block Gaussian elimination technique (7.2) seems to be highly efficient in terms of memory and computational cost. A closer look on its numerical stability, however, shows that the method becomes sufficiently robust only with the addition of some special iterative refinement called iterative refinement sweeps. We will briefly sketch this technique and work out its consequences for the construction of Newton and Gauss-Newton methods—for details see the original paper [70] by P. Deuflhard and G. Bader.

Let $\nu = 0, 1, \ldots$ be the indices of the iterative refinement steps. In lieu of the exact Newton corrections Δx_j the block Gaussian elimination will supply certain error carrying corrections $\Delta \tilde{x}_j^\nu$ so that iterative refinement will produce nonvanishing differences

$$dx_j^\nu \approx \Delta \tilde{x}_j^{\nu+1} - \Delta \tilde{x}_j^\nu \,.$$

In the present framework we might first consider the following algorithm:

(a) $du^\nu = dr^\nu + B\left[dF_{m-1}^\nu + G_{m-1}dF_{m-2}^\nu + \cdots + G_{m-1}\cdots G_2 dF_1^\nu \right]$,

(b) $E dx_1^\nu = -du^\nu$,

(c) $dx_{j+1}^\nu = G_j dx_j^\nu + dF_j^\nu \quad j = 1, \ldots, m.$

However, as shown by the detailed componentwise round-off error analysis in [70], this type of iterative refinement is only guaranteed to converge under the sufficient condition

$$\varepsilon(m-1)(2n+m-1)\kappa[a,b] \ll 1 \,,$$

wherein ε denotes the relative machine precision and $\kappa[a,b]$ the IVP condition number over the whole interval $[a,b]$—to be associated with single instead of multiple shooting. This too restrictive error growth can be avoided by a modification called *iterative refinement sweeps*. As before, this modification also begins with an implementation of iterative refinement for the 'condensed' linear system

$$E\Delta\tilde{x}_1 + u \approx 0\,.$$

Suppose, for the time being, that

$$\|d\tilde{x}_1\| \leq \text{eps},$$

where eps is the relative tolerance prescribed for the Newton iteration. Then some *sweep-index* $j_\nu \geq 1$ can be defined such that

$$\|d\tilde{x}_j^\nu\| \leq \text{eps}, \quad j = 1, \ldots, j_\nu\,.$$

If we now set part of the residuals deliberately to machine-zero, say,

$$dF_j^\nu = 0, \quad j = 1, \ldots, j_\nu - 1\,,$$

then this modified iterative refinements process can be shown to converge under the less restrictive sufficient condition

$$\varepsilon(m-1)(2n+m-1)\kappa_\Delta[a,b] < 1\,,$$

wherein now the quantity $\kappa_\Delta[a,b]$ enters, which denotes the maximum of the IVP condition numbers on each of the subintervals of the partitioning Δ. Obviously, this quantity reflects the IVP condition number to be naturally associated with multiple shooting. Under this condition it can be shown that

$$j_{\nu+1} \geq j_\nu + 1\,,$$

hence the process terminates, at the latest, after $m-1$ refinement sweeps.

Whenever the above excluded case $j_0 = 0$ occurs, the iterative refinement cannot even start. This occurrence does not necessarily imply that the BVP as such is ill-conditioned—for a detailed discussion of this aspect see again the textbook [71].

Rank reduction. The iterative refinement sweeps cheaply supply a condition number estimate for the sensitivity matrix E via

$$\text{cd}(E) = \frac{\|d\tilde{x}_1^0\|}{\|\Delta\tilde{x}_1^0\|\varepsilon} \leq \text{cond}(E)\,.$$

Even without iterative refinement sweeps a cheap condition number estimate may be available: Assume that separable boundary conditions have been split off via the above described projection. Let E denote the remaining part of the sensitivity matrix which is then treated by QR-decomposition with column pivoting—for details see, e.g., [77, Section 3.2.2]. In this setting the *subcondition number*

$$sc(E) \leq cond(E)$$

is easily computable.

If either

$$\varepsilon \, cd(E) \geq \tfrac{1}{2} \quad \Longleftrightarrow \quad \|d\tilde{x}_1^0\| \geq \tfrac{1}{2}\|\Delta\tilde{x}_1^0\|,$$

which is equivalent to $j_0 = 0$ or

$$\varepsilon \, sc(E) \geq \tfrac{1}{2},$$

then

$$\varepsilon \, cond(E) \geq \tfrac{1}{2}.$$

In other words: in either case E is *rank-deficient* and the condensed system is ill-conditioned. In this situation we may replace the condensed equation

$$E\Delta x_1 = -u$$

by the *underdetermined* linear least squares problem

$$\|E\Delta x_1 + u\|_2 = min .$$

This linear system may be 'solved' by means of the Moore-Penrose pseudo-inverse as

$$\Delta x_1 = -E^+ u .$$

Upon leaving the remaining part of the condensing algorithm unaltered, the thus modified elimination process can be formally described by some generalized inverse

$$J^- = RS^+ L \qquad (7.5)$$

with R,S,L as defined in Lemma 7.1 and

$$S^+ = \begin{bmatrix} E^+ & & & \\ & I & & \\ & & \ddots & \\ & & & I \end{bmatrix}.$$

As can be easily verified (see Exercise 4.7), this generalized inverse is an *outer inverse* and can be uniquely defined by the set of four axioms

$$\begin{aligned}
(J^-J)^T &= (RR^T)^{-1}J^-J(RR^T), \\
(JJ^-)^T &= (L^TL)JJ^-(L^TL)^{-1}, \\
J^-JJ^- &= J^-, \\
JJ^-J &= J.
\end{aligned} \qquad (7.6)$$

This type of generalized inverse plays a role in a variety of more general BVPs, some of which are given in the subsequent Sections 7.2 and 7.3.

7.1.2 Realization of Newton methods

On the basis of the preceding sections we are now ready to discuss the actual realization of global Newton methods within multiple shooting techniques for timelike BVPs.

Jacobian matrix approximations. In order to establish the total Jacobian J, we must approximate the boundary derivative matrices A, B and the propagation matrices $G_1, \ldots, G_{m-1}$.

Boundary derivatives. Either an analytic derivation of r (not too rare case) or a *finite difference approximation*

$$A \doteq \frac{\delta r}{\delta x_1}, \quad B \doteq \frac{\delta r}{\delta x_m}$$

will be realized.

Propagation matrices. The propagation matrices G_j are also called Wronskian matrices. Whenever the derivative matrix $f_y(y)$ of the right side is analytically available, then numerical integration of the n variational equations

$$G'_j = f_y(y(t))G_j, \quad G_j(t_j) = I_n \tag{7.7}$$

might be the method of choice to compute them. If f_y is not available analytically, then some *internal* differentiation as suggested by H.G. Bock [31, 32] should be applied—see also [71, Section 8.2.1]. Its essence is a numerical differencing of the form

$$f_y(y) \doteq \frac{\delta f(y)}{\delta y},$$

which then enters into the numerical solution of *discrete variational equations* instead of (7.7). The actual realization of this idea requires special variants of standard integration software [112, 31, 32]. Note that any such approach involves, of course, the simultaneous numerical integration of $y' = f(y(t))$ to obtain the argument $y(t)$ in f_y.

Scaling. Formally speaking, each variable x_j will be transformed as

$$x_j \longrightarrow D_j^{-1} x_j,$$

wherein the diagonal matrices $D_j > 0$ represent some carefully chosen scaling. Formal consequences are then

$$
\begin{aligned}
F_j &\longrightarrow D_{j+1}^{-1} F_j, \\
G_j &\longrightarrow D_{j+1}^{-1} G_j D_j =: \hat{G}_j.
\end{aligned}
\tag{7.8}
$$

In actual computation, this means replacing

$$
\begin{aligned}
\|F_j\| &\longrightarrow \|D_{j+1}^{-1} F_j\|, \\
\|\Delta x_j\| &\longrightarrow \|D_j^{-1} \Delta x_j\|.
\end{aligned}
$$

Inner products and norms. In view of the underlying BVP we may want to modify the Euclidean inner product and norm by including information about the mesh $\Delta = \{t_1, \ldots, t_m\}$. For example, let $(\cdot, \cdot)$ denote some (possibly scaled) Euclidean inner product for the vectors $u = (u_1, \ldots, u_m)$, $v = (u_1, \ldots, v_m)$, $u_j, v_j \in \mathbb{R}^n$. Then we may define some *discrete L^2-product* by virtue of

$$
\begin{aligned}
|b - a|(u, v)_\Delta \;=\; & (u_1, v_1)|t_2 - t_1| + (u_m, v_m)|t_m - t_{m-1}| \\
& + \textstyle\sum_{j=2}^{m-1}(u_j, v_j)|t_{j+1} - t_{j-1}|
\end{aligned}
\tag{7.9}
$$

and its induced *discrete L^2-norm* as

$$
(u, u)_\Delta \equiv \|u\|_\Delta^2 .
$$

Quasi-Newton updates. Any approximation of the Wronskian matrices G_j requires a computational cost of $\sim n$ trajectory evaluations. In order to save computing time per Wronskian evaluation, we may apply rank-1 updates as long as the iterates remain within the Kantorovich domain around the solution point—i.e., when the damping strategies in Section 2.1.4 supply

$$
\lambda_k = \lambda_{k-1} = 1 .
$$

Of course, the sparse structure of the total Jacobian J must be taken into account. We assume the boundary derivative approximations A and B as fixed. Then the *secant condition* (1.17) for the total Jacobian

$$
(J_{k+1} - J_k)\Delta x^k = F(x^{k+1})
$$

splits into the separate block secant conditions

$$
(G_j^{k+1} - G_j^k)\Delta x_j^k = F_j(x_j^{k+1}, x_{j+1}^{k+1}) , \quad j = 1, \ldots, m - 1 .
$$

Upon applying the ideas of Section 2.1.4, we arrive at the following rank-1 update formula:

$$
G_j^{k+1} = G_j^k + F_j(x_j^{k+1}, x_{j+1}^{k+1})\frac{(\Delta x_j^k)^T}{\|\Delta x_j^k\|_2^2} , \quad j = 1, \ldots, m - 1 .
\tag{7.10}
$$

In a *scaled* version of the update formula (7.10), we will either update the $\hat{G}_j$ from (7.8) directly or, equivalently, update G_j replacing

$$
\frac{\Delta x_j^T}{\|\Delta x_j\|_2^2} \;\longrightarrow\; \frac{(D_j^{-2}\Delta x_j)^T}{\|D_j^{-1}\Delta x_j\|_2^2}
$$

in the representation (7.10). As worked out in detail in Section 2.1.4 above, scaling definitely influences the convergence of the corresponding quasi-Newton iteration (compare also [59, Section 4.2]).

Adaptive rank strategy. Assume that the condensed matrix E has been indicated as being 'rank-deficient'. In this case we need not terminate the Newton iteration, but may continue by an intermediate Gauss-Newton step with a correction of the form

$$\Delta x^k = -F'(x^k)^- F(x^k) \, .$$

Upon recalling Section 4.1.1, the generalized inverse $F'(x)^-$ can be seen to be an *outer* inverse. Therefore Theorem 4.7 guarantees that the thus defined ordinary Gauss-Newton iteration converges locally to a solution of the system

$$F_j(x_j, x_{j+1}) = 0 \, , \quad j = 1, \ldots, m - 1 \, ,$$
$$\|r(x_1, x_m)\|_2 = \min \, . \tag{7.11}$$

The modified trust region strategies of Section 4.3.5 can be adapted for the special projector

$$P := J^- J \, .$$

Note, however, that intermediate rank reductions in this context will *not* guarantee an increase of the feasible damping factors (compare Lemma 4.17 or Lemma 4.18), since P is generically *not* orthogonal (for $m > 2$). Nevertheless a significant increase of the damping factors has been observed in numerical experiments.

Obviously, the thus constructed Gauss-Newton method is associated with an underdetermined *least squares* BVP of the kind

$$y' = f(y) \, ,$$
$$\|r(y(a), y(b))\|_2 = \min \, .$$

Level functions. In the *rank-deficient* case, the *residual* level function

$$T(x|I) = \|F(x)\|^2 = \sum_{j=1}^{m} \|F_j(x)\|^2$$

no longer has the Gauss-Newton correction $\Delta x = -F'(x)^- F(x)$ as a descent direction. Among the practically interesting level functions, this property still holds for the above used *natural* level function $T(x|J^-)$ or for the *hybrid* level function

$$T(x|R^{-1}J^-) = \|R^{-1}J^- F(x)\|^2 = \|\Delta x_1\|^2 + \sum_{j=1}^{m-1} \|F_j(x)\|^2 \, .$$

The proof of these statements is left as Exercise 7.4.

BIBLIOGRAPHICAL NOTE. The affine covariant Newton method as described
here is realized, e.g., in the multiple shooting code BVPSOL due to P. Deuflhard
and G. Bader [70] and the optimal control code BOUNDSCO due to J. Oberle
[162]. Among these only BVPSOL realizes the Gaussian block elimination (Section 7.1.1) including the rank-deficient option with possible intermediate
Gauss-Newton steps. Global sparse solution of the cyclic linear Newton systems is implemented in the code BOUNDSCO and as one of two options in
BVPSOL; a rank-strategy is not incorporated within global elimination.

7.1.3 Realization of continuation methods

Throughout this section we consider parameter dependent two-point boundary value problems of the kind

$$y' = f(y, \tau),$$

$$r(y(a), y(b), \tau) = 0,$$

which give rise to some parameter dependent cyclic system of nonlinear equations

$$F(x, \tau) = 0.$$

Typical situations are that either the τ-family of BVP solutions needs to
be studied or a continuation method is applied to globalize a local Newton method. Generally speaking, the parameter dependent mapping F is
exactly the case treated in Section 5. Hence, any of the continuation methods described there can be transferred—including the automatic control of
the parameter stepsizes $\Delta\tau$.

Newton continuation methods. Assume the BVP under consideration
has no turning or bifurcation points—known either from external insight
into the given scientific problem or from an a-priori analysis. Then Newton
continuation methods as presented in Section 5.1 are applicable.

Classical continuation method. This algorithm (of order $p = 1$) deserves no
further explanation. All the details of Section 5.1 carry over immediately.

Tangent continuation method. For this algorithm (of order $p = 2$) we need to
solve the linear system

$$F_x(x, \tau)\dot{x}(\tau) = F_\tau(x, \tau),$$

which is the same type of block cyclic linear system as for the Newton corrections. The above right hand term $F_\tau(x, \tau)$ can be computed by numerical
integration of the associated variational equations or by *internal* numerical
differentiation (cf. [31, 32] or [71, Section 8.2.1]).

Continuation via trivial BVP extension. A rather popular trick is to just extend the standard BVP such that

$$y' = f(y, \tau), \quad \tau' = 0,$$
$$r(y(0), y(T), \tau) = 0, \quad h(\tau) = 0 \tag{7.12}$$

with $h'(\tau) \neq 0$. Any BVP solver applied to this extended BVP then defines some extended mapping

$$F(x, \tau) = 0, \quad h(\tau) = 0. \tag{7.13}$$

Let Δx^k denote the Newton correction for the equations with fixed τ. Then the Newton correction $(\Delta z^k, \Delta \tau)$ for (7.13) turns out to be

$$\Delta z^k = \Delta x^k + \Delta \tau^k \dot{\bar{x}}(\tau^k), \quad \Delta \tau^k = -\frac{h(\tau^k)}{h'(\tau^k)}. \tag{7.14}$$

The proof of this connection is left as Exercise 7.1. If one selects τ^0 such that $h(\tau^0) \neq 0$, then the extension (7.12) realizes a mixture of continuation methods of order $p = 1$ and $p = 2$. An adaptive control of the stepsizes $\Delta \tau$ here arises indirectly via the damping strategy of Newton's method.

Example 7.1 *Space shuttle problem.* This optimal control problem stands for a class of highly sensitive BVPs from space flight engineering. The underlying physical model (very close to realistic) is due to E.D. Dickmanns [87]. The full mathematical model has been documented in [81]. The stated mathematical problem is to find an optimal trajectory of the second stage of a Space Shuttle such that a prescribed maximum permitted skin temperature of the front shield is not exceeded. The real problem of interest is a study with respect to the temperature parameter, say τ. For technological reasons, the aim is to drive down the temperature as far as possible. This problem gave rise to a well-documented success story for error oriented Newton methods (earlier called affine 'invariant' instead of affine covariant).

The unconstrained trajectory goes with a temperature level of $2850°F$ (equivalent to $\tau = 0.072$). The original technological objective of NASA had been optimal flight trajectories at temperature level $2000°F$ (equivalent to $\tau = 0.$). However, the applied continuation methods just failed to continue to temperatures lower than $2850°F$! One reason for that failure can already be seen in the sensitivity matrix: the early optimal control code OPTSOL of R. Bulirsch [42], improved 1972 by P. Deuflhard [59] (essentially in the direction of error oriented Newton methods), revealed a subcondition number

$$\mathrm{sc}(E) = 0.2 \cdot 10^{10}$$

at that temperature. As a consequence, any traditional residual based Newton methods, which had actually been used at NASA within the frame of classical continuation, are bound to fail. The reasons for such an expectation have been discussed in Sections 3.3.1 and 3.3.2.

In 1973, H.-J. Pesch attacked this problem by means of OPTSOL, which in those days still contained classical continuation with *empirical* stepsize control, but already error oriented Newton methods [59] with *empirical* damping strategy. With these techniques at hand, the first successful continuation steps to $\tau < 0.072$ were at all possible—nicely illustrating the geometric insight from Section 3.3.2. However, computing times had been above any tolerable level, so that H.-J. Pesch eventually terminated the continuation process at $\tau = 0.0080$ with a final empirical stepsize $\Delta\tau = -0.0005$. Further improvements were possible by replacing

- the classical Newton damping strategy by an adaptive one [63], similar to the adaptive trust region predictor given in Section 3.3.3 and
- the classical continuation method with empirical stepsize selection by their adaptive counterparts as presented in Section 5.1.

For the last continuation step performed by H.-J. Pesch, Table 7.1 shows the comparative computational amount for different continuation methods (counting full trajectories to be computed within the multiple shooting approach).

Continuation method	Newton method	work
classical	residual based	failure
classical	error oriented, empirical damping	~ 340
classical	error oriented, adaptive trust region	114
trivial BVP extension	error oriented, adaptive trust region	48
tangent	error oriented, adaptive trust region	18

Table 7.1. Space Shuttle problem: Fixed continuation step from $\tau = 0.0085$ to $\tau = 0.0080$. Comparative computational amount for different Newton continuation methods.

In 1975, an adaptive error oriented Newton method [81] in connection with the trivial BVP extension made it, for the first time, possible to solve the original NASA problem for temperature level $2000°F$ ($\tau = 0.$). Results of technical interest have been published by E.D. Dickmanns and H.-J. Pesch [88]. The performance of this computational technique for temperatures even below the NASA objective value is documented in Table 7.2. As can be seen, this kind of continuation technique, even though it succeeds to solve the problem, still performs a bit rough.

Continuation sequence			work		Remarks
0.0	→	-0.0050	60	fail	switching structure totally disturbed
0.0	→	-0.0010	60	fail	negative argument in log-function
0.0	→	-0.0005	80		
-0.0005	→	-0.0010	63		
-0.0010	→	-0.0020	50		
-0.0020	→	-0.0050	65		
-0.0050	→	-0.0200	30	fail	switching structure totally disturbed
-0.0050	→	-0.0100	30	fail	as above
-0.0050	→	-0.0100	67	fail	Newton method fails to converge
0.0050	→	-0.0080	66		prescribed final parameter
0.0	→	-0.0080	571		overall amount

Table 7.2. Space Shuttle problem: Adaptive continuation method [81] via trivial BVP extension (7.12).

A much smoother and faster behavior occurs when adaptive tangent continuation as worked out in Section 5.1 is applied—just see Table 7.3. With this method the temperature could be lowered even down to $1700°F$. Starting from these data, H.G. Bock [30] computed an achievable temperature of only $890°C$ from the multiple shooting solution of a Chebyshev problem assuming that all state constraints of the problem are to be observed.

Continuation sequence			work	Remarks
0.0	→	-0.0035	47	ordinary Newton method
-0.0035	→	-0.0057	32	throughout the computation;
-0.0057	→	-0.0080[a]	31	switching structure never disturbed
0.0	→	-0.0080	110	overall amount
[a] stepsize cut off to prescribed final value $\tau = -0.0080$.				

Table 7.3. Space Shuttle problem: Adaptive tangent continuation [61]. See also Section 5.1 here.

Remark 7.1 It may be interesting to hear that none of these 'cooler' space shuttle trajectories has been realized up to now. In fact, the author of this book has presented the optimal $2000°F$ trajectories in 1977 within a seminar at NASA, Johnson Space Flight Center, Houston; the response there had been that the countdown for the launching of the first space shuttle (several years ahead) had already gone too far to make any substantial changes. The second chance came when Europe thought about launching its own space shuttle HERMES; in fact, the maximum skin temperature assumed therein turned out to be the same as for the present NASA flights! Sooner or later, a newcomer in the space flight business (and this is big business!) will exploit this kind of knowledge which permits him (or her) to build a space shuttle in a much cheaper technology.

Gauss-Newton continuation method. As soon as *turning* or *bifurcation points* might arise, any Newton continuation method is known to be inefficient and Gauss-Newton techniques come into play—compare Section 5.2. The basic idea behind these techniques is to treat the parameter dependent nonlinear equations as an *underdetermined* system in terms of the extended variable $z = (x, \tau) = (x_1, \ldots, x_m, \tau)$. In addition to the Wronskian approximations G_j we therefore need the derivatives

$$g_j := \frac{\partial \Phi^{t_j+1, t_j}(\tau) x_j}{\partial \tau}, \quad j = 1, \ldots, m.$$

With these definitions the Jacobian $(nm, nm + 1)$-matrix now has the block structure

$$J = \begin{bmatrix} G_1 & -I & & & g_1 \\ & \ddots & \ddots & & \vdots \\ & & G_{m-1} & -I & g_{m-1} \\ A & & & B & g_m \end{bmatrix}.$$

Based on this structure, *Gaussian block elimination* offers a convenient way to compute a Gauss-Newton correction

$$\widehat{\Delta z} = -J^- F$$

in terms of the outer inverse J^- already introduced in (7.5). The computation of $\widehat{\Delta z}$ can be conveniently based on a QR-decomposition of the $(n, n + 1)$-matrix $[E, g]$, where

$$\begin{aligned} E &:= A + B G_{m-1} \cdots \cdots G_1, \\ g &:= g_m + B(g_{m-1} + \cdots + G_{m-1} \cdots \cdots G_2 g_1). \end{aligned}$$

With this we obtain a variant of the condensing algorithm

$$\begin{pmatrix} \widehat{\Delta x_1} \\ \widehat{\Delta \tau} \end{pmatrix} = -[E, g]^+ u, \tag{7.15}$$

$$\widehat{\Delta x_{j+1}} = G_j \widehat{\Delta x_j} + g_j \widehat{\Delta \tau} + F_j, \quad j = 1, \ldots, m - 1.$$

Of course, iterative refinement sweeps must be properly added, see Section 7.1.1. This kind of Gauss-Newton continuation would need to be coupled by some extra step-size control in the continuation parameter τ—which is not worked out here.

Instead we advocate a realization of the standard Gauss-Newton continuation method as developed in Section 5.2. In the spirit of (5.28) and (5.29), a Gauss-Newton correction in terms of the Moore-Penrose pseudo-inverse can be easily computed via

$$\Delta z = -J^+ F = \widehat{\Delta z} - \frac{(t, \widehat{\Delta z})}{(t, t)} t, \qquad (7.16)$$

wherein t now denotes any *kernel vector* satisfying $Jt = 0$. Let $t = (t_1, \ldots, t_m, t_\tau)$ denote the partitioning of a kernel vector. Then components (t_1, t_τ) can be computed from the $(n, n+1)$-system

$$Et_1 + gt_\tau = 0$$

again via the QR-decomposition. The remaining components are once more obtained via explicit recursion as

$$t_{j+1} = G_j t_j + g_j t_\tau, \quad j = 1, \ldots, m-1.$$

Insertion of the particular correction $\widehat{\Delta z}$ and the kernel vector t finally yields the Gauss-Newton correction Δz. The local convergence analysis as well as the corresponding step-size control in τ can then be copied from Section 5.2 without further modification.

As for the inner products arising in the above formula, the discrete L^2-product $(\cdot, \cdot)_\Delta$ defined in (7.9) looks most promising, since it implicitly reflects the structure of the BVP.

Detection of critical points. Just as in Section 5.2.3, certain determinant pairs need to be computed. This is especially simple in the context of the QR-decomposition of the matrix $[E, g]$ in (7.15).

BIBLIOGRAPHICAL NOTE. More details are given in the original paper [73] by P. Deuflhard, B. Fiedler, and P. Kunkel. There also a performance comparison of MULCON, a multiple shooting code with Gauss-Newton continuation as presented here, and AUTO, a collocation code with pseudo-arclength continuation due to E. Doedel [89] is presented: the given numerical example nicely shows that the *empirical* pseudo-arclength continuation is robust and reliable, but too cautious and therefore slower, whereas the *adaptive* Gauss-Newton continuation is also robust and reliable, but much faster. Moreover, continuous analogs of the augmented system of G. Moore for the characterization of bifurcations are worked out therein.

7.2 Parameter Identification in ODEs

This section deals with the *inverse problem* in ordinary differential equations (ODEs): given a system of n nonlinear ODEs

$$y' = f(y,p), \quad y(0) \text{ given}, \quad p \in \mathbb{R}^q,$$

determine the unknown parameter vector p such that the solution $y(t,p)$ 'fits' given experimental data

$$(\tau_j, z_j) \qquad z_j \in \mathbb{R}^n, \quad j = 1, \ldots, M.$$

Let

$$\delta y(\tau_j, p) := y(\tau_j, p) - z_j, \quad j = 1, \ldots, M.$$

denote the *pointwise deviations* between model and data with prescribed statistical tolerances δz_j, $j = 1, \ldots, M$. If some of the components of z_j are not available, this formally means that the corresponding components of δz_j are infinite. In nonlinear least squares, the deviations are measured via a discrete (weighted) l_2-product $(\cdot, \cdot)$, which leads to the problem

$$(\delta y, \delta y) = \frac{1}{M} \sum_{j=1}^{M} \|D_j^{-1} \delta y(\tau_j, p)\|_2^2 = \min$$

with diagonal weighting matrices

$$D_j := \operatorname{diag}(\delta z_{j1}, \ldots, \delta z_{jn}), \quad j = 1, \ldots, M.$$

If we define some nonlinear mapping F by

$$F(p) := \begin{bmatrix} D_1^{-1} \delta y(\tau_1, p) \\ \vdots \\ D_M^{-1} \delta y(\tau_M, p) \end{bmatrix},$$

then our least squares problem reads

$$\|F(p)\|_2^2 \equiv (\delta y, \delta y) = \min.$$

If *all* components at every data point τ_j have been measured (rare case), then $F : \mathbb{R}^q \longrightarrow \mathbb{R}^L$ with $L = nM$. Otherwise some $L < nM$ occurs.

We are thus guided to some *constrained nonlinear least squares problem*

$$y' = f(y,p), \quad (\delta y, \delta y) = \min,$$

where the ODEs represent the equality constraints. This problem type leads to a modification of the standard multiple shooting method.

The associated Jacobian (L, q)-matrix $F'(p)$ must also be computed exploiting its structure numerically

$$F'(p) = \begin{bmatrix} D_1^{-1} y_p(\tau_1) \\ \vdots \\ D_l^{-1} y_p(\tau_l) \end{bmatrix}. \tag{7.17}$$

Herein the *sensitivity matrices* y_p each satisfy the *variational equation*

$$y_p' = f_y(y, p) y_p + f_p(y, p)$$

with initial values $y_p(t_j) = 0, t_j \in \Delta$. Of course, we will naturally pick m multiple shooting nodes out of the set of M measurement nodes, which means that

$$\Delta := \{t_1, \dots, t_m\} \subseteq \{\tau_1, \dots, \tau_M\}$$

with, in general, $m \ll M$. As before, sub-trajectories $\Phi^{t,t_j}(p) x_j$ are defined per each subinterval $t \in [t_j, t_{j+1}]$ via the initial value problem

$$y' = f(y, p), \quad y(t_j) = x_j.$$

Figure 7.3 gives a graphical illustration of the situation for the special case $M = 13$, $m = 4$.

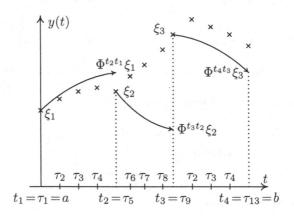

Fig. 7.3. Multiple shooting for parameter identification ($M = 13, m = 4$).

Unknowns to be determined are $(x, p) = (x_1, \dots, x_m, p)$. If we introduce the convenient notation

$$r(x_1, \dots, x_m, p) := \begin{bmatrix} D_1^{-1}(\Phi^{\tau_1, t_1}(p) x_1 - z_1) \\ \vdots \\ D_{M-1}^{-1}(\Phi^{\tau_{M-1}, t_{m-1}}(p) x_{m-1} - z_{M-1}) \\ D_M^{-1}(x_m - z_M) \end{bmatrix},$$

we arrive at the parameter identification problem in its multiple shooting version

$$F_j(x_j, x_{j+1}, p) := \Phi^{t_{j+1}, t_j}(p)x_j - x_{j+1} = 0 \,, \quad j = 1, \ldots, m-1 \,,$$

$$\|r(x_1, \ldots, x_m, p)\|_2^2 = \min \,.$$

Obviously, this is a constrained nonlinear least squares problem with the continuity equations as nonlinear constraints, an overdetermined extension of (7.11).

For ease of notation we write the whole mapping as

$$F(x, p) = \begin{bmatrix} F_1(x_1, x_1, p) \\ \vdots \\ F_{m-1}(x_{m-1}, x_m, p) \\ r(x_1, \ldots, x_m, p) \end{bmatrix}$$

and its block structured Jacobian matrix (ignoring weighting matrices for simplicity) as

$$J = \begin{bmatrix} G_1 & -I & & & P_1 \\ & \ddots & \ddots & & \vdots \\ & & G_{m-1} & -I & P_{m-1} \\ B_1 & \cdots & B_{m-1} & B_m & P_m \end{bmatrix}.$$

The above matrices P_j, $j = 1, \ldots, m$ represent the parameter derivatives of the mapping F consisting just as in (7.17) of sensitivity matrices; their length and initial values depend on the available measurement data and on the selection of the multiple shooting nodes out of the measurement modes— details are omitted here, since they require clumsy notation.

Upon recalling (7.11) and (7.5), the corresponding *constrained Gauss-Newton* corrections are defined as

$$(\Delta x^k, \Delta p^k) = -J(x^k, p^k)^- F(x^k, p^k)$$

or, more explicitly, via the block system

$$G_j \Delta x_j - \Delta x_{j+1} + P_j \Delta p = -F_j \,, \quad j = 1, \ldots, m-1 \,,$$

$$\|B_1 \Delta x_1 + \cdots + B_m \Delta x_m + P_m \Delta p + r\|_2^2 = \min \,.$$

Gaussian block elimination. Proceeding as in the simpler BVP case, we here obtain

$$\begin{pmatrix} \Delta x_1 \\ \Delta p \end{pmatrix} = -[E, P]^+ u \,,$$

wherein the quantities E, P, u are computed recursively from

$$\bar{P}_m := P_m, \bar{B}_m := B_m,$$

$$j = m-1,\ldots,1: \bar{P}_j := \bar{P}_{j+1} + \bar{B}_{j+1}P_j, \quad \bar{B}_j := B_j + \bar{B}_{j+1}G_j,$$

$$E := \bar{B}_1, \quad P := \bar{P}_1,$$

$$u := r + B_m[F_{m-1} + \cdots + G_{m-1} \cdots \cdots G_2 F_1].$$

The remaining correction components follow from

$$\Delta x_{j+1} = G_j\Delta x_j + P_j\Delta p + F_j, \quad j = 1,\ldots,m-1.$$

In analogy with (7.5) and with the notation for the block matrices L, R introduced there, the generalized inverse J^- can be formally written as

$$J^- = RS^-L, \quad S^- = \mathrm{diag}\,([E,P]^+, I, \cdots, I).$$

BIBLIOGRAPHICAL NOTE. Since 1981, this version of the multiple shooting method for parameter identification in differential equations has been suggested and driven to impressive perfection by H.G. Bock [29, 31, 32] and his coworkers. It is implemented in the program PARFIT. A single shooting variant especially designed for parameter identification in large chemical reaction kinetic networks has been worked out in detail by U. Nowak and the author [158, 159] in the code PARKIN.

Iterative refinement sweeps. In order to start the iterative refinement sweeps, we require some iterative correction of the above condensed least-squares system. A naive iterative correction approach, however, would not be suitable for large residuals

$$\bar{r} = E\Delta x_1 + P\Delta p + u.$$

Thus we recommend an algorithm proposed by Å. Björck [25] to be adapted to the present situation. In this approach the above linear least-squares problem is first written in the form of the augmented linear system of equations

$$-\bar{r} + E\Delta x_1 + P\Delta p + u = 0,$$
$$[E,P]^T\bar{r} = 0$$

in the variables Δx_1, Δp, $\bar{r}$. The iterative correction is then applied to this system. If it converges—which means that the condensed linear least-squares problem is regarded as well-posed—then, without any changes, the iterative refinement sweeps for the explicit recursion can be applied in the same form as presented in the standard situation treated in Section 7.1.1.

7.3 Periodic Orbit Computation

In this section we are interested in continuous solutions of *periodic boundary value problems:*

$$y' = f(y) ,$$
$$r(y(0), y(T)) := y(T) - y(0) = 0 \tag{7.18}$$

with (hidden) time variable t and (unknown) period T. Here we treat only the situation that f is *autonomous*, the nonautonomous case is essentially standard. In this case f satisfies the *variational equation*

$$f' = f_y \cdot f ,$$

which can be formally solved as

$$f(y(t)) = W(t, 0) \, f(y(0)) , \tag{7.19}$$

wherein $W(\cdot, \cdot)$ is once more the propagation matrix of the variational equation. Insertion of the periodicity condition $y(T) = y(0)$ then yields

$$f(y(T)) = f(y(0)) = W(T, 0) f(y(0)) ,$$

or, equivalently, with $E = E(0) = W(T, 0) - I$ inserted:

$$Ef(y(0)) = 0 .$$

Obviously, the sensitivity matrix E is *singular* and $f(y(0))$ is a right eigenvector associated with eigenvalue zero, if only $f(y(0)) \neq 0$. The singularity of E reflects the fact that the phase or time origin is undetermined, causing a special nonuniqueness of solutions: whenever $y(t)$, $t \in [0, T]$ is a periodic solution, then $y(t + t_0)$, $t \in [0, T]$, $t_0 \neq 0$, is a different periodic solution, even though it is represented by the same *orbit*. Obviously, there exists a continuous solution set generated by S^1-symmetry. In this situation, we will naturally aim at computing the *orbit* directly, which means computing *any* trajectory $y(t + t_0)$, $t \in [0, T]$ without fixing the phase t_0.

In what follows we will assume that the eigenvalue zero of E is *simple*, which then implies that

$$\mathrm{rank}[E, f(y(0))] = n$$

and

$$\ker[E, f] = (f, 0) . \tag{7.20}$$

7.3.1 Single orbit computation

First we treat the case when a single periodic orbit is wanted. In order to convey the main idea, we start with the derivation of a special Gauss-Newton method in the framework of single shooting ($m = 2$).

Single Shooting. In this approach one obtains the *underdetermined* system of n equations

$$F(z) = \Phi^T x - x = 0$$

in the $n+1$ unknowns $z = (x, T)$ is generated. The corresponding Jacobian $(n, n+1)$-matrix has the form

$$F'(z) = [r_x, r_T] = [E, f(x(T))],$$

or, after substituting a periodic solution,

$$F'(z) = [E, f(x(0))].$$

Under the assumption made above the Jacobian matrix has full row rank and its Moore-Penrose pseudoinverse $F'(z)^+$ has full column rank. Hence, instead of a Newton method we can construct the *Gauss-Newton* iteration

$$\Delta z^k = -F'(z^k)^+ F(z), \quad z^{k+1} = z^k + \Delta z^k, \quad k = 0, 1, \ldots .$$

Also under the full rank assumption this iteration will converge *locally quadratically* to *some* solution z^* in the 'neighborhood' of a starting point z^0, i.e., to an arbitrary point on the orbit. This point determines the whole orbit uniquely.

Multiple shooting. In multiple shooting we need to have fixed nodes. So we introduce the dimensionless independent variable

$$s := \frac{t}{T} \in [0, 1]. \tag{7.21}$$

Let

$$\Delta := \{0 = s_1 < s_2 < \cdots < s_m = 1\}$$

denote the given partitioning with mesh sizes

$$\Delta s_j := s_{j+1} - s_j, \; j = 1, \ldots, m-1.$$

Then the following conditions must hold

$$F_j(x_j, x_{j+1}, T) \; := \; \Phi^{T \Delta s_j} x_j - x_{j+1} = 0, \; j = 1, \ldots, m-1,$$
$$r(x_1, x_m) \; := \; x_m - x_1 = 0.$$

The subtrajectories can be formally represented by

$$\Phi^{T \Delta s_j} x_j = x_j + \int\limits_{s=0}^{T \Delta s_j} f(y(s)) ds.$$

With the Wronskian (n, n)-matrices

$$G_j \quad := \quad \frac{\partial \Phi^{T \Delta s_j} x_j}{\partial x_j} = W(s_{j+1}, s_j) \Big|_{\Phi^{T(s-s_j)} x_j}$$

and the n-vectors

$$g_j \quad := \quad \frac{\partial \Phi^{T \Delta s_j} x_j}{\partial T} = \Delta s_j f(\Phi^{T \Delta s_j} x_j)$$

the underdetermined linear system to be solved in each Gauss-Newton step has the form

$$
\begin{array}{rcl}
G_1 \Delta x_1 - \Delta x_2 \qquad\qquad\qquad +g_1 \Delta T &=& -F_1 \\
\ddots \qquad \ddots \qquad\qquad\qquad \vdots &&\ \ \vdots \\
G_{m-1} \Delta x_{m-1} - \Delta x_m + g_{m-1} \Delta T &=& -F_{m-1} \\
-\Delta x_1 \qquad\qquad +\Delta x_m &=& 0\,.
\end{array}
$$

Gaussian block elimination. The 'condensing' algorithm as described in Section 7.1.1 will here lead to the small underdetermined linear system

$$E \Delta x_1 + g \Delta T + u = 0\,,$$

where

$$E = G_{m-1} \cdots \cdots G_1 - I\,,$$

$$g := g_{m-1} + \cdots + G_{m-1} \cdots \cdots G_2 g_1\,,$$

$$u := F_{m-1} + \cdots + G_{m-1} \cdots \cdots G_2 F_1\,.$$

At a solution point $z^* = (x_1^*, \ldots, x_m^*, T^*)$ we obtain

$$E^* = W(1,0) - I\,, \quad g^* = f(x_1^*)\,,$$

just recalling that

$$f(x_m^*) = W(s_m, s_j) f(x_j^*) = f(x_1^*),$$
$$\Delta s_1 + \cdots + \Delta s_{m-1} = 1\,.$$

Hence, the $(n, n+1)$-matrix

$$[E^*, g^*] = [E(0), f(y(0))]$$

has again full row rank n. Consequently, an extension of the single shooting rank-deficient Gauss-Newton method realizing the Jacobian outer inverse J^- can be envisioned.

Better convergence properties, however, are expected by the standard Gauss-Newton method which requires the Moore-Penrose pseudoinverse J^+ of the total block Jacobian. As in the case of parameter dependent BVPs, we again compute a kernel vector $t = (t_1, \ldots, t_m, t_T)$, here according to

$$[E, g] \begin{pmatrix} t_1 \\ t_T \end{pmatrix} = 0,$$

$$t_{j+1} = G_j t_j + g_j t_T, \quad j = 1, \ldots, m-1$$

and combine it with iterative refinement sweeps, of course. The computation of the actual Gauss-Newton correction then follows from

$$\Delta z = \widehat{\Delta z} - \frac{(t, \widehat{\Delta z})}{(t, t)} t,$$

where $\widehat{\Delta z}$ is an arbitrary particular correction vector satisfying

$$F'(z^k)\widehat{\Delta z} + F(z^k) = 0.$$

Simplified Gauss-Newton method. At the solution point z^*, the above equations lead (up to some normalization factor) to the known solution

$$(t_1, t_T) = (f(x_1^*), 0).$$

Upon inserting this result into (7.19), we immediately arrive at

$$t_j = f(x_j^*), \quad j = 2, \ldots, m.$$

This inspires an intriguing modification of the above Gauss-Newton method: we may insert the 'iterative' kernel vector

$$t_T^k = 0, \; t_j^k = f(x_j^k) \; j = 1, \ldots, m$$

into the expression (7.16). In this way we again obtain some pseudo-inverse and, in turn, thus define some associated Gauss-Newton method.

BIBLIOGRAPHICAL NOTE. The multiple shooting version realizing the Jacobian outer inverse J^- has been suggested in 1984 by P. Deuflhard [64] and implemented in the code PERIOD. The improvement realizing J^+ has been proposed in 1994 by C. Wulff, A. Hohmann, and P. Deuflhard[199] and realized in the orbit continuation code PERHOM—for details see the subsequent Section 7.3.2. The same paper also contains a possible exploitation of symmetry for equivariant orbit problems, following up the work of K. Gatermann and A. Hohmann [95] for equivariant steady state problems.

7.3.2 Orbit continuation methods

In this section we consider the computation of families of orbits for the parameter dependent periodic boundary value problem

$$y' = f(y, \lambda),$$

$$r(y(0), y(T)) := y(T) - y(0) = 0,$$

where λ is the embedding parameter and T the unknown period. Note that the S^1-symmetry now only holds for *fixed* λ, so that the orbits can be explicitly parametrized with respect to λ. Throughout this section, let $\{\lambda_\nu\}$ denote the parameter sequence and

$$\Delta\lambda_\nu := \lambda_{\nu+1} - \lambda_\nu$$

the corresponding continuation step sizes to be automatically selected.

Single shooting. In order to convey the main geometrical idea, we again start with this simpler case. There we must solve the sequence of problems

$$F(x,T,\lambda) := \Phi^T(\lambda)x - x = 0$$

for the parameters $\lambda \in \{\lambda_\nu\}$.

Classical continuation method. This continuation method, where the previous orbit just serves as starting guess for the Gauss-Newton iteration to compute the next orbit, can be implemented without any further discussion, essentially as described in Section 5.1.

Tangent continuation method. The realization of this method deserves some special consideration. The Jacobian $(n, n+2)$-matrix has the following substructure

$$[F_x, F_T, F_\lambda] = [E, f, p]$$

with E, f as introduced above and p defined as

$$p := \frac{\partial \Phi^T(\lambda)x}{\partial\lambda}.$$

Let $t = (t_x, t_T, t_\lambda)$ denote any kernel vector satisfying

$$Et_x + f \cdot t_T + p \cdot t_\lambda = 0.$$

Under the above assumption that $[E, f]$ has full row rank n, we here know that

$$\dim \ker[E, f, p] = 2.$$

A natural basis for the kernel will be $\{t^1, t^2\}$ such that

$$t^1 := \ker[E, f], \quad t^2 \perp t^1, \tag{7.22}$$

wherein

$$t^i = (t_x^i, t_T^i, t_\lambda^i), \quad i = 1, 2.$$

From (7.20) and (7.22) we are directly led to the representations (ignoring any normalization)

$$t_x^1 = f, \quad t_T^1 = 0, \quad t_\lambda^1 = 0$$

and

$$\begin{pmatrix} t_x^2 \\ t_T^2 \end{pmatrix} = -[E,f]^+ p \,, \quad t_\lambda^2 := 1 \,.$$

Upon recalling that t^1 reflects the S^1-symmetry of each orbit, tangent continuation will mean to continue along t^2. In the notation of Section 5 this means that guesses $(\hat{x}, \hat{T})$ can be predicted as

$$\begin{pmatrix} \hat{x}(\lambda_{\nu+1}) - \bar{x}(\lambda_\nu) \\ \hat{T}(\lambda_{\nu+1}) - t(\lambda_\nu) \end{pmatrix} = \begin{pmatrix} t_x^2 \\ t_T^2 \end{pmatrix}\bigg|_{\lambda_\nu} \cdot \Delta\lambda_\nu \,.$$

These guesses are used as starting points for the Gauss-Newton iteration as described in Section 7.3.1. Since $t_T^1 = t_\lambda^1 = 0$, the property $t^1 \perp t^2$ also implies $t_x^1 \perp t_x^2$, which means that

$$\hat{x}(\lambda_{\nu+1}) - \bar{x}(\lambda_\nu) \perp f(\bar{x}(\lambda_\nu)) \,. \tag{7.23}$$

The geometric situation is represented schematically in Figure 7.4.

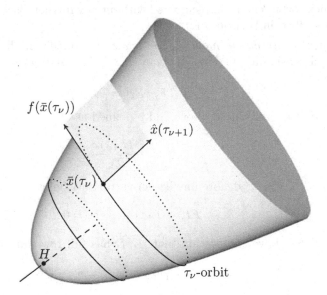

Fig. 7.4. Orbit continuation: H Hopf bifurcation point, — stable steady states, --- unstable steady states.

Multiple Shooting. As in (7.21) we again deal with fixed nodes

$$\Delta := \{0 = s_1 < s_2 < \cdots < s_m = 1\}$$

instead of variable nodes $t_j = s_j T$, $j = 1, \ldots, m$.

Switching notation, let now $t = (t_1, \ldots, t_m, t_T, t_\lambda)$ denote a selected kernel vector satisfying

$$G_j t_j - t_{j+1} + g_j \cdot t_T + p_j t_\lambda = 0 , \quad j = 1, \ldots, m-1 ,$$

$$t_m = t_1 ,$$

wherein G_j, g_j, and p_j are essentially defined as in single shooting $(m = 2)$. For $m > 2$, the Jacobian nullspace is still two-dimensional. For continuation we again choose the tangent vector

$$t = (f(x_1), \ldots, f(x_m), 0, 0), \quad x = \bar{x}(\lambda_\nu) .$$

Gaussian block elimination. In this setting, we will compute

$$\begin{pmatrix} \widehat{\Delta x_1} \\ \widehat{\Delta T} \end{pmatrix} := -[E, f]^+ p \cdot \Delta \lambda_\nu$$

with

$$p := p_{m-1} + \cdots + G_{m-1} \cdots \cdot G_2 p_1$$

and, recursively, for $j = 1, \ldots, m-2$:

$$\widehat{\Delta x_{j+1}} = G_j \widehat{\Delta x_j} + g_j \widehat{\Delta T} + p_j \Delta \lambda_\nu .$$

As a straightforward consequence, we may verify that

$$\widehat{\Delta x_m} = \widehat{\Delta x_1} .$$

The thus defined continuation

$$\hat{x}_j(\lambda_{\nu+1}) - \bar{x}_j(\lambda_\nu) = \widehat{\Delta x_j} , \quad j = 1, \ldots, m, \quad \hat{T}(\lambda_{\nu+1}) - t(\lambda_\nu) = \widehat{\Delta T}$$

clearly satisfies the *local* orthogonality property

$$\hat{x}_1(\lambda_{\nu+1}) - \bar{x}_1(\lambda_\nu) \perp f(\bar{x}_1(\lambda_\nu)) ,$$

which is *biased* towards the node $s_1 = 0$. Therefore, already from a geometrical point of view, the above continuation method should be modified such that the *global* orthogonality

$$\hat{x}(\lambda_{\nu+1}) - \bar{x}(\lambda_\nu) \perp f(\bar{x}(\lambda_\nu))$$

holds as a natural extension of (7.23). This directly leads us to the following orbit continuation method

$$\hat{x}(\lambda_{\nu+1}) - \bar{x}(\lambda_\nu) = \widehat{\Delta x} - \frac{(f_\nu, \widehat{\Delta x})}{(f_\nu, f_\nu)} f_\nu$$

with $f_\nu = f(\bar{x}(\lambda_\nu))$. Once more, the periodicity condition

$$\hat{x}_m(\lambda_{\nu+1}) = \hat{x}_1(\lambda_{\nu+1})$$

can be shown to hold.

Hopf bifurcations. The *detection* of Hopf bifurcation points H is an easy computational task. First, orbit continuation beyond H is impossible, as long as we come from the side of the periodic orbits, which we here do—see Figure 7.4. Second, at H, the condition $f = 0$ leads to rank$[E, f] < n$—a behavior that already shows up in a neighborhood of H. Third, the automatic stepsize control will lead to a significant reduction of the stepsizes $\Delta\lambda_\nu$. As soon as a Hopf bifurcation point seems to close by, its precise computation can be done switching to the augmented system suggested by A.D. Jepson [124].

BIBLIOGRAPHICAL NOTE. The above orbit continuation has been worked out in 1994 by C. Wulff, A. Hohmann, and P. Deuflhard [199] and realized in the code PERHOM. The same paper also covers the detection and computation of Hopf bifurcations and period doublings. Particular attention is paid to the computational exploitation of symmetries following up work of M. Dellnitz and B. Werner [50] and of K. Gatermann and A. Hohmann [95], the latter for steady state problems only, the former including Hopf bifurcations as well.

7.3.3 Fourier collocation method

In quite a number of application fields the desired periodic solution y to period $T = 2\pi/\omega$ is just expanded into a Fourier series according to

$$y(t) = \tfrac{1}{2}a_0 + \sum_j (a_j \cos(j\omega t) + b_j \sin(j\omega t)) \,.$$

Such a solution living in the infinite dimensional function space L_2 cannot be directly computed from the periodic BVP (7.18). Instead one aims at computing some *Fourier-Galerkin approximation* y_m out of the finite dimensional subspace $U_m \subset L_2$ according to the finite Fourier series ansatz

$$y_m(t) = \tfrac{1}{2}a_0 + \sum_{j=1}^{m} (a_j \cos(j\omega t) + b_j \sin(j\omega t)) \,, \tag{7.24}$$

where $a_j, b_j \in \mathbb{R}^n$. Insertion into the *approximate periodic* BVP

$$y_m' = f(y_m)\,, \qquad y_m(T) = y_m(0)$$

will require the coefficients of the derivative defined via

$$y_m'(t) = \sum_{j=1}^{m} (a_j' \cos(j\omega t) + b_j' \sin(j\omega t))$$

with

$$a_j' = j\omega b_j\,, \qquad b_j' = -j\omega a_j$$

as well as those of the right side defined via

$$f(y_m(t)) = \sum_{j=1}^{m} (\alpha_j \cos(j\omega t) + \beta_j \sin(j\omega t)) . \qquad (7.25)$$

Upon inserting the two expansions into the BVP (7.24), we arrive at the system of $N = 2m + 1$ relations

$$j\omega b_j = \alpha_j , \quad j = 0, \ldots, m , \qquad -j\omega a_j = \beta_j , \quad j = 1, \ldots, m .$$

From the theory of Fourier transforms we know that the right side coefficients can be computed via

$$\alpha_j = \frac{2}{T} \int_{t=0}^{T} f(y_m(t)) \cos(j\omega t)\, dt , \qquad \beta_j = \frac{2}{T} \int_{t=0}^{T} f(y_m(t)) \sin(j\omega t)\, dt . \qquad (7.26)$$

Obviously, this Galerkin approach involves a continuous Fourier transform which inhibits the construction of an approximation scheme for y_m.

Therefore, already in 1965, M. Urabe [188] suggested to replace the continuous Fourier transform by a *discrete Fourier transform*, i.e., by trigonometric interpolation over equidistant nodes

$$t_k = T\frac{k}{N} , \quad k = 0, 1, \ldots, N .$$

Formally speaking, the integrals in (7.26) are then approximated by their trapezoidal sums defined over the selected set of nodes. As a consequence, we now substitute the Galerkin approximation y_m by a *Galerkin-Urabe approximation* defined via the modified Fourier series expansion

$$\hat{y}_m(t) = \tfrac{1}{2}\hat{a}_0 + \sum_{j=1}^{m} \left(\hat{a}_j \cos(j\omega t) + \hat{b}_j \sin(j\omega t) \right) \qquad (7.27)$$

and the corresponding expansion for its derivative $\hat{y}'_m(t)$. Instead of (7.25) we now have a modified expansion

$$f(\hat{y}_m(t)) = \sum_{j=1}^{m} \left(\hat{\alpha}_j \cos(j\omega t) + \hat{\beta}_j \sin(j\omega t) \right)$$

and instead of the representation (7.26) we obtain the well-known trigonometric expressions

$$\hat{\alpha}_j = \frac{2}{N} \sum_{k=0}^{N-1} f(\hat{y}_m(t_k)) \cos(j\omega t_k) , \qquad \hat{\beta}_j = \frac{2}{N} \sum_{k=0}^{N-1} f(\hat{y}_m(t_k)) \sin(j\omega t_k) .$$

$$(7.28)$$

We again require the $N = 2m + 1$ relations

$$j\omega\hat{b}_j = \hat{\alpha}_j\,,\quad j = 0,\ldots,m\,,\qquad -j\omega\hat{a}_j = \hat{\beta}_j\,,\quad j = 1,\ldots,m\,.\qquad (7.29)$$

As before, we insert the expansion for $\hat{y}_m$ into the formal representation (7.28) so that the coefficients $\hat{\alpha}_j, \hat{\beta}_j$ drop out and we arrive at a system of nN equations, in general nonlinear. Originally, Urabe had suggested this approach, also called *harmonic balance method*, for nonautonomous periodic BVPs where T (or ω, respectively) is given in the problem so that the nN unknown coefficients $(\hat{a}_0, \hat{a}_1, \hat{b}_1, \ldots, \hat{a}_m, \hat{b}_m)$ can, in principle, be computed.

For *autonomous* periodic BVPs as treated here, system (7.29) turns out to be underdetermined with $nN + 1$ unknowns $(\hat{a}_0, \hat{a}_1, \hat{b}_1, \ldots, \hat{a}_m, \hat{b}_m, \omega)$ to be computed. We observe that in this kind of approximation local nonuniqueness shows up just as in the stated original problem: given a solution with computed coefficients $\hat{a}_j, \hat{b}_j$, then any trajectory defined by the modified coefficients

$$\tilde{a}_0 = \hat{a}_0\,,\quad \tilde{a}_j = \hat{a}_j \cos(j\omega\tau) + \hat{b}_j \sin(j\omega\tau)\,,\quad \tilde{b}_j = \hat{b}_j \cos(j\omega\tau) - \hat{a}_j \sin(j\omega\tau)\,,$$

is also a solution, shifted by τ. Therefore we may simply transfer the Gauss-Newton methods for single orbit computation or for orbit continuation (see the preceding sections) to the nonlinear mapping as just defined.

Numerical realization of Gauss-Newton method. For ease of writing, we here ignore the difference between y_m and $\hat{y}_m$ and skip all 'hats' in the coefficients. Then the underdetermined system has the form

$$F(z) = \begin{pmatrix} \dfrac{2}{N}\sum_{k=0}^{N-1} f(y_m(t_k)) \\ \vdots \\ \dfrac{2}{N}\sum_{k=0}^{N-1} f(y_m(t_k))\cos(j\omega t_k) + j\omega b_j \\ \dfrac{2}{N}\sum_{k=0}^{N-1} f(y_m(t_k))\sin(j\omega t_k) - j\omega a_j \\ \vdots \end{pmatrix} = 0\,,\quad z = \begin{pmatrix} a_0 \\ \vdots \\ a_j \\ b_j \\ \vdots \\ \omega \end{pmatrix}$$

in terms of a mapping $F : \mathbb{R}^{nN+1} \longrightarrow \mathbb{R}^{nN}$. Herein z additionally enters via the expression (7.27) understood to be inserted for y_m. For the corresponding Jacobian $(nN, nN + 1)$-matrix we may write block columnwise

$$F'(z) = (F_{a_0}(z), F_{a_1}(z), F_{b_1}(z), \ldots, F_{a_m}(z), F_{b_m}(z), F_\omega(z))\,.$$

Assume we have already computed the Fourier series expansion of the (n, n)-matrix $f_y(y_m)$, say with coefficients

$$A_j = \frac{2}{N} \sum_{k=0}^{N-1} f_y(y_m(t_k)) \cos(j\omega t_k), \quad B_j = \frac{2}{N} \sum_{k=0}^{N-1} f_y(y_m(t_k)) \sin(j\omega t_k).$$

Then a straightforward calculation reveals that

$$F_\omega(z) = \begin{pmatrix} 0 \\ \vdots \\ ja_j \\ -jb_j \\ \vdots \end{pmatrix}, \quad F_{a_0}(z) = \begin{pmatrix} \frac{1}{2}A_0 \\ \vdots \\ \frac{1}{2}B_j \\ \frac{1}{2}A_j \\ \vdots \end{pmatrix},$$

and, for $l = 1, \ldots, m$:

$$F_{a_l}(z) = \begin{pmatrix} A_l \\ \vdots \\ \frac{1}{2}(B_{l+j} - B_{l-j}) + j\omega\delta_{jl}I_n \\ \frac{1}{2}(A_{l+j} + A_{l-j}) \\ \vdots \end{pmatrix},$$

and

$$F_{b_l}(z) = \begin{pmatrix} B_l \\ \vdots \\ \frac{1}{2}(-BA_{l+j} + A_{l-j}) \\ \frac{1}{2}(B_{l+j} + B_{l-j}) - j\omega\delta_{jl}I_n \\ \vdots \end{pmatrix}.$$

In this representation, certain indices run out of the permitted index set: whenever an index $l > m$ appears, then replace A_l by A_{N-l} and B_l by $-B_{N-l}$; whenever $l < 0$, then replace A_l by A_{-l} and B_l by $-B_{-l}$. In this way all computations can be performed using the FFT algorithm for f and f_y, assuming that an iterate y_m is at hand—which it is during the Gauss-Newton iterations for single orbit computation or orbit continuation (see preceding sections).

Adaptivity device. Up to now, we have not discussed the number m of terms necessary to obtain an approximations to prescribed accuracy. Let $|\cdot|$ denote the $L_2[0, T]$-norm then we have

$$\varepsilon_m = |y - y_m| = \left(\sum_{j=m+1}^{\infty} (a_j^2 + b_j^2) \right)^{\frac{1}{2}}.$$

If we repeat the Galerkin-Urabe procedure with m replaced by $M \gg m$ then we may choose the computationally available term

$$[\varepsilon_m] = |y_M - y_m| = \left(\sum_{j=m+1}^{M} (a_j^2 + b_j^2) \right)^{\frac{1}{2}} \leq \varepsilon_m$$

as a reasonable error estimate. Upon again ignoring the difference between the Fourier coefficients y_m and $\hat{y}_m$, we may apply a well-known approximation result from Fourier analysis: assume that the unknown function y and all its approximations y_m are analytic, then the coefficients obey some exponential decay law, which we write in the form

$$\varepsilon_m \doteq Ce^{-\gamma m},$$

where the coefficients C, γ are unknown a-priori and need to be estimated. For this purpose, let the optimal number m^* be such that

$$\varepsilon_{m^*} \doteq \text{TOL}$$

and assume that this is not yet achieved for the actual index m. Then a short calculation (see also Exercise 7.6 for more details) shows that m^* can be estimated by the adaptive rule (with some further index $l \ll m$)

$$m^* \doteq m + (m - l)\frac{\log([\varepsilon_m]/\text{TOL})}{\log([\varepsilon_l]/[\varepsilon_m])}. \tag{7.30}$$

Only with such an adaptivity device added, the Galerkin-Urabe (also: harmonic balance) method can be expected to supply reliable computational results.

From (7.28) and (7.29) we may readily observe that in the Galerkin-Urabe approach the BVP (7.24) has been tacitly replaced by the *discrete boundary value problem*

$$\hat{y}_m'(t_k) = f(\hat{y}_m(t_k)) , \quad k = 0, 1, \ldots, N , \quad \hat{y}_m(T) = \hat{y}_m(0) ,$$

wherein $t_0 = 0, t_N = T$ by definition. The boundary conditions are implicitly taken into account by the Fourier ansatz. One of the conditions, at $t = 0$ or at $t = T$, can be dropped due to periodicity so that there are nN so-called *collocation conditions* left. By construction, the method inherits the symmetry of the BVP with respect of an interchange of the boundaries $t = 0$ and $t = T$—indicating that this computational approach treats periodic BVPs as *spacelike* BVPs—as opposed to the preceding Sections 7.3.1 and 7.3.2 where multiple shooting approaches for timelike BVPs have been discussed. The following Section 7.4 is fully devoted to collocation methods for spacelike BVPs—there, however, in connection with polynomial approximation.

7.4 Polynomial Collocation for Spacelike BVPs

In the collocation approach the interval $[a, b]$ is subdivided into a partition

$$\Delta = \{a = t_1 < t_2 < \cdots < t_m = b\}, \quad m > 2,$$

where each subinterval $I_j = [t_j, t_{j+1}]$ of length $\tau_j = t_{j+1} - t_j$ is further subdivided by s internal nodes, the so-called *collocation points*

$$t_{ji} = t_j + c_i \tau_j, \quad i = 1, \ldots, s, \quad 0 \le c_1 < \cdots < c_s \le 1$$

corresponding to some quadrature rule of order p with nodes c_i. Let Δ_* denote the union of all collocation points—to be distinguished from the above defined coarse mesh Δ.

Let u denote the collocation polynomial to be computed for the given BVP. The collocation polynomial is defined via the n boundary conditions

$$F_m = r(u(a), u(b)) = 0$$

typically assumed to be linear separated (cf. [9, 71])

$$F_m = Au(a) + Bu(b) - d = 0, \tag{7.31}$$

so that all components arising therein can be fixed. At the 'internal' nodes we require the $(m-1)sn$ 'local' collocation conditions

$$F_{ji} = u'(t_{ji}) - f(u(t_{ji})) = 0, \qquad t_{ji} \in \Delta_* \tag{7.32}$$

often in the scaling invariant form (i.e. invariant under rescaling of the variable t)

$$F_{ji} = \tau_j \left(u'(t_{ji}) - f(u(t_{ji})) \right) = 0, \qquad t_{ji} \in \Delta_*.$$

Finally, the $(m-1)n$ 'global' collocation conditions

$$F_j = u(t_{j+1}) - u(t_j) - \tau_j \sum_{l=1}^{s} b_l f(u(t_{jl})) = 0, \quad t_j \in \Delta \tag{7.33}$$

must hold, wherein the quadrature rule implies the relation

$$\sum_{l=1}^{s} b_l = 1.$$

By construction, the collocation approach is invariant under $a \leftrightarrow b$ whenever a *symmetric* quadrature rule with

$$c_i = 1 - c_{s+1-i}, \quad b_i = b_{s+1-i}, \qquad i = 1, \ldots, s$$

is selected. In fact, symmetric collocation methods are realized in nearly all public domain codes, since they permit the highest possible convergence orders p by one out of the following two options:

- Gauss methods. Here collocation points are selected as the nodes of Gauss-Legendre quadrature. This leads to the highest possible order $p = 2s$. The simplest case with $s = 1$ is just the implicit midpoint rule. Since $c_0 > 0$ and $c_s < 1$, the nodes of the coarse mesh are not collocation points, i.e., $\Delta_* \cap \Delta = \emptyset$, and hence only $u \in C^0[a, b]$ is obtained.

- Lobatto methods. Here the collocation points are selected as the nodes of Lobatto quadrature. The attainable order is $p = 2s - 2$. The simplest case is the implicit trapezoidal rule with $p = s = 2$. Since $c_0 = 0$ and $c_s = 1$, the nodes of the coarse mesh are included in the set of collocation points, i.e., $\Delta_* \cap \Delta = \Delta$. The lower order (compared with the Gauss methods) comes with better global smoothness, since here $u \in C^1[a, b]$.

In algorithmic implementations, Gauss methods are usually preferred due to their more robust behavior in nonsmooth BVPs.

BIBLIOGRAPHICAL NOTE. Efficient collocation methods have been implemented in the classical code COLSYS of U.M. Ascher, J. Christiansen, and R.D. Russell and its more recent variant COLNEW by G. Bader and U.M. Ascher [16]. An adaptive Gauss-Newton continuation method (as described in Section 5.2) has been implemented in the code COLCON by G. Bader and P. Kunkel [17]. An advanced *residual based inexact* Gauss-Newton continuation method has been designed for collocation by A. Hohmann [120] and realized in the rather robust research code COCON. Unfortunately, that line of development has not continued toward a fully satisfactory general purpose collocation code. We will resume the topic, in Section 7.4.2 below, in the frame of *error oriented inexact* Newton methods. Recently, this concept has regained importance in a novel multilevel algorithm for optimal control problems based on function space complementarity methods—a topic beyond the present scope, for details see M. Weiser and P. Deuflhard [197].

7.4.1 Discrete versus continuous solutions

From multiple shooting techniques we are accustomed to the fact that, whenever the underlying BVP has a locally unique solution, the discrete system also has a locally unique solution—just look up Lemma 7.1 and the interpretation thereafter. *This need not be the case for global discretization methods.* Here additional 'spurious' discrete solutions may occur that have nothing to do with the unique continuous BVP solution. In [71, Section 8.4.1] this situation has been analyzed for the special method based on the implicit trapezoidal rule, the simplest Lobatto method already mentioned above. In what follows we want to give the associated analysis for the whole class of collocation methods. We will mainly focus on Gauss collocation methods, which are the ones actually realized in the most efficient collocation codes. Our results do, however, also apply to other collocation schemes.

Throughout this section we assume—without proof—that the BVP has a *unique* solution $y \in C^{p+1}[a, b]$ and that the collocation polynomial u is globally continuous, i.e., $u \in C^0[a, b]$, but only piecewise sufficiently differentiable, $u \in C^{p+1}[t_j, t_{j+1}]$, which we denote by $u \in C_\Delta^{p+1}[a, b]$. In what follows we restrict our attention to Gauss methods, which means $p = 2s$. Consequently, the discretization error $\epsilon(t) = u(t) - y(t), t \in [a, b]$ satisfies $\epsilon \in C_\Delta^{p+1}[a, b]$. In order to study its behavior, we introduce norms over the grids Δ, Δ_*, for example

$$|\epsilon|_\Delta = \max_{t \in \Delta} \|\epsilon(t)\|$$

in terms of some vector norm $\| \cdot \|$. Let

$$\tau = \max_{j=1,\ldots,m-1} \tau_j$$

denote the maximum mesh size on the coarse grid Δ. With these preparations we are now ready to state a convergence theorem for Gauss collocation methods.

Theorem 7.2 *Notation as just introduced. Consider a BVP on $[a, b]$ with linear separated boundary conditions and a right side f that is p-times differentiable with respect to its argument. Assume that the BVP has a unique solution $y \in C^{p+1}[a, b]$. Let this solution be well-conditioned with bounded interval condition number $\bar{\rho}$. Define a global Lipschitz constant ω via*

$$\|f_y(v) - f_y(w)\| \leq \omega \|v - w\| \, . \tag{7.34}$$

Consider a Gauss collocation scheme based on a quadrature rule of order $p = 2s$. Let the discrete BVP have a collocation solution $u \in C_\Delta^{p+1}[a, b]$ that is consistent with the BVP solution y. Let γ, γ^* denote error coefficients corresponding to the Gauss quadrature rule and depending on the smoothness of the right hand side f. Then, for

$$\tau \leq \left(2\omega\gamma^*(\bar{\rho}|b - a|)^2\right)^{-\frac{1}{s+1}} , \tag{7.35}$$

the following results hold:

(I) *At the local (internal) nodes the pointwise approximation satisfies*

$$|u - y|_{\Delta_*} \leq 2\bar{\rho}|b - a|\gamma^* \tau^{s+1} \, . \tag{7.36}$$

(II) *At the global nodes superconvergence holds in the sense that*

$$|u - y|_\Delta \leq \bar{\rho}|b - a| \left(2\omega(\bar{\rho}|b - a|\gamma^* \tau)^2 + \gamma\right) \tau^{2s} \, . \tag{7.37}$$

Proof. I. We begin with deriving a perturbed variational equation for the discretization error $\epsilon \in C_\Delta^{p+1}[a, b]$. For $t \in I_j = [t_j, t_{j+1}]$ we may write

$$\epsilon'(t) - f_y(y(t))\epsilon(t) = \delta f(t) + \delta\varphi(t)\,, \tag{7.38}$$

where

$$\delta f(t) = u'(t) - f(u(t))$$

and

$$\begin{aligned}
\delta\varphi(t) &= f(u(t)) - f(y(t)) - f_y(y(t))\epsilon(t) \\
&= \int_{\Theta=0}^{1} \big(f_y(y(t) + \Theta\epsilon(t)) - f_y(y(t))\big)\epsilon(t)\, d\Theta\,.
\end{aligned}$$

The above first term δf vanishes at the collocation points and gives rise to the local upper bound for the interpolation error (see, e.g., [71, Thm. 7.16])

$$\max_{\sigma\in I_j}\|\delta f(\sigma)\| \le \overline{\gamma}\tau_j^s\,. \tag{7.39}$$

The second term $\delta\varphi$ contains the nonlinear contribution and satisfies the pointwise estimate

$$\|\delta\varphi(t)\| \le \tfrac{1}{2}\omega\|\epsilon(t)\|^2\,. \tag{7.40}$$

By variation of constants (see Exercise 7.7) the differential equation (7.38) can be formally solved for $t \in [t_j, t_{j+1}]$ to yield

$$\epsilon(t) = W(t,t_j)\epsilon(t_j) + \int_{\sigma=t_j}^{t} W(t,\sigma)\,(\delta f(\sigma) + \delta\varphi(\sigma))\, d\sigma\,,$$

where $W(\cdot,\cdot)$ denotes the (Wronskian) propagation matrix, the solution of the unperturbed variational equation—see [71, Section 3.1.1]. This, however, is just a representation of the solution of the IVP on each subinterval. Therefore, any estimates based on this formula would bring in the IVP condition number, which we want to avoid for spacelike BVPs.

For this reason, we need to include the boundary conditions, known to be linear separated, so that

$$\delta r = A\epsilon(t_1) + B\epsilon(t_m) = 0\,.$$

Upon combining these results, we obtain the formal global representation

$$\epsilon(t) = \int_{\sigma=a}^{b} G(t,\sigma)\,(\delta f(\sigma) + \delta\varphi(\sigma))\, d\sigma\,,$$

where $G(\cdot,\cdot)$ denotes the *Green's function* of the (linear) variational BVP. Its global upper bound is the *condition number* of the (nonlinear) BVP (as defined in [71, Section 8.1.2]):

$$\bar\rho = \max_{t,\bar t \in [a,b]} \|G(t,\bar t)\|.$$

Since the BVP is assumed to be well-conditioned, the above condition number is bounded. Note that, by definition, the Green's function has a jump at $t = \bar t$ such that

$$G(t,t^+) - G(t,t^-) = I.$$

For the subsequent derivation we decompose the integral into subintegrals over each of the subintervals such that

$$\epsilon(t) = \sum_{k=1}^{m-1} \int_{\sigma=t_k}^{t_{k+1}} G(t,\sigma) \left(\delta f(\sigma) + \delta\varphi(\sigma)\right) d\sigma.$$

II. We are now ready to derive upper bounds for the discretization error. For $t_j \in \Delta$, we may directly apply the corresponding Gauss quadrature rule, since the above jumps occur only at the boundaries of each of the subintegrals. Along this line we obtain

$$\epsilon(t_j) = \sum_{k=1}^{m-1} \tau_k \left(\sum_{l=1}^{s} b_l G(t_j, t_{kl})\delta\varphi(t_{kl}) + \Gamma_k(\cdot)\tau_k^{2s} \right).$$

The argument in the remainder term $\Gamma_k(\cdot)$ is dropped, since we are only interested in its global upper bound, say

$$\|\Gamma_k(\cdot)\| \le \bar\rho\gamma.$$

Introducing pointwise norms on the coarse grid Δ, and exploiting (7.40) and (7.36), we are then led to

$$\|\epsilon(t_j)\| \le \bar\rho|b-a| \left(|\delta\varphi|_{\Delta_*} + \gamma\tau^{2s}\right),$$

which yields

$$|\epsilon|_\Delta \le \bar\rho|b-a| \left(\tfrac{1}{2}\omega|\epsilon|^2_{\Delta_*} + \gamma\tau^{2s}\right). \tag{7.41}$$

Next we consider arguments $t = t_{ji} \in \Delta_*$. In this case the jumps do occur inside one of the subintegrals. Hence, we have to be more careful in our estimate. We start with

$$\|\epsilon(t)\| \le \sum_{k=1}^{m-1} \left\| \int_{\sigma=t_k}^{t_{k+1}} G(t,\sigma) \left(\delta f(\sigma) + \delta\varphi(\sigma)\right) d\sigma \right\|$$

from which we immediately see that the integrals over I_k for $k \ne j$ can be treated as before

$$\left\| \int_{\sigma=t_k}^{t_{k+1}} G(t,\sigma) \left(\delta f(\sigma) + \delta\varphi(\sigma)\right) d\sigma \right\| \le \bar\rho\tau_k \left(\gamma\tau_k^{2s} + \tfrac{1}{2}\omega|\epsilon|^2_{\Delta_*}\right).$$

For $k = j$, however, we just obtain

$$\left\| \int_{\sigma=t_j}^{t_{j+1}} G(t,\sigma)\,(\delta f(\sigma) + \delta\varphi(\sigma))\,\,d\sigma \right\| \le \bar{\rho}\tau_j \left(\max_{\sigma \in I_j} \|\delta f(\sigma)\| + \tfrac{1}{2}\omega|\epsilon|_{\Delta_*}^2 \right).$$

If we recall (7.39) and define the quantity

$$\gamma^*\tau = \left(\gamma\tau^s + \bar{\gamma}\frac{\tau}{|b-a|}\right),$$

we end up with the quadratic inequality

$$|\epsilon|_{\Delta_*} \le \bar{\rho}|b-a|(\tfrac{1}{2}\omega|\epsilon|_{\Delta_*}^2 + \gamma^*\tau^{s+1}). \qquad (7.42)$$

For the solution of this inequality, we introduce the majorant $|\epsilon|_{\Delta_*} \le \bar{\epsilon}$ generating the quadratic equation

$$\bar{\epsilon} = \bar{\rho}|b-a|(\tfrac{1}{2}\omega\bar{\epsilon}^2 + \gamma^*\tau^{s+1}).$$

For the discriminant to be nonnegative we need to require

$$\tau^{s+1} \le \frac{1}{2\omega\gamma^*(\bar{\rho}|b-a|)^2},$$

which is just statement (7.35) of the theorem. This situation is represented graphically in Figure 7.5.

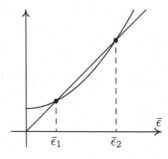

Fig. 7.5. Left and right side of quadratic inequality (7.42).

With the notations

$$\alpha = \frac{1}{\bar{\rho}|b-a|\omega}, \quad \tau_c = \left(2\omega\gamma^*(\bar{\rho}|b-a|)^2\right)^{-\frac{1}{s+1}}$$

we obtain the two majorant roots

$$\bar{\epsilon}_1 = \frac{\alpha(\tau/\tau_c)^{s+1}}{1 + \sqrt{1 - (\tau/\tau_c)^{s+1}}}, \quad \bar{\epsilon}_2 = \alpha\left(1 + \sqrt{1 - (\tau/\tau_c)^{s+1}}\right) = 2\alpha - \bar{\epsilon}_1.$$

For $\tau \in [0, \tau_c[$ we obtain the bounds

$$0 \leq \bar{\epsilon}_1 \leq \alpha^-, \quad 2\alpha \geq \bar{\epsilon}_2 \geq \alpha^+.$$

The situation is depicted in Figure 7.6.

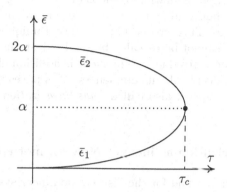

Fig. 7.6. Error bounds for consistent discrete solution ($\bar{\epsilon} \leq \bar{\epsilon}_1$) and spurious or 'ghost' solutions ($\bar{\epsilon} \geq \bar{\epsilon}_2$).

First we pick $\bar{\epsilon}_1$, the root consistent with the continuous BVP, and are led to the approximation result

$$|\epsilon|_{\Delta_*} \leq \bar{\epsilon}_1 \leq 2\bar{\rho}|b - a|\gamma^* \tau^{s+1},$$

which verifies statement (7.36). Second we study the root $\bar{\epsilon}_2$. Again under the meshsize constraint (7.35), the quadratic inequality can be seen to characterize a further *discrete solution* branch by

$$|\epsilon|_{\Delta_*} \geq \bar{\epsilon}_2.$$

Obviously, the proof permits the existence of inconsistent discrete solutions. However, if such solutions exist, they are well-separated from the consistent ones as long as $\tau < \tau_c$.

As a final step of the proof, we may just insert the upper bound (7.36) into (7.41) and arrive at

$$|\epsilon|_\Delta \leq \bar{\rho}|b - a| \left(\tfrac{1}{2}\omega(\bar{\rho}|b - a|\gamma^* \tau^{s+1})^2 + \gamma\tau^{2s}\right)$$
$$\leq \bar{\rho}|b - a| \left(\tfrac{1}{2}\omega(\bar{\rho}|b - a|\gamma^* \tau)^2 + \gamma\right)\tau^{2s}.$$

This is the desired superconvergence result (7.37) and thus completes the proof.

Of course, the theorem does not state anything about the situation when the
meshsize restriction (7.35) does *not* hold, i.e. for $\tau > \tau_c$.   □

The above discussed possible occurrence of *'ghost' solutions*, i.e. of *inconsistent discrete solutions*, which have nothing to do with the continuous solution, has been experienced by computational scientists, both in ODEs and
in PDEs. In words, the above theorem states that a computed collocation
solution u is a valid approximation of the BVP solution y only, if the applied
mesh is 'sufficiently fine' and if this solution 'essentially is preserved' on successively finer meshes. This means that the actual uniqueness structure of a
collocation solution cannot be revealed by solving just one finite-dimensional
problem. Rather, successive mesh refinement is additionally needed as an algorithmic device to decide about uniqueness. This paves the way to Newton
methods in function space, also called *quasilinearization*—to be treated in
the next section.

7.4.2 Quasilinearization as inexact Newton method

Instead of a Newton method for the discrete nonlinear system (7.31), (7.32),
and (7.33), the popular collocation codes realize some *quasilinearization* technique, i.e., a Newton method in function space. Of course, approximation
errors are unavoidable, which is why *inexact* Newton methods in function
space are the correct conceptual frame.

We start with the *exact* ordinary Newton iteration in *function space*

$$y_{k+1}(t) = y_k(t) + \delta y_k(t) , \quad k = 0, 1, \dots .$$

Herein the Newton corrections δy_k satisfy the linearized BVP, which is the
perturbed variational equation with linear separated boundary conditions:

$$\delta y_k' - f_y(y_k(t))\delta y_k = -(y_k'(t) - f(y_k(t))) , \quad t \in [a, b],$$
$$A\delta y_k(a) + B\delta y_k(b) = 0 .$$

Newton's method in the infinite dimensional function space can rarely be
realized, apart from toy problems. Instead we study here the corresponding
exact Newton method in *finite dimensional space*, i.e., the space spanned by
the collocation polynomials on the grids Δ and Δ_* (as introduced in the
preceding section). Formally, this method replaces $y, \delta y$ by their polynomial
representations $u, \delta u$, where

$$\delta u_k'(t_{ji}) - f_y(u_k(t_{ji}))\delta u_k(t_{ji}) = -\delta f_k(t_{ji}) = -(u_k'(t_{ji}) - f(u_k(t_{ji}))) ,$$
$$A\delta u_k(a) + B\delta u_k(b) = 0 .$$

$$(7.43)$$

If we include a damping factor λ_k to expand the local domain of convergence
of the ordinary Newton method, we arrive at

$$u_{k+1}(t) = u_k(t) + \lambda_k \delta u_k(t) \,. \tag{7.44}$$

Note that here Theorem 7.2 applies with global Lipschitz constant $\omega = 0$. As a consequence there are *no spurious solutions* δu to be expected—which clearly justifies the use of quasilinearization. Nevertheless, the chosen mesh might be not 'fine enough' to represent the solution correctly—see the discussion on mesh selection below.

Linear band system. Within each iteration, dropping the index k, we have to solve the finite-dimensional linear system consisting of the (scaling invariant) local collocation conditions

$$F_{ji} = \tau_j \left(\delta u'(t_{ji}) - f_y(u(t_{ji}))\delta u(t_{ji}) + \delta f(t_{ji}) \right) = 0 \,,$$

the global collocation conditions

$$F_j = \delta u(t_{j+1}) - \delta u(t_j) - \tau_j \sum_{l=1}^{s} b_l \left(f_y(u(t_{jl}))\delta u(t_{jl}) - \delta f(t_{jl}) \right) = 0 \tag{7.45}$$

and the boundary conditions

$$F_m = A\delta x_1 + B\delta x_m = 0 \,.$$

In most implementations, the local collocation conditions are realized in the equivalent initial value problem form

$$F_{ji} = \delta u(t_{ji}) - \delta u(t_j) - \tau_j \sum_{l=1}^{s} a_{il} \left(f_y(u(t_{jl}))\delta u(t_{jl}) - \delta f(t_{ji}) \right) \,. \tag{7.46}$$

In this case, the local variables $u(t_{j1}), \ldots, u(t_{js})$ can be condensed—which means expressed in terms of the global variables $u(t_j)$. However, unlike the equations (7.45) and (7.43), the part (7.46) is not symmetric with respect to $a \leftrightarrow b$. The separated boundary conditions can be dropped by fixing the proper components of the boundary values.

The remaining linear system has block tridiagonal structure. It is usually solved by some global direct elimination method such as the modified band solver due to J.M. Varah [191].

In the already mentioned Gauss-Newton continuation method for parameter dependent BVPs (cf. [17]), the discretized linear system is just enhanced by a further column, which requires a slight modification of the elimination method; the corresponding rank-deficient Moore-Penrose pseudoinverse is then simply realized via the representation (5.29) as presented in Section 5.2.

Norms. As in Section 7.4.1 above we here also write $\| \cdot \|$ for a local vector norm. With $| \cdot |$ we will mean the canonical C^0-norm defined as

$$|\epsilon|_{0,[a,b]} = \max_{t \in [a,b]} \|\epsilon(t)\|,$$

whose discretization is

$$|\epsilon|_{0,\Delta} = \max_{t \in \Delta} \|\epsilon(t)\|.$$

If we scale the L_2-norm appropriately, we find that

$$|\epsilon|_{2,[a,b]} = \left(\frac{1}{b-a} \int_a^b \|\epsilon(t)\|^2 dt \right)^{\frac{1}{2}} \leq |\epsilon|_{0,[a,b]}.$$

From this, we may turn to the corresponding discretization

$$|\epsilon|_{2,\Delta} = \left(\frac{1}{b-a} \sum_{j=1}^{m-1} \tau_j \sum_{l=1}^{s} b_l \|\epsilon(t_{jl})\|^2 \right)^{\frac{1}{2}}$$

and obtain the corresponding relation

$$|\epsilon|_{2,\Delta} \leq |\epsilon|_{0,\Delta^*}.$$

Note that, due to Gaussian quadrature, we have the approximation property

$$|\epsilon|_{2,\Delta} = |\epsilon|_{2,[a,b]} + O(\tau^{2s}) \leq |\epsilon|_{0,[a,b]} + O(\tau^{2s}).$$

Below we will not distinguish between any of these essentially equivalent norms and write subscripts only where necessary.

Discretization error estimates. The above exact finite dimensional Newton method can also be viewed as an *inexact* Newton method in *function space*, if we include the discretization errors into a unified mathematical frame. In fact, any *adaptive* collocation method will need to control the arising discretization error

$$\epsilon(t) = \delta u(t) - \delta y(t)$$

at least via some estimate of it, in some suitable norm (see above). For convenience, we again use the notation $I_j = [t_j, t_{j+1}]$ for the subintervals of the coarse mesh Δ.

Before we start, let us draw a useful consequence of Theorem 7.2: we interpret the Gaussian collocation method as an implicit Runge-Kutta method (compare, e.g., [115, 71]) and thus immediately obtain (for $k < s$)

$$\delta u^{(k)}(t) - \delta y^{(k)}(t) = \delta y^{(s+1)}(t_j) \tau_j^{s+1} P_s^{(k)} \left(\frac{t - t_j}{\tau_j} \right) (1 + O(\tau_j)) + O(\tau_j^{2s}), \quad (7.47)$$

where P is a polynomial of degree s—see, e.g., [9, Section 9.3]. Since the local collocation polynomials are of order s, their derivative $\delta u^{(s)}$ is piecewise constant, i.e.,

$$\delta u^{(s)}(t) = \text{const}_j, \ , \quad \delta u^{(s+1)}(t) = 0 \ , \quad t \in I_j \ .$$

At first glance, we do not seem to have a reasonable approximation of $\delta y^{(s+1)}$ that could be used to estimate the error bound (7.47). However, we may overcome this lack of information by defining piecewise linear functions δz_s via the interpolation conditions at the subinterval midpoints $\bar{t}_j = t_j + \frac{1}{2}\tau_j$ such that

$$\delta z_s(\bar{t}_j) = \delta u^{(s)}(\bar{t}_j) \ .$$

The situation is depicted schematically in Figure 7.7. The derivative $\delta z_s'$ is piecewise constant with jumps at the subinterval midpoints $\bar{t}_j$.

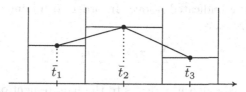

Fig. 7.7. Piecewise linear approximation δz_s of piecewise constant function $\delta u^{(s)}$

Since $P_s^{(s)}(\frac{1}{2}) = 0$, (7.47) implies the superconvergence result

$$\delta u^{(s)}(\bar{t}_j) - \delta y^{(s)}(\bar{t}_j) = O(\tau_j^{2s}) \ ,$$

from which we obtain the approximation property

$$\delta z_s'(t) = \delta y^{(s+1)}(t_j) + O(\tau_j) \ , \quad t \in I_j \ .$$

Hence, in first order of the local mesh size τ_j, we obtain the componentwise estimate

$$|\delta u(t) - \delta y(t)| \doteq |\delta z_s'(t_j)|\tau_j^{s+1} \ , \quad t \in I_j \ .$$

If we average according to the rule

$$|\delta z_s'(\bar{t}_1)| = |\delta z_s'(t_1)| \ , \quad |\delta z_s'(\bar{t}_m)| = |\delta z_s'(t_m)| \ ,$$
$$|\delta z_s'(\bar{t}_j)| = \tfrac{1}{2}\left(|\delta z_s'(t_j)| + |\delta z_s'(t_{j+1})|\right) \ , \quad j = 2, \ldots, m-1 \ ,$$

we arrive at the cheaply computable local discretization error estimates

$$\max_{t \in I_j} |\delta u(t) - \delta y(t)| \doteq \epsilon_j = |\delta z_s'(\bar{t}_j)|\tau_j^{s+1} \ . \tag{7.48}$$

In [120], A. Hohmann has suggested a realization with variable order $p_j = 2s_j$ in different subintervals I_j; for such an $h-p$-strategy, the above estimation technique will no longer be applicable. Instead, the Gauss interpolation polynomial is extended by the additional boundary nodes t_j, t_{j+1}. This kind of collocation polynomial is of order $s_j + 2$. Its difference to the computed solution can then serve as an estimate of order $s_j + 1$ replacing (7.48).

Global mesh selection. This issue is crucial in every global discretization method, at least when the numerical solution of really challenging spacelike BVPs is envisioned, including the possible occurrence of internal or boundary layers. Typically, a starting mesh Δ will be defined at the beginning of the discretization process, which may already contain some information about the expected behavior of the solution. Within each quasilinearization step the componentwise global discretization error estimate

$$|\epsilon| = \max_{j=1,..,m-1} \epsilon_j$$

can be computed as indicated above. In order to minimize this maximum subject to the constraint

$$\sum_{j=1}^{m-1} \tau_j = b - a \,,$$

the well-known greedy algorithm leads to the requirement of *equidistribution* of the local discretization errors. Therefore, any reasonable mesh selection device will aim, at least asymptotically, at

$$\epsilon_j = C_j \tau_j^{s+1} \approx \text{const} \,, \quad j = 1, \ldots, m - 1$$

with coefficients C_j as defined above; an equivalent formulation is

$$\epsilon_j \approx \bar{\epsilon} = \frac{1}{m-1} \sum_{l=1}^{m-1} \epsilon_l \,.$$

There are various options to realize this equidistribution principle. In the code family COLSYS the number m of nodes is adapted such that the global error estimate eventually satisfies

$$|\epsilon| \le \text{TOL}$$

in terms of the user prescribed error tolerance TOL. Some codes also permit nonnested successive meshes.

In view of the theoretical approximation results (as given, e.g., in the previous section), COLSYS uses a further *global* mesh refinement criterion based on an affine covariant Newton method including a damping strategy [60, 63, 9]. Whenever the damping factor turns out to be 'too small', i.e., whenever $\lambda_k < \lambda_{\min}$ occurs, for some prescribed threshold value $\lambda_{\min} \ll 1$, then a new mesh with precisely halved local stepsizes is generated. Note that in Newton methods, which are *not* affine covariant, such a criterion may be activated in the wrong situation: not caused by the nonlinearity of the problem, but by the ill-conditioning of the discrete equations—which automatically comes with the fact that the BVP operator is noncompact.

Local mesh refinement. Instead of the mesh selection devices realized within COLSYS we here want to work out an alternative option in the spirit of *adaptive multilevel methods* for partial differential equations (to be treated in Section 8.3 below). Let

$$\Delta_0 \subset \Delta_1 \subset \cdots \subset \Delta_d$$

denote a sequence of *nested* meshes. By construction, the mesh Δ_0 is just the given initial mesh, while the mesh Δ_1 is obtained by halving of all subintervals. All further meshes can be obtained using an adaptivity device based on *local extrapolation*—a technique that has been suggested by I. Babuška and W.C. Rheinboldt [12] already in 1978, there for finite element methods in partial differential equations. For details we here recur to Section 9.7.1 of the elementary textbook [77], where this technique is explained in the simple context of numerical quadrature. Assume we have already computed local error estimates on two consecutive meshes, on the given mesh Δ and its coarser predecessor Δ^-. Let $I := (t_l, t_m, t_r) \in \Delta^-$ denote an interval bisected into subintervals $I_l, I_r \in \Delta$, where

$$I_l := \left(t_l, \tfrac{1}{2}(t_l + t_m), t_m\right) \quad \text{and} \quad I_r := \left(t_m, \tfrac{1}{2}(t_r + t_m), t_r\right) .$$

Repeated refinement leads to a recursive *binary tree*—see Figure 7.8.

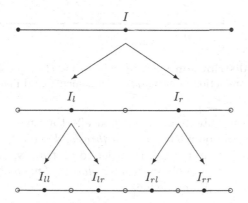

Fig. 7.8. Double refinement of subinterval $I := (t_l, t_m, t_r)$

In contrast to our above approximation results, we make the following more general assumption for the local discretization error

$$\epsilon(I_j) \doteq C\tau_j^\gamma \qquad (7.49)$$

with a local order γ and a local problem dependent constant C to be roughly identified; note that above we derived $\gamma = s$ in the upper bounds, which may be unrealistic in a specific example. Let I be a subinterval obtained by refinement; then we denote the starting interval from the previous mesh by I^-, i.e., $I_r^- = I_l^- = I$. Dropping the local index j, assumption (7.49) then implies

$$\epsilon(I^-) \doteq C(2\tau)^\gamma = 2^\gamma C\tau^\gamma \doteq 2^\gamma \epsilon(I) \,,$$

from which we conclude that

$$\epsilon(I_l) \doteq C\tau^\gamma 2^{-\gamma} \doteq \epsilon(I)\epsilon(I)/\epsilon(I^-) \,.$$

Thus, through *local extrapolation*, we have obtained a local *error prediction*

$$\epsilon^+(I) := \frac{\epsilon^2(I)}{\epsilon(I^-)} \approx \varepsilon(I_l) \,.$$

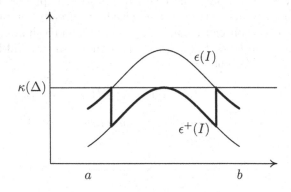

Fig. 7.9. Error distributions: before refinement: $\epsilon(I)$, prediction after *global* refinement: $\epsilon^+(I)$, prediction after *adaptive* refinement: bold line.

We can therefore estimate in advance, what effect a refinement of an interval $I \in \Delta$ would have. We only have to fix a *threshold value* for the local errors, above which we refine an interval. In order to do this, we take the maximal local error, which we would obtain from a *global refinement*, i.e., refinement of *all* subintervals $I \in \Delta$, and define

$$\kappa(\Delta) := \max_{I \in \Delta} \epsilon^+(I) \,.$$

In order to illustrate the situation, we plot the computed estimated errors $\epsilon(I)$ together with the predicted errors $\epsilon^+(I)$ in a smoothed histogram, see Figure 7.9.

Following the aim of local error equidistribution, we do not need to refine near the left and right boundary; rather, refinement will pay off only in the center region. We thus arrive at the following *refinement rule*: Refine only those intervals $I \in \Delta$, for which

$$\epsilon(I) \geq \kappa(\Delta).$$

This yields the error distribution displayed in bold line in Figure 7.9, which is narrower than both the original distribution and the one to be expected from global refinement.

Clearly, repeated refinement will ultimately generate some rough error equidistribution, provided the local error estimation technique is reliable, which here means that the finest meshes need to be 'fine enough' to activate the superconvergence properties of Gauss collocation methods.

Error matching. If we view quasilinearization as an inexact Newton method in function space, we need to match the Newton corrections δu_k and the discretization errors $\delta u_k - \delta y_k$. In his dissertation [120], A. Hohmann worked out a residual based inexact Newton method using a Fredholm basis for the local representation of the collocation polynomials and obtained some rather robust algorithm. Here, however, we want to realize the *error oriented* local (Section 2.1.5) and global Newton methods (Section 3.3.4) within the collocation code COLSYS—transferring the finite dimensional case therein to the present infinite dimensional one.

At each iteration index k, identify $u_k = y_k$—as a common starting point, say—and define the two Newton corrections in function space

$$F'(u_k)\delta u_k = -F(u_k) + r_k , \quad F'(u_k)\delta y_k = -F(u_k). \tag{7.50}$$

In contrast to the setting in finite dimensional inexact Newton methods (see Chapters 2 and 3), here the residual r_k is not generated by some inner iteration, but by the discretization error such that

$$F'(u_k)(\delta u_k - \delta y_k) = r_k .$$

In collocation, the situation is characterized by the fact that we have cheap computational estimates of the relative *discretization error*

$$\delta_k = \frac{|\delta u_k - \delta y_k|}{|\delta u_k|}$$

available in some (approximate) norm $|\cdot|$—compare (3.50). This quantity essentially depends on the selected mesh. For successively fine meshes, this quantity will approach zero, which is part of the asymptotic mesh independence to be discussed in detail in Section 8.1 below (see also Exercise 8.3).

The iteration (7.44) is realized as a finite dimensional Newton iteration with adaptive trust region (or damping) strategy—as long as the mesh is kept unchanged. However, in order to really solve the BVP and not just some discrete substitute system with unclear approximation quality, the mesh should be adapted along with the iteration: the idea advocated here is to aim at some asymptotic confluence with an exact Newton iteration based on the correction δy_k. In view of (3.55) and the analysis in Section 3.3.4, we will require that, for some $\rho \leq 1$,

$$\delta_k \leq \frac{\rho}{2(1+\rho)} \leq \tfrac{1}{4} \text{ for } \overline{\lambda}_k < 1. \tag{7.51}$$

As soon as the iteration swivels in the *ordinary* Newton phase with $\lambda = 1$ throughout, then a more careful consideration is needed, which can be based on the following theoretical estimates.

Theorem 7.3 *Let δu_k and δy_k denote the inexact and exact Newton corrections as defined in (7.50). Let ω be the affine covariant Lipschitz constant defined via*

$$\left| F'(u)^{-1} \left(F'(v) - F'(w) \right) \right| \leq \omega |v - w| .$$

Then, with the notation of the present section, we obtain

(I) *for the ordinary exact Newton method in finite dimension the quadratic convergence result*

$$|\delta u_{k+1}| \leq \tfrac{1}{2} h_k^\delta |\delta u_k|$$

with $h_k^\delta = \omega |\delta u_k|$,

(II) *for the stepwise 'parallel' ordinary inexact Newton method in function space, as defined in (7.50), the mixed convergence results*

$$|\delta y_{k+1}| \leq \frac{\tfrac{1}{2} h_k^\delta + (1 + h_k^\delta)\delta_k}{1 - \delta_k} |\delta y_k|$$

and, under the additional matching assumption for some safety factor $\rho \leq 1$,

$$\delta_k \leq \tfrac{1}{2}\rho \frac{h_k^\delta}{1 + h_k^\delta} , \tag{7.52}$$

the modified quadratic convergence results

$$|\delta y_{k+1}| \leq \tfrac{1}{2}(1 + \rho) \frac{h_k^\delta}{1 - \delta_k} |\delta y_k| ,$$

where contraction is realized, if $h_0^\delta < \dfrac{2(1 - \delta_0)}{1 + \rho}$.

Proof. Part (I) of the theorem is standard. Part (II) is a slight modification of Theorem 2.11 in Section 2.1.5, there derived for the inner iterative solver GBIT, which does not satisfy any orthogonality properties. The definition of

h_k^δ is different here; it can be directly inserted into the intermediate result (2.54). □

Clearly, we will select the same value of ρ in both (7.51) for the damped Newton method and (7.52) for the ordinary Newton method.

Recall that the derivation of (7.51) required that λ_k needs to be chosen according to an adaptive trust region strategy based on computationally available Kantorovich estimates $[h_k^\delta] \leq h_k^\delta$. We are still left with the construction of such estimates. Since the right hand upper bound in (7.52) is a monotone increasing function of h_k^δ, we may then construct an adaptive matching strategy just replacing h_k^δ by its lower bound $[h_k^\delta]$. From the above lemma we directly obtain the a-posteriori estimates

$$[h_k^\delta]_1 = 2\Theta_k \leq h_k^\delta \, , \quad \text{where } \Theta_k = \frac{|\delta u_{k+1}|}{|\delta u_k|}$$

and the a-priori estimate

$$[h_k^\delta] = 2\Theta_{k-1}^2 \leq h_k^\delta \, .$$

With these preparations we are led to the following informal

Error matching algorithm. As long as the finite dimensional global Newton method is still damped, we realize $\delta_k \leq 1/4$ via appropriate mesh selection as given above. Let the index $k = 0$ characterize the beginning of the local Newton method with $\lambda_0 = 1$. Then the following steps are required (skipping emergency exits to avoid infinite loops):

1. $k = 0$: Given u_0, compute δu_0 and its norm $|\delta u_0|$. Compute the discretization error estimate δ_0, e.g., via the suggestion (7.48).

2. **If** $\delta_0 > \frac{1}{4}$, then refine the mesh and goto 1,
   else $u_1 = u_0 + \delta u_0$.

3. $k \geq 1$: Given u_k, compute δu_k, its norm $|\delta u_k|$ and the contraction factor

$$\Theta_{k-1} = \frac{|\delta u_k|}{|\delta u_{k-1}|} \, .$$

   If $\Theta_{k-1} > 1$, then realize an adaptive trust region strategy with $\lambda < 1$ as described in Section 3.3.4,
   else compute the discretization error estimate δ_k.

4. **If**

$$\delta_k > \min\left(\frac{\rho}{2(1+\rho)}, \frac{\rho\Theta_{k-1}^2}{1+\Theta_{k-1}^2}\right) \, ,$$

   then refine the mesh and goto 3,
   else $u_{k+1} = u_k + \delta u_k \to u_k$ and goto 3.

If the adaptive collocation algorithm is performed including the matching strategy as described here, the amount of work will clearly increase from step to step, since the required discretization error estimate will ask for finer and finer (adaptive) meshes. Of course, the process may be modified such that at some iterate the value of δ_k is frozen—with the effect of eventually obtaining unreasonably accurate results u_k.

Remark 7.2   An adaptive collocation method as sketched here has not been implemented so far. However, this kind of ideas has entered into the adaptive multilevel collocation method used within a function space complementarity approach to constrained optimal control problems that has recently been suggested and worked out by M. Weiser and P. Deuflhard [197].

Exercises

Exercise 7.1   Given the extended mapping (7.12), derive the expressions (7.14) for the associated Newton corrections. Sketch details of the corresponding adaptive trust region method. In which way is a control of the actual parameter stepsize performed? Why should the initial iterate τ^0 satisfy $h(\tau^0) \neq 0$?

Exercise 7.2   In order to compute a periodic orbit, one may apply a gradient method [156] as an alternative to the Gauss-Newton method described in Section 7.3. Let, in a single shooting approach, the functional

$$\varphi := \tfrac{1}{2}\|r\|_2^2$$

be minimized. Then

$$\operatorname{grad}\varphi = [E(0),\ f(y(T))]^T r$$

in the autonomous case. Show that

$$E^T r = u(0) - u(T)$$

for some u satisfying

$$u' = -f_y(y)^T u,\quad u(T) = r.$$

Discuss the computational consequences of this relation. Why is, nevertheless, such a gradient method unsatisfactory?

Exercise 7.3   In multiple shooting techniques, assume that local rank reduction leads to a replacement of the Jacobian inverse J^{-1} by the generalized inverse J^- defined in (7.5).

a) Verify the four axioms (7.6) using the Penrose axioms for the Moore-Penrose pseudo-inverse.

b) In the notation of Section 4.3.5, let $\lambda_k \sim [\omega_k]^{-1}$ denote a damping factor estimate associated with the Jacobian inverse J_k^{-1} and $\lambda_k' \sim [\omega_k']^{-1}$ the analog estimate associated with the generalized inverse J_k^-. Give a computationally economic expression for λ_k'. Show that, in general, the relation $\lambda_k' > \lambda_k$ need not hold, unless $m = 2$.

Exercise 7.4   Consider a local rank reduction in multiple shooting techniques as defined by J^- in (7.5) and the axioms (7.6). Show that for the Gauss-Newton direction

$$\Delta x = -J^- F.$$

the residual level function

$$T(x|I) := \|r\|_2^2 + \sum_{j=1}^{m-1} \|F_j\|_2^2$$

is, in general, no longer an appropriate *descent function* in the sense of Lemma 3.11, whereas both the natural level function $T(x|J^-)$ and the hybrid level function $T(x|R^{-1}J^-)$ still are.

Exercise 7.5   Consider a singular perturbation problem of the type

$$\varepsilon y'' + f(t)y' + g(t)y = h(t), \quad y(0) = y_0, \quad y(T) = y_T$$

having one internal layer at some $\tau \in]0, T[$ with $f(\tau) = 0$. Assume that the initial value problem is well-conditioned from τ to 0 and from τ to T so that numerical integration in these directions can be conveniently performed. Consider a multiple shooting approach with $m = 3$ nodes $\{0, \tau, T\}$ that involves numerical integration in the well-conditioned directions. Study the associated Newton method in detail with respect to necessary Jacobian approximations, condensed linear system solver, and iterative refinement sweeps. Discuss extensions for $m > 3$, where the above 3 nodes are among the selected multiple shooting nodes. Interpret this approach in the light of Lemma 7.1.

Exercise 7.6   Consider the Fourier collocation method (also: Urabe or harmonic balance method) as presented in Section 7.3.3.

a) Given the asymptotic decay law

$$\varepsilon_m \doteq Ce^{-\gamma m},$$

verify the computationally available estimate (7.30) for the optimal number m^* of terms needed in the Fourier series expansion (7.24). How many different Galerkin-Urabe approximations are at least required for this estimate?

b) In the derivation of Section 7.3.3 we ignored the difference between the Galerkin approximation y_m with Fourier coefficients a_j, b_j and the actually computed Galerkin-Urabe approximation with coefficients a_j^m, b_j^m depending on the truncation index m. Modify the error estimates so that this feature is taken into account.

Exercise 7.7   Given a perturbed variational equation in the form

$$\epsilon'(t) - f_y(y(t))\epsilon(t) = \delta f(t), \quad t \in [a, b],$$

prove the closed analytic expression

$$\epsilon(t) = W(t, a)\epsilon(a) + \int_{\sigma=a}^{t} W(t, \sigma)\delta f(t)d\sigma.$$

Hint: Apply the variation of constants method recalling that the propagation matrix $W(t, a)$ is a solution of the unperturbed variational equation.

8 PDE Boundary Value Problems

This chapter deals with Newton methods for boundary value problems (BVPs) in nonlinear partial differential equations (PDEs). There are two principal approaches: (a) finite dimensional Newton methods applied to given systems of already discretized PDEs, also called *discrete Newton methods*, and (b) function space oriented inexact Newton methods directly applied to continuous PDEs, at best in the form of *inexact Newton multilevel methods*.

Before we discuss the two principal approaches in detail, we study the underlying feature of *asymptotic mesh independence* that connects the finite dimensional and the infinite dimensional Newton methods, see Section 8.1. In Section 8.2, we assume the standard situation in industrial technology software, where the grid generation module is strictly separated from the solution module. Consequently, nonlinear PDEs there arise as discrete systems of nonlinear equations with fixed finite, but usually high dimension n and large sparse ill-conditioned Jacobian (n, n)-matrix. This is the domain of applicability of finite dimensional inexact Newton methods. More advanced, but typically less convenient in a general industrial environment, are *function space* oriented inexact Newton methods, which additionally include the adaptive manipulation of discretization meshes within a multilevel or multigrid solution process. This situation is treated in Section 8.3 and compared there with *finite dimensional* inexact Newton techniques.

We will *not* treat 'multilevel Newton methods' here (often also called 'Newton multilevel methods'), which are in between discrete Newton methods and inexact Newton methods in function space; they have been extensively treated in the classical textbook [113] by W. Hackbusch or in the synoptic study [135] by R. Kornhuber, who uses an affine conjugate Lipschitz condition.

8.1 Asymptotic Mesh Independence

The term 'mesh independence' characterizes the observation that finite dimensional Newton methods, when applied to a nonlinear PDE on successively finer discretizations with comparable initial guesses, show roughly the same convergence behavior on all sufficiently fine discretizations. In this section, we want to analyze this experimental evidence from an abstract point of view.

Let a general nonlinear operator equation be denoted by

$$F(x) = 0 \,, \tag{8.1}$$

where $F : D \to Y$ is defined on a convex domain $D \subset X$ of a Banach space X with values in a Banach space Y. We assume the existence of a unique solution x^* of this operator equation. The corresponding ordinary Newton method in Banach space may then be written as

$$F'(x^k)\Delta x^k = -F(x^k) \,, \quad x^{k+1} = x^k + \Delta x^k \,, \quad k = 0, 1, \dots \,. \tag{8.2}$$

In each Newton step, the linearized operator equation must be solved, which is why this approach is often also called *quasilinearization*. We assume that an affine covariant version of the classical Newton-Mysovskikh theorem holds— like Theorem 2.2 for the finite dimensional case. Let ω denote the affine covariant Lipschitz constant characterizing the mapping F. Then the quadratic convergence of Newton's method is governed by the relation

$$|x^{k+1} - x^k| \le \tfrac{1}{2}\omega|x^k - x^{k-1}|^2 \,,$$

where $|\cdot|$ is a norm in the domain space X.

In actual computation, we can only solve discretized nonlinear equations of finite dimension, at best on a sequence of successively finer mesh levels, say

$$F_j(x_j) = 0 \,, \quad j = 0, 1, \dots \,,$$

where $F_j : D_j \to Y_j$ denotes a nonlinear mapping defined on a convex domain $D_j \subset X_j$ of a finite-dimensional subspace $X_j \subset X$ with values in a finite dimensional subspace $Y_j \subset X$. The corresponding finite dimensional ordinary Newton method reads

$$F_j'(x_j^k)\Delta x_j^k = -F_j(x_j^k) \,, \quad x_j^{k+1} = x_j^k + \Delta x_j^k \,, \quad k = 0, 1, \dots \,.$$

In each Newton step, a system of linear equations must be solved, which may be a quite challenging task of its own in discretized PDEs. The above Newton system can be interpreted as a discretization of the linearized operator equation (8.2) and, at the same time, as a linearization of the discretization of the nonlinear operator equation (8.1). Again we assume that Theorem 2.2 holds, this time for the finite dimensional mapping F_j. Let ω_j denote the corresponding affine covariant Lipschitz constant. Then the quadratic convergence of this Newton method is governed by the relation

$$\|x_j^{k+1} - x_j^k\| \le \tfrac{1}{2}\omega_j\|x_j^k - x_j^{k-1}\|^2 \,, \tag{8.3}$$

where $\|\cdot\|$ is a norm in the finite dimensional space X_j.

Under the assumptions of Theorem 2.2 there exist unique discrete solutions x_j^* on each level j. Of course, we want to choose appropriate discretization schemes such that

$$\lim_{j \to \infty} x_j^* = x^* .$$

From the synopsis of discrete and continuous Newton method, we immediately see that any comparison of the convergence behavior on different discretization levels j will direct us toward a comparison of the affine covariant Lipschitz constants ω_j. Of particular interest is the connection with the Lipschitz constant ω of the underlying operator equation.

Consistent norms. An important issue for any comparison of affine covariant Lipschitz constants ω_j on different discretization levels j is the choice of *consistent* norms. In the mathematical treatment of Galerkin methods, we will identify the norm $| \cdot |$ in X with the norm $\| \cdot \|$ in $X_j \subset X$. Moreover, the needs of algorithmic adaptivity strongly advise to choose *smooth* norms. These considerations bring us to Sobolev H^p-norms to be properly selected in each particular problem.

For non-Galerkin methods such as finite difference methods, the easiest way to construct consistent norms is to discretize the function space norm $| \cdot |$ appropriately, which directs us toward *discrete H^p-norms*. For example, in one-dimensional BVPs we may naturally use discrete L^2-norms (7.9) to treat highly nonuniform meshes—see also their application in (7.16). For uniform one-dimensional meshes, the discrete L^2-norms on level j differ from the Euclidean vector norms in $\mathbb{R}^{n_j}$ only by some dividing factor $\sqrt{n_j}$. Insertion of the discrete L^2-norm instead of the Euclidean vector norm into the Lipschitz condition (8.3) shows that this same factor would now multiply ω_j. As long as merely a single finite dimensional system were to be analyzed, this change would not make a substantial difference, but only affect the interpretation. A synoptic analysis of a *sequence* of nonlinear mappings, however, will be reasonable only, if consistent discrete norms are used.

In what follows we will consider the phenomenon of mesh independence of Newton's method along two lines. First, we will show that the discrete Newton sequence tracks the continuous Newton sequence closely, with a maximal distance bounded in terms of the mesh size; both of the Newton sequences behave nearly identically until, eventually, a small neighborhood of the solution is reached. Second, we prove the existence of affine covariant Lipschitz constants ω_j for the discretized problems, which approach the Lipschitz constant ω of the continuous problem in the limit $j \longrightarrow \infty$; again, the distance can be bounded in terms of the mesh size. Upon combining these two lines, we finally establish the existence of locally unique discrete solutions x_j^* in a vicinity of the continuous solution x^*.

To begin with, we prove the following nonlinear perturbation lemma.

Lemma 8.1 *Consider two Newton sequences* $\{x^k\}, \{y^k\}$ *starting at initial guesses* x^0, y^0 *and continuing as*

$$x^{k+1} = x^k + \Delta x^k , \quad y^{k+1} = y^k + \Delta y^k ,$$

where $\Delta x^k, \Delta y^k$ *are the corresponding ordinary Newton corrections. Assume that an affine covariant Lipschitz condition with Lipschitz constant* ω *holds. Then the following propagation result holds:*

$$\|x^{k+1} - y^{k+1}\| \le \omega \left(\frac{1}{2}\|x^k - y^k\| + \|\Delta x^k\| \right) \|x^k - y^k\| . \qquad (8.4)$$

Proof. Dropping the iteration index k we start with

$$x + \Delta x - y - \Delta y$$
$$= x - F'(x)^{-1}F(x) - y + F'(y)^{-1}F(y)$$
$$= x - F'(x)^{-1}F(x) + F'(x)^{-1}F(y) - F'(x)^{-1}F(y) - y + F'(y)^{-1}F(y)$$
$$= x - y - F'(x)^{-1}(F(x) - F(y)) + F'(x)^{-1}(F'(y) - F'(x))F'(y)^{-1}F(y)$$
$$= F'(x)^{-1} \left(F'(x)(x - y) - \int_{t=0}^{1} F'(y + t(x - y))(x - y)\, dt \right)$$
$$+ F'(x)^{-1}(F'(y) - F'(x))\Delta y.$$

Upon using the affine covariant Lipschitz condition, we arrive at

$$\|x^{k+1} - y^{k+1}\| \le \int_{t=0}^{1} \|F'(x^k)^{-1}\big(F'(x^k) - F'(y^k + t(x^k - y^k))\big)(x^k - y^k)\|\, dt$$
$$+ \|F'(x^k)^{-1}(F'(y^k) - F'(x^k))\Delta y^k\|$$
$$\le \frac{\omega}{2}\|x^k - y^k\|^2 + \omega\|x^k - y^k\|\,\|\Delta y^k\|,$$

which confirms (8.4). $\qquad\qquad\qquad\qquad\qquad\qquad\qquad\qquad\qquad \square$

With the above auxiliary result, we are now ready to study the relative behavior of discrete versus continuous Newton sequences.

Theorem 8.2 *Notation as introduced. Let* $x^0 \in \bigcap X_j \subset X$ *denote a given starting value such that the assumptions of Theorem 2.2 hold including*

$$h_0 = \omega\|\Delta x^0\| < 2 .$$

For the discrete mappings F_j *and all arguments* $x_j \in S(x^0, \rho + \frac{2}{\omega}) \cap X_j$ *define*

$$F_j'(x_j)\Delta x_j = -F_j(x_j) , \quad F'(x_j)\Delta x = -F(x_j) .$$

Assume that the discretization is fine enough such that

$$\|\Delta x_j - \Delta x\| \le \delta_j \le \frac{1}{2\omega} . \tag{8.5}$$

Then the following cases occur:

I. *If* $h_0 \le 1 - \sqrt{1 - 2\omega\delta_j}$, *then*

$$\|x_j^k - x^k\| < 2\delta_j \le \frac{1}{\omega} , \quad k = 0, 1, \dots .$$

II. *If* $1 - \sqrt{1 - 2\omega\delta_j} < h_0 \le 1 + \sqrt{1 - 2\omega\delta_j}$, *then*

$$\|x_j^k - x^k\| \le \frac{1}{\omega}(1 + \sqrt{1 - 2\omega\delta_j}) < \frac{2}{\omega} , \quad k = 0, 1, \dots .$$

In both cases I and II, the asymptotic result

$$\limsup_{k\to\infty} \|x_j^k - x^k\| \le \frac{1}{\omega}(1 - \sqrt{1 - 2\omega\delta_j}) < 2\delta_j \le \frac{1}{\omega}$$

can be shown to hold.

Proof. In [114, pp. 99, 160], E. Hairer and G. Wanner introduced 'Lady Windermere's fan' as a tool to prove discretization error results for evolution problems based on some linear perturbation lemma. We may copy this idea and exploit our nonlinear perturbation Lemma 8.1 in the present case. The situation is represented graphically in Figure 8.1.

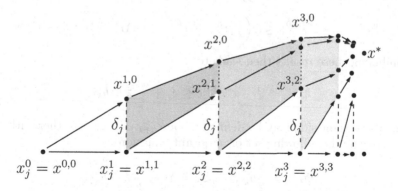

Fig. 8.1. Lady Windermere's fan: continuous versus discrete Newton iterates.

The discrete Newton sequence starting at the given initial point $x_j^0 = x^{0,0}$ is written as $\{x^{k,k}\}$. The continuous Newton sequence, written as $\{x^{k,0}\}$, starts at the same initial point $x^0 = x^{0,0}$ and runs toward the solution point x^*. In between we define further continuous Newton sequences, written as $\{x^{i,k}\}, k = i, i+1, \ldots$, which start at the discrete Newton iterates $x_j^i = x^{i,i}$ and also run toward x^*. Note that the existence or even uniqueness of a discrete solution point x_j^* is not clear yet.

For the purpose of repeated induction, we assume that

$$\|x_j^{k-1} - x^0\| < \rho + \frac{2}{\omega},$$

which certainly holds for $k = 1$. In order to characterize the deviation between discrete and continuous Newton sequences, we introduce the two majorants

$$\omega\|\Delta x^k\| \le h_k , \quad \|x_j^k - x^{k,0}\| \le \epsilon_k .$$

Recall from Theorem 2.2 that

$$h_{k+1} = \tfrac{1}{2}h_k^2 . \tag{8.6}$$

For the derivation of a second majorant recursion, we apply the triangle inequality in the form

$$\|x^{k+1,k+1} - x^{k+1,0}\| \le \|x^{k+1,k+1} - x^{k+1,k}\| + \|x^{k+1,k} - x^{k+1,0}\|.$$

The first term can be treated using assumption (8.5) so that

$$\|x^{k+1,k+1} - x^{k+1,k}\| = \|x_j^k + \Delta x_j^k - \left(x^{k,k} + \Delta x^{k,k}\right)\| = \|\Delta x_j^k - \Delta x^{k,k}\| \le \delta_j .$$

For the second term, we may apply our nonlinear perturbation Lemma 8.1 (see the shaded regions in Fig. 8.1) to obtain

$$\|x^{k+1,k} - x^{k+1,0}\| \le \omega\left(\tfrac{1}{2}\|x^{k,k} - x^{k,0}\| + \|\Delta x^{k,0}\|\right)\|x^{k,k} - x^{k,0}\| .$$

Combining these results then leads to

$$\|x^{k+1,k+1} - x^{k+1,0}\| \le \delta_j + \frac{\omega}{2}\epsilon_k^2 + h_k\epsilon_k .$$

The above right side may be defined to be ϵ_{k+1}. Hence, together with (8.6), we arrive at the following set of majorant equations

$$h_{k+1} = \frac{1}{2}h_k^2 , \quad \epsilon_{k+1} = \delta_j + \frac{\omega}{2}\epsilon_k^2 + h_k\epsilon_k .$$

If we introduce the quantities $\alpha_k = \omega\epsilon_k + h_k$ and $\delta = \omega\delta_j$, we may obtain the decoupled recursion

$$\alpha_{k+1} = \delta + \tfrac{1}{2}\alpha_k^2 , \tag{8.7}$$

which can be started with $\alpha_0 = h_0$, since $\epsilon_0 = 0$. Upon solving the equation

$$\delta = \hat{\alpha} - \tfrac{1}{2}\hat{\alpha}^2 \, ,$$

we get the two equilibrium points

$$\hat{\alpha}_1 = 1 - \sqrt{1 - 2\delta} < 1 \, , \quad \hat{\alpha}_2 = 1 + \sqrt{1 - 2\delta} > 1 \, .$$

Insertion into the recursion (8.7) then leads to the form

$$\alpha_{k+1} - \hat{\alpha} = \tfrac{1}{2}(\alpha_k - \hat{\alpha})(\alpha_k + \hat{\alpha}) \, .$$

For $\alpha_k < \hat{\alpha}_2$ we see that

$$\tfrac{1}{2}(\alpha_k + \hat{\alpha}_1) < \tfrac{1}{2}(\hat{\alpha}_2 + \hat{\alpha}_1) = 1 \, ,$$

which implies that

$$|\alpha_{k+1} - \hat{\alpha}_1| < |\alpha_k - \hat{\alpha}_1| \, . \tag{8.8}$$

Hence, the fixed point $\hat{\alpha}_1$ is attractive, whereas $\hat{\alpha}_2$ is repelling. Moreover, since $\alpha_k + \hat{\alpha}_1 > 0$, we immediately obtain the result

$$\text{sign}(\alpha_{k+1} - \hat{\alpha}) = \text{sign}(\alpha_k - \hat{\alpha}) \, .$$

Therefore, we have the following cases:

I. $\alpha_0 \leq \hat{\alpha}_1 \implies \alpha_k \leq \hat{\alpha}_1 \, ,$
II. $\hat{\alpha}_1 < \alpha_0 < \hat{\alpha}_2 \implies \hat{\alpha}_1 \leq \alpha_k < \hat{\alpha}_2 \, .$

Insertion of the expressions for the used quantities then shows that cases I,II directly correspond to cases I,II of the theorem. Its last asymptotic result is an immediate consequence of (8.8). Finally, with application of the triangle inequality

$$\|x_j^{k+1} - x^0\| \leq \epsilon_{k+1} + \|x^{k+1} - x^0\| < \frac{2}{\omega} + \rho$$

the induction and therefore the whole proof is completed.   □

We are, of course, interested whether a discrete solution point x_j^* exists. The above tracking theorem, however, only supplies the following result.

Corollary 8.3 *Under the assumptions of Theorem 8.2, there exists at least one accumulation point*

$$\hat{x}_j \in S\left(x^*, 2\delta_j\right) \cap X_j \subset S\left(x^*, \frac{1}{\omega}\right) \cap X_j \, ,$$

which need not be a solution point of the discrete equations $F_j(x_j) = 0$.

Proof. This is just the last asymptotic result of Theorem 8.2.   □

In order to prove more, Theorem 2.2 directs us to study the question of whether an affine covariant Lipschitz condition holds for the finite dimensional mapping F_j, too.

Lemma 8.4 *Let Theorem* 2.2 *hold for the mapping* $F : X \to Y$. *For collinear* $x_j, y_j, y_j + v_j \in X_j$, *define quantities* $w_j \in X_j$ *and* $w \in X$ *according to*

$$F_j'(x_j)w_j = \left(F_j'(y_j + v_j) - F_j'(y_j)\right) v_j \,, \ F'(x_j)w = \left(F'(y_j + v_j) - F'(y_j)\right) v_j \,.$$

Assume that the discretization method satisfies

$$\|w - w_j\| \le \sigma_j \|v_j\|^2 \,. \tag{8.9}$$

Then there exist constants

$$\omega_j \le \omega + \sigma_j \tag{8.10}$$

such that the affine invariant Lipschitz condition

$$\|w_j\| \le \omega_j \|v_j\|^2$$

holds for the discrete Newton process .

Proof. The proof is a simple application of the triangle inequality

$$\|w_j\| \le \|w\| + \|w_j - w\| \le \omega\|v_j\|^2 + \sigma_j\|v_j\|^2 = (\omega + \sigma_j)\,\|v_j\|^2.$$

□

Finally, the existence of a unique solution x_j^* is now an immediate consequence.

Corollary 8.5 *Under the assumptions of Theorem* 8.2 *and Lemma* 8.4 *the discrete Newton sequence* $\{x_j^k\}, k = 0, 1, \dots$ *converges quadratically to a unique discrete solution point*

$$x_j^* \in S\left(x^*, 2\delta_j\right) \cap X_j \subset S\left(x^*, \frac{1}{\omega}\right) \cap X_j \,.$$

Proof. We just need to apply Theorem 2.2 to the finite dimensional mapping F_j with the starting value $x_j^0 = x^0$ and the affine invariant Lipschitz constant ω_j from (8.10).   □

Remark 8.1   In the earlier papers [3, 4] two assumptions of the kind

$$\|F_j'(x_j)^{-1}\| \le \beta_j \,, \quad \|F_j'(x_j + v_j) - F_j'(x_j)\| \le \gamma_j \|v_j\|$$

have been made in combination with the *uniformity* requirements

$$\beta_j \leq \beta, \quad \gamma_j \leq \gamma. \tag{8.11}$$

Obviously, these assumptions lack any affine invariance. More important, however, and as a consequence of the noninvariance, these conditions are phrased in terms of *operator norms*, which, in turn, depend on the relation of norms in the domain and the image space of the mappings F_j and F, respectively. For typical PDEs and norms we would obtain

$$\lim_{j \to \infty} \beta_j \to \infty,$$

which clearly contradicts the uniformity assumption (8.11). Consequently, an analysis in terms of β_j and γ_j would not be applicable to this important case.

The situation is different with the affine covariant Lipschitz constants ω_j: they only depend on the choice of *norms in the domain space*. It is easy to verify that

$$\omega_j \leq \beta_j \gamma_j.$$

From the above Lemma 8.4 we see that the ω_j remain bounded in the limit $j \longrightarrow \infty$, as long as ω is bounded—even if β_j or γ_j blow up. Moreover, even when the product $\beta_j \gamma_j$ remains bounded, the Lipschitz constant ω_j may be considerably lower, i.e.

$$\omega_j \ll \beta_j \gamma_j.$$

For illustration, just compare the simple $\mathbb{R}^2$-example in Exercise 2.3.

Summarizing, we come to the following conclusion, at least in terms of upper bounds: If the asymptotic properties

$$\lim_{j \to \infty} \delta_j = 0, \quad \lim_{j \to \infty} \sigma_j = 0$$

can be shown to hold, then the convergence speed of the discrete ordinary Newton method is asymptotically just the one for the continuous ordinary Newton method—compare Exercises 8.3 and 8.4. Moreover, if related initial guesses x^0 and x_j^0 and a common termination criterion are chosen, then even the number of iterations will be nearly the same.

BIBLIOGRAPHICAL REMARK. The 'mesh independence' principle has been reported and even exploited for mesh design in papers by E.L. Allgower and K. Böhmer [3] and S.F. McCormick [148]. Further theoretical investigations of the phenomenon have been given in the paper [4] by E.L. Allgower, K. Böhmer, F.A. Potra, and W.C. Rheinboldt; that paper, however, lacked certain important features, which have been discussed in Remark 8.1 above. A first affine covariant theoretical study has later been worked out by P. Deuflhard and F.A. Potra in [82]; from that analysis, the modified term 'asymptotic mesh independence' naturally emerged. The presentation here follows the much simpler and more intuitive treatment [198] of the topic as given recently by M. Weiser, A. Schiela, and the author.

8.2 Global Discrete Newton Methods

In the present section we regard BVPs for nonlinear PDEs as given in already discretized form on a fixed mesh, to be briefly called *discrete PDEs* here. In what follows, we report about the comparative performance of exact and inexact Newton methods in solving such problems. Part of the results are from a recent paper by P. Deuflhard, U. Nowak, and M. Weiser [80].

In the exact Newton methods, we use either band mode LU-decomposition or a sparse solver provided by MATLAB. Failure exits in the various numerical tests are characterized by

- OUTMAX: the Newton iteration (outer iteration in inexact Newton methods) does not converge within 75 iterations,
- ITMAX: the inner iteration per inexact Newton step does not converge within ITMAX iterations,
- λ-fail: the adaptive damping strategy suggests some 'too small' damping factor $\lambda_k < 10^{-4}$.

8.2.1 General PDEs

This section documents the comparative performance of residual based (or affine contravariant) Newton methods versus error oriented (or affine covariant) Newton methods, both for the exact and the inexact versions, at a common set of discrete PDE test problems.

Common test set. We consider a subset of the discrete PDE problems given in [160]. In order to be able to compare exact and inexact methods, we selected examples in only two space dimensions. This choice leads to moderate system dimensions n that still permit a direct solution of the arising linear equations. Throughout the examples, we use the usual second order, centered finite differences on tensor product grids. Neumann boundary conditions are included by simple one-sided differences, as usual.

Example 8.1 *Artificial test problem (atp1).* This problem comprises the simple scalar PDE

$$\Delta u - (0.9 \exp(-q) + 0.1u)(4x^2 + 4y^2 - 4) - g = 0,$$

where

$$g = \exp(u) - \exp(\exp(-q)) \text{ and } q = x^2 + y^2,$$

with boundary conditions $u|_{\partial\Omega} = 0$ on the domain $\Omega = [-3,3]^2$. The analytical solution is known to be $u(x,y) = \exp(-q)$.

Example 8.2 *Driven cavity problems (dcp1000, dcp5000).* This problem involves the steady stream-function/vorticity equations

$$\Delta\omega + \mathrm{Re}(\psi_x\omega_y - \psi_y\omega_x) = 0 , \quad \Delta\psi + \omega = 0 ,$$

where ψ is the stream-function and ω the vorticity. For the domain $\Omega = [0,1]^2$ the following discrete boundary conditions are imposed (with $\Delta x, \Delta y$ the mesh sizes in x,y-direction)

$$\frac{\partial\psi}{\partial y}(x,1) = -16x^2(1-x)^2 ,$$
$$\omega(x,0) = -\tfrac{2}{\Delta y^2}\psi(x,\Delta y) ,$$
$$\omega(x,1) = -\tfrac{2}{\Delta y^2}[\psi(x,1-\Delta y) + \Delta y\tfrac{\partial\psi}{\partial y}(x,1)] ,$$
$$\omega(0,y) = -\tfrac{2}{\Delta x^2}\psi(\Delta x,y) ,$$
$$\omega(1,y) = -\tfrac{2}{\Delta x^2}\psi(1-\Delta x,y) .$$

Problems *dcp1000, dcp5000* correspond to Reynolds numbers $\mathrm{Re}=1000,5000$, respectively. For both cases the default initial guess is $\psi^0 = \omega^0 = 0$.

As will be seen below, the residual based Newton strategy was unable to solve problems *dcp1000* and *dcp5000* with this initial guess. That is why we added problems *dcp1000a* and *dcp5000a* with the better initial guesses $\omega^0 = y^2\sin(\pi x)$, $\psi^0 = 0.1\sin(\pi x)\sin(\pi y)$.

Example 8.3 *Supersonic transport problem (sst2).* The four model equations for the chemical species O, O_3, NO, NO_2, represented by the unknown functions (u_1, u_2, u_3, u_4), are

$$0 = D\Delta u_1 + k_{1,1} - k_{1,2}u_1 + k_{1,3}u_2 + k_{1,4}u_4 - k_{1,5}u_1u_2 - k_{1,6}u_1u_4 ,$$
$$0 = D\Delta u_2 + k_{2,1}u_1 - k_{2,2}u_2 + k_{2,3}u_1u_2 - k_{2,4}u_2u_3 ,$$
$$0 = D\Delta u_3 - k_{3,1}u_3 + k_{3,2}u_4 + k_{3,3}u_1u_4 - k_{3,4}u_2u_3 + 800.0 + SST ,$$
$$0 = D\Delta u_4 - k_{4,1}u_4 + k_{4,2}u_2u_3 - k_{4,3}u_1u_4 + 800.0 ,$$

where $D = 0.5\cdot 10^{-9}$,
$k_{1,1},\ldots,k_{1,6} = 4\cdot 10^5, 272.443800016, 10^{-4}, 0.007, 3.67\cdot 10^{-16}, 4.13\cdot 10^{-12}$,
$k_{2,1},\ldots,k_{2,4} = 272.4438, 1.00016\cdot 10^{-4}, 3.67\cdot 10^{-16}, 3.57\cdot 10^{-15}$,
$k_{3,1},\ldots,k_{3,4} = 1.6\cdot 10^{-8}, 0.007, 4.1283\cdot 10^{-12}, 3.57\cdot 10^{-15}$,
$k_{4,1},\ldots,k_{4,3} = 7.000016\cdot 10^{-3}, 3.57\cdot 10^{-15}, 4.1283\cdot 10^{-12}$, and

$$SST = \begin{cases} 3250 & \text{if } (x,y) \in [0.5, 0.6]^2 \\ 360 & \text{otherwise.} \end{cases}$$

Homogeneous Neumann boundary conditions are imposed on the unit square. For the initial guess we take

$$u_1^0(x,y) = 10^9 \,, u_2^0(x,y) = 10^9 \,, u_3^0(x,y) = 10^{13} \,, u_4^0(x,y) = 10^7 \,.$$

Again we consider better initial guesses to allow for convergence in the residual based Newton methods:

$$u_i^0 \to (1 + 100(\sin(\pi x)\sin(\pi y))^2)u_i^0 \,.$$

Name	Grid	Dim n	OrdNew
atp1	31×31	961	4
dcp1000	31×31	1922	OUTMAX
dcp1000a	31×31	1922	9
dcp5000	63×63	7983	OUTMAX
dcp5000a	63×63	7983	OUTMAX
sst2	51×51	10404	OUTMAX
sst2a	51×51	10404	OUTMAX

Table 8.1. Test set characteristics.

Characteristics of the selected test set are arranged in Table 8.1. In order to give some idea about the complexity of the individual problems, we first applied an exact ordinary Newton method (uncontrolled)—see the last column of the table. All of its failures are due to 'too many' Newton (outer) iterations (recall OUTMAX= 75).

Exact Newton methods. Recall that *exact* Newton methods require the direct solution of the arising linear subsystems for the Newton corrections. Hence, adaptivity only shows up through affine invariant trust region (or damping) strategies. From the code family NLEQ we compare the following variants:

- NLEQ-RES requiring monotonicity in the residual norm $\|F\|_2$, as discussed in Section 3.2.2,

- NLEQ-RES/L requiring monotonicity in the preconditioned residual norm $\|C_L F\|_2$, also discussed in Section 3.2.2; the preconditioner C_L comes from incomplete LU-decomposition with fill-in only accepted within the block pentadiagonal structure (compare, e.g., [184]), and

- NLEQ-ERR requiring monotonicity in the natural level function, as discussed in Section 3.3.3.

The residual based methods realize the restricted monotonicity test (3.32). For termination, the (possibly preconditioned) criterion (2.70) with FTOL = 10^{-8} has been taken, except for the badly scaled problems *sst*, which required FTOL = 10^{-5} to terminate within a tolerable computing time. The error oriented methods realize the restricted monotonicity test (3.47) and the (scaled) termination criterion (2.14) with XTOL = 10^{-8}.

Name	NLEQ-RES	NLEQ-RES/L	NLEQ-ERR
atp1	4 (0)	4 (0)	4 (0)
dcp1000	OUTMAX	10 (5)	8 (4)
dcp1000a	21 (17)	8 (2)	8 (2)
dcp5000	OUTMAX	OUTMAX	11 (7)
dcp5000a	42 (39)	λ-fail	8 (2)
sst2	λ-fail	12 (11)	13 (8)
sst2a	38 (33)	15 (13)	19 (14)

Table 8.2. Exact Newton codes: adaptive control via residual norm (NLEQ-RES), preconditioned residual norm (NLEQ-RES/L), and error norm (NLEQ-ERR).

In Table 8.2 we compare the residual based versus the error oriented exact Newton codes in terms of Newton steps (in parentheses: damped). As can be seen, there is striking evidence that the error oriented adaptive Newton methods are clearly preferable to the residual based ones, at least for the problem class tested here.

The main reason for this phenomenon is certainly that the arising discrete Jacobian matrices are bound to be ill-conditioned, the more significant the finer the mesh is. For this situation, the limitation of residual monotonicity has been described at the end of Section 3.3.1. Example 3.1 has given an illustration representative for a class of ODE boundary value problem. The experimental evidence here seems to indicate that the limitation carries over to PDE boundary value problems as well.

Inexact Newton methods. Finite dimensional *inexact* Newton methods contain some inner iterative solver, which induces the necessity of an accuracy matching between inner and outer iteration. The implemented ILU-preconditioning [184] is the same as in the exact Newton codes above. In the code family GIANT, various affine invariant damping and accuracy matching strategies are realized—according to the selected affine invariance class. The failure exit ITMAX was activated at 2000 inner iterations.

Residual based methods. For this type of inexact Newton method, we chose the codes GIANT-GMRES/R and GIANT-GMRES/L with right (R) or left (L) preconditioning.

As a first test, we selected the standard convergence mode from Sections 2.2.4 and 3.2.3, prescribing $\eta_k \le \bar{\eta}$ with threshold values $\bar{\eta} = 0.1$ and $\bar{\eta} = 0.001$.

Name	NLEQ-RES	GIANT-GMRES/R/0.001		GIANT-GMRES/R/0.1	
atp1	4 (0)	4 (0)	34	4 (1)	28
dcp1000	OUTMAX	OUTMAX		OUTMAX	
dcp1000a	21 (17)	21 (17)	788	28 (23)	605
dcp5000	OUTMAX	OUTMAX		OUTMAX	
dcp5000a	42 (39)	43 (39)	5021	58 (53)	3208
sst2	λ-fail	15 (10)	1376	ITMAX	
sst2a	38 (33)	ITMAX		ITMAX	

Table 8.3. Residual based Newton codes: exact version NLEQ-RES versus inexact version GIANT-GMRES/R for threshold values $\bar{\eta} = 0.001$ and $\bar{\eta} = 0.1$.

Name	NLEQ-RES/L	GIANT-GMRES/L/0.001		GIANT-GMRES/L/0.1	
atp1	4 (0)	4 (0)	31	4 (1)	25
dcp1000	10 (5)	10 (5)	380	16 (10)	309
dcp1000a	8 (2)	8 (1)	279	12 (3)	229
dcp5000	OUTMAX	ITMAX		OUTMAX	
dcp5000a	λ-fail	24 (15)	1700	OUTMAX	
sst2	12 (10)	15 (12)	252	OUTMAX	
sst2a	15 (13)	18 (15)	465	OUTMAX	

Table 8.4. Preconditioned residual based Newton codes: exact version NLEQ-RES/L versus inexact version GIANT-GMRES/L for threshold values $\bar{\eta} = 0.001$ and $\bar{\eta} = 0.1$.

In Table 8.3, we compare exact versus inexact Newton methods, again at the common test set, in terms of Newton steps (in parentheses: damped Newton steps) and inner iterations. For comparison, the first column is identical to the first one from Table 8.2. In Table 8.4, the performance of two GIANT-GMRES/L versions is documented. This time, the first column is the second one from Table 8.2.

As can be seen from both tables, the inexact Newton codes behave very much like their exact counterparts in terms of outer iterations, with erratic discrepancies now and then. In view of the anyway poor behavior of the residual based Newton methods in this problem class, we did not realize the fully adaptive accuracy matching strategy (linear or quadratic convergence mode) in the frame of residual based inexact Newton methods.

Error oriented Newton methods. For this type of inexact Newton method, we chose the codes GIANT-CGNE/L and GIANT-GBIT/L, both with left (L) pre-conditioning. Adaptive matching strategies as worked out in Sections 2.1.5 and 3.3.4 have been realized. Initial values for the arising inner iterations were chosen according to the nested suggestions (3.59) and (3.60). Note that these inexact codes realize a damping strategy and a termination criterion slightly different from those in NLEQ-ERR. In view of a strict comparison, we constructed an exact variant NLEQ-ERR/I, which realizes just these modifications, i.e. which is an inexact Newton-ERR code with exact inner solution.

Name	NLEQ-ERR/I	GIANT-CGNE/L		GIANT-GBIT/L	
atp1	5 (0)	5 (0)	237	5 (0)	122
dcp1000	10 (5)	ITMAX		10 (5)	1388
dcp1000a	9 (3)	ITMAX		9 (3)	2241
dcp5000	13 (8)	ITMAX		14 (8)	5943
dcp5000a	10 (3)	ITMAX		10 (3)	9504
sst2	15 (11)	16 (11)	23084	16 (11)	1549
sst2a	20 (15)	20 (15)	39889	20 (15)	2399

Table 8.5. Error oriented Newton codes: exact version NLEQ-ERR/I versus inexact versions GIANT-CGNE/L and GIANT-GBIT/L for threshold values $\bar{\delta} = 10^{-3}$.

In Table 8.5, we give results for the 'standard convergence mode', imposing the (scaled) condition $\delta_k \leq \bar{\delta}$ for the threshold value $\bar{\delta} = 10^{-3}$ throughout, as defined in (2.61) for the local Newton method and (3.55) for the global Newton method, the latter via $\rho = 2\bar{\delta}/(1 - 2\bar{\delta})$. As can be seen, the first column for NLEQ-ERR/I and the third column of Table 8.2 for NLEQ-ERR differ only marginally.

From these numerical experiments, we may keep the following information:

- The error estimator (1.31) for CGNE is more reliable than (1.37) for GBIT.

- Nevertheless CGNE is less efficient than GBIT—compare Remark 1.2.

- The code GIANT-GBIT/L essentially reproduces the outer iteration pattern of the exact Newton code NLEQ-ERR*.

More insight into GIANT-GBIT/L can be gained from Table 8.6 where we compare the 'standard convergence mode' (SM) with the 'quadratic convergence mode' (QM), again in terms of outer (damped) and inner iterations. Different sets of control parameters are applied. The parameter ρ defines $\bar{\delta} = \frac{1}{2}\rho/(1+\rho)$ via (3.55). As a default, the parameter $\tilde{\rho}$ is fixed to $\tilde{\rho} = \frac{1}{2}\rho$, which, in turn, defines $\bar{\rho}_{max} = \tilde{\rho}/(1 + \tilde{\rho})$ via (3.70).

The first column, SM(.025, .05), presents results obtained over our common test set, when the accuracy matching strategies (3.66) with (3.71) and (3.50) with (3.55) are implemented; the values $(\tilde{\rho}, \rho) = (.025, .05)$ represent the largest values, for which all problems from the common test set were still solvable. This column should be compared with the third column in Table 8.5, where GIANT-GBIT/L has been realized roughly in an SM(.001, .002) variant: considerable savings are visible.

Detailed examination of the numerical results has revealed that the weakest point of this algorithm is the rather poor error estimator (1.37) realized

Name	SM(.025, .05)		SM*(1/16, 1/8)		QM(.025, .05)	
atp1	5 (0)	91	5 (0)	66	5 (0)	97
dcp1000	11 (5)	904	12 (5)	817	10 (5)	852
dcp1000a	11 (3)	1272	12 (4)	1458	9 (3)	1180
dcp5000	19 (11)	3952	16 (8)	3417	16 (11)	3802
dcp5000a	16 (3)	4304	11 (1)	3475	10 (3)	3539
sst2	22 (13)	963	19 (12)	1037	18 (13)	842
sst2a	25 (16)	1429	24 (17)	1597	22 (16)	1365

Table 8.6. Comparison of different variants of error oriented inexact Newton code `GIANT-GBIT/L`. Accuracy matching strategies $\mathrm{SM}(\tilde{\rho}, \rho)$ and $\mathrm{QM}(\tilde{\rho}, \rho)$ for control parameters $(\tilde{\rho}, \rho)$. SM* realizes an exact computation of the inner iteration error.

within `GBIT`, which is often quite off scale. For an illustration of this effect, the second column presents results for version SM*, which realizes a *precise* error estimator

$$\epsilon_i = \|\delta x_i^k - \Delta x^k\|$$

instead of the unsatisfactory estimator (1.37); in this case, the relaxed choice $(\tilde{\rho}, \rho) = (1/16, 1/8)$ appeared to be possible. Consequently, savings of inner iterations can be observed.

These savings, however, are not too dramatic when compared with the version QM shown in the third column; here the quadratic accuracy matching rule (3.56) is realized, which does not differ too much from the rule (2.62). For the control parameters we again selected $(\tilde{\rho}, \rho) = (.025, .05)$ to allow for a comparison with SM in the first column. Obviously, this column shows the best comparative numbers.

Summary. From our restricted set of numerical experiments, we may nevertheless draw certain practical conclusions:

- Among the inner iterations for an inexact Newton-ERR method, the algorithm `GBIT` is the clear 'winner'—despite the rather poor computational estimator for the inner iteration error, which is presently realized.

- Linear preconditioning also plays a role in nonlinear preconditioning as realized in the inexact Newton-ERR codes; in particular, the better the linear preconditioner, the better the inner iteration error estimator, the better the performance of the whole inexact Newton-ERR method.

- The 'quadratic convergence mode' in the local convergence phase can save a considerable amount of computing time over the 'standard convergence mode'.

8.2.2 Elliptic PDEs

This section documents the comparative performance of exact versus inexact affine conjugate Newton methods at a common set of nonlinear BVPs for elliptic discrete PDEs. Recall that elliptic PDEs are associated with underlying convex optimization problems—see Sections 2.3 and 3.4.

Test set. Below three discretized nonlinear elliptic PDE BVPs in two space dimensions are given. Of course, their corresponding discretized functional is also at hand. All discretizations are simple finite difference schemes on uniform meshes.

Example 8.4 *Simple elastomechanics problem (elas).* For $\Omega =]0, 1[^2$, minimize the functional (total energy in Ogden material law)

$$\int_\Omega \left(\|F\|^2 + (\det F)^{-1} - M(1/2, -1)u \right) dx \quad \text{with } F = I + \nabla u .$$

Homogeneous Dirichlet conditions on the boundary part $\{0\} \times [0, 1]$ and natural boundary conditions on the remaining boundary part are imposed. Physically speaking, $u(x) \in \mathbb{R}^2$ is the displacement of an elastic body. The volume force $(1/2, -1)^T$ acting on the body is scaled by M, which can be used to weight the 'nonlinearity' of the problem. As initial value we chose $u^0 = 0$ in agreement with the Dirichlet conditions.

Detailed examination reveals that the above functional is not globally convex on the whole domain of definition, but only in a neighborhood of the solution. Fortunately, for the given initial guesses, our Newton codes did not encounter any nonpositive second derivatives. The locally unique solution is depicted in Figure 8.2, right.

Example 8.5 *Minimal surface problem over convex domain (msc).* Given $\Omega =]0, 1[^2$, minimize the surface area

$$\int_\Omega (1 + |\nabla u|^2)^{\frac{1}{2}} dx$$

subject to the Dirichlet boundary conditions

$$u(x_1, x_2) = M(x_1 + (1 - 2x_1)x_2) \quad \text{on } \partial\Omega .$$

Here $u(x) \in \mathbb{R}$ is the vertical position of the surface parametrized over Ω. The scaling parameter M of the boundary conditions allows to weight the 'nonlinearity' of the problem. The initial value u^0 is chosen as the bilinear interpolation of the boundary conditions. This problem has a unique well-defined solution depicted in Figure 8.2, left.

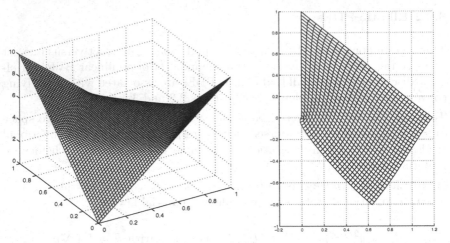

Fig. 8.2. *Left:* solution of problem *msc* ($M = 10, h = 1/63$). *Right:* solution of problem *elas* ($M = 2, h = 1/31$).

Example 8.6 *Minimal surface problem over nonconvex domain (msnc).* Given the domain $\Omega =]0, 2[^2 \backslash]1, 2[^2$, minimize the surface area

$$\int_{\Omega} (1 + |\nabla u|^2)^{\frac{1}{2}} \, dx$$

subject to the Dirichlet boundary conditions

$$u = 0 \text{ on } [0, 2] \times \{0\} \cup \{0\} \times [0, 2] \,, \quad u = M \text{ on } [1, 2] \times \{1\} \cup \{1\} \times [1, 2] \,.$$

On the remaining boundary parts, $[0, 1] \times \{2\} \cup \{2\} \times [0, 1]$, homogeneous Neumann boundary conditions $\partial_n u = 0$ are imposed. Here $u(x) \in \mathbb{R}$ is the vertical position of the surface parameterized over Ω. The scaling parameter M plays the same role as in problem *msc*. The initial value u^0 is chosen to be the linear interpolation of the Dirichlet boundary conditions on $[0, 1] \times [1, 2] \cup [1, 2] \times [0, 1]$ and the bilinear interpolation of the thus defined boundary values on $[0, 1]^2$.

This problem has been deliberately constructed such that the underlying PDE does *not* have a unique *continuous* solution. Indeed, function space Newton multilevel methods (to be presented in Section 8.3 below) are able to detect this nonexistence: even though there exists a finite dimensional 'pseudosolution' on each mesh with size h, the local convergence domain of Newton's method shrinks when $h \to 0$. In the present setting of discrete PDEs, however, Newton's method will just supply a discrete solution on each of the meshes. As shown in Figure 8.3, these discrete solutions exhibit an interior 'discrete discontinuity', which is the sharper, the finer the mesh is.

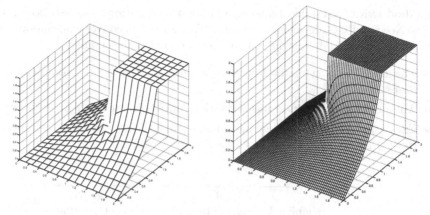

Fig. 8.3. Discrete solutions of problem *msnc*. *Left:* $M = 2, h = 1/8$, *right:* $M = 2, h = 1/32$.

Table 8.7 gives some 'measure' of the problem complexity for selected problem sizes of low dimension n: the value $M_{\max}$ in the last column indicates the maximal nonlinearity weight factor, for which the ordinary Newton method (uncontrolled) had still converged in our tests.

Name	Grid	Dim n	$M_{\max}$
msc	32×32	1024	6.2
elas	32×32	2048	1.0
msnc	32×32	3072	1.9

Table 8.7. Test set characteristics for special 2D grid.

Incidentally, below we also treat much larger problems, where dimensions up to $n \approx 200.000$ arise.

Exact versus inexact Newton methods. For the exact as well as the inexact Newton iteration, the energy error termination criterion (2.110) with ETOL $= 10^{-8}$ is taken. For Newton-PCG methods, we use the inner iteration termination criterion (1.25) and the accuracy matching strategy as worked out in Sections 2.3.3 and 3.4.3. As preconditioners we tested both the Jacobi and the incomplete Cholesky preconditioner (ICC) provided by MATLAB (with droptol $= 10^{-3}$). The failure exit ITMAX was activated at more than 500 inner iterations.

Local versus global Newton methods. In Table 8.8, we give comparative results for varying weight factor M at problem *msc*. Among the local Newton methods, we deliberately included the rather popular simplified Newton

method (with initial Jacobian throughout the iteration)—see Section 2.3.2.
The Newton-PCG algorithms are run in the *quadratic* convergence mode—see
Section 2.3.3. The following features can be clearly observed:

- The local Newton methods converge only for the mildly nonlinear case
  (here: small M).
- Among the local Newton methods, the simplified variant behaves poorest.
- Exact and inexact Newton methods realize nearly the same number of
  (outer) Newton iterations, both local and global.

M		local			global	
	simplified	exact	inexact		exact	inexact
2	21	5	5		9	9
5	DIV	7	7		10	9
10	DIV	DIV	DIV		10	10

Table 8.8. Problem *msc*: comparative Newton steps (DIV: divergence).

Asymptotic mesh independence. In Table 8.9, we test different discrete Newton algorithms over the whole test set for decreasing mesh sizes. Asymptotic
mesh independence as studied in Section 8.1 (see also Exercise 8.4) is clearly
visible in problems *msc* and *elas*, but not in problem *msnc*, which does not
have a unique continuous solution (see also Table 8.11 in Section 8.3.2 below). In the latter problem failures occur on the finest meshes—in agreement
with the subsequent Example 8.9. The missing entries indicate the fact that
the inexact codes were able to tackle much larger problems than the exact
ones—both due to time and, even more pronounced, memory requirements
of the direct solver on the finer meshes.

N	$msc(M = 10)$		$elas(M = 2)$		$msnc(M = 2)$	
	exact	inexact	exact	inexact	exact	inexact
4	9	8	10	9	9	8
8	10	9	10	10	9	9
16	10	9	10	10	10	10
32	10	10	10	10	10	11
64	10	10		11		13
128		10				λ-fail
256		10				ITMAX

Table 8.9. Test set: Newton steps for decreasing mesh sizes $h = 1/N$.

Different preconditioners. In Table 8.10, the Jacobi preconditioner (Jac) and the incomplete Cholesky preconditioner (ICC) are compared for the quadratic and the linear convergence mode. For the linear convergence mode, $\bar{\Theta} = 0.5$ has been chosen. As can be seen, Jac is insufficient for fine meshes. ICC is more effective, at least for small up to moderate size meshes. The linear convergence mode is comparable to the quadratic convergence mode—as opposed to the behavior in the function space Newton method presented in Section 8.3 below.

	n	quadratic		linear	
		ICC	Jac	ICC	Jac
msc	4	7 (16)	7 (39)	7 (14)	8 (30)
(M=3.5)	8	6 (15)	6 (134)	6 (12)	15 (120)
	16	6 (19)	7 (385)	6 (12)	$\Theta \geq 1$
	32	6 (25)	8 (921)	7 (16)	$\Theta \geq 1$
	64	6 (35)	ITMAX	9 (24)	$\Theta \geq 1$
	128	6 (57)	ITMAX	12 (52)	$\Theta \geq 1$
	256	6 (103)	ITMAX	15 (96)	$\Theta \geq 1$
	512	6 (210)	ITMAX	19 (235)	$\Theta \geq 1$
clas	4	6 (18)	6 (174)	6 (12)	$\Theta \geq 1$
(M=0.2)	8	5 (19)	6 (479)	6 (12)	$\Theta \geq 1$
	16	5 (29)	ITMAX	7 (18)	$\Theta \geq 1$
	32	5 (44)	ITMAX	9 (36)	$\Theta \geq 1$
	64	5 (80)	ITMAX	11 (67)	$\Theta \geq 1$
	128	6 (176)	ITMAX	14 (144)	ITMAX

Table 8.10. Local inexact Newton-PCG method: comparative outer (inner) iterations. Quadratic versus linear convergence mode, Jacobi (Jac) versus incomplete Cholesky (ICC) preconditioning.

Summary. For elliptic discrete nonlinear PDEs both the exact and the inexact affine conjugate Newton methods perform efficiently and reliably, in close connection with the associated convergence theory. The inexact Newton code GIANT-PCG with ICC preconditioning seems to be a real competitor to so-called nonlinear PCG methods (for references see Section 2.3.3).

8.3 Inexact Newton Multilevel FEM for Elliptic PDEs

In this section we consider minimization problems of the kind

$$f(x) = \min,$$

wherein $f : D \subset X \to \mathbb{R}$ is assumed to be a *strictly convex* C^2-functional defined on an open *convex* subset D of a Banach space X. Let X be endowed with a norm $\| \cdot \|$. In order to assure the *existence* of a minimum point $x^* \in D$, we assume that X is *reflexive*; in view of the subsequent finite element method (FEM), we choose $X = W^{1,p}$ for $1 < p < \infty$. Moreover, for given initial guess $x^0 \in D$, we assume that the level set $\mathcal{L}_0 := \{x \in D | f(x) \leq f(x^0)\}$ is nonempty, closed, and bounded. Under these assumptions the existence of a *unique* minimum point x^* is guaranteed. In this case the nonlinear minimization problem is equivalent to the nonlinear operator equation

$$F(x) := f'(x) = 0 , \quad x \in D . \qquad (8.12)$$

In the present section, this equation is understood to be a nonlinear elliptic PDE problem. In order to guarantee the feasibility of Newton's method, we further assume that the PDE problem is *strictly* elliptic, which means that its symmetric Frechét-derivative $F'(x) = f''(x)$ is *strictly positive*.

In abstract notation, the ordinary Newton method for the mapping F reads $(k = 0, 1, \dots)$

$$F'(x^k)\Delta x^k = -F(x^k) , \quad x^{k+1} = x^k + \Delta x^k ,$$

which just describes the successive linearization, often also called *quasilinearization*, of the above nonlinear operator equation. Since equation (8.12) is a nonlinear elliptic PDE in the Banach space $W^{1,p}$, the above Newton sequence consists of solutions of linear elliptic PDEs in some Hilbert space, say H_k, associated with each iterate $x^k \in W^{1,p}$. For reasonable arguments x, there exist energy products $\langle \cdot, F'(x) \cdot \rangle$, which induce energy norms $\langle \cdot, F'(x) \cdot \rangle^{1/2}$. The question of whether these energy norms are bounded for all arguments of interest needs to be discussed inside the proofs of the theorems to be stated below.

In the subsequent analysis, we will 'lift' these energy products and energy norms from H to $W^{1,p}$ (in the sense of dual pairing) defining the corresponding *local energy products* $\langle \cdot, F'(x) \cdot \rangle$ as symmetric bilinear forms and the induced *local energy norms* $\langle \cdot, F'(x) \cdot \rangle^{1/2}$ for arguments x in appropriate subsets of $W^{1,p}$. Moreover, motivated by the notation in Hilbert space, where the operator $F'(x)^{1/2}$ is readily defined, we also adopt the shorthand notation

$$\|F'(x)^{1/2} \cdot \| \equiv \langle \cdot, F'(x) \cdot \rangle^{1/2}$$

to be only used in connection with the local energy norms.

As already mentioned for space-like ODE BVPs in Section 7.4.2, *quasilinearization*, here for BVPs in more than one space dimension, cannot be realized without approximation errors. This means that we need to study *inexact* Newton methods

$$F'(x^k)\,\delta x^k = -F(x^k) + r^k ,$$

equivalently written as

$$F'(x^k)\left(\delta x^k - \Delta x^k\right) = r^k.$$

Here the discretization errors show up either as residuals r^k or as the discrepancy between the inexact Newton corrections δx^k and the exact Newton corrections Δx^k. Among the discretization methods, we will focus on Galerkin methods, known to satisfy the *Galerkin condition*

$$\langle \delta x^k, F'(x^k)(\delta x^k - \Delta x_k) \rangle = \langle \delta x^k, r^k \rangle = 0. \tag{8.13}$$

Such a condition also holds in the finite dimensional inexact Newton-PCG method, where the residuals originate from the use of PCG as inner iterative solver—just look up condition (2.98) in Section 2.3.3. Recall that Newton-PCG methods are relevant for the discrete PDE situation, as presented in Section 8.2.2. Here, however, we want to treat the infinite dimensional exact Newton method approximated by an *adaptive* finite dimensional inexact Newton method. The benefit to be gained from adaptivity will become apparent in the following.

8.3.1 Local Newton-Galerkin methods

In this section we study the *ordinary* Newton-Galerkin method

$$x^{k+1} = x^k + \delta x^k,$$

where the iterates x^k are in $W^{1,p}$ and the inexact Newton corrections δx^k satisfy (8.13). With the above theoretical considerations we are ready to just modify the local convergence theorem for Newton-PCG methods (Theorem 2.20) in such a way that it covers the present infinite dimensional setting.

Theorem 8.6 *Let $f : D \to \mathbb{R}$ be a strictly convex C^2-functional to be minimized over some open convex domain $D \subset W^{1,p}$ endowed with the norm $\|\cdot\|$. Let $F'(x) = f''(x)$ be strictly positive. For collinear x, y, $z \in D$, assume the affine conjugate Lipschitz condition*

$$\left\| F'(z)^{-1/2}(F'(y) - F'(x))v \right\| \le \omega \left\| F'(x)^{1/2}(y-x) \right\| \cdot \left\| F'(x)^{1/2}v \right\|$$

for some $0 \le \omega < \infty$. Consider an ordinary Newton-Galerkin method satisfying (8.13) with approximation errors bounded by

$$\delta_k := \frac{\| F'(x^k)^{1/2}(\delta x^k - \Delta x^k) \|}{\| F'(x^k)^{1/2}\delta x^k \|}.$$

At any well-defined iterate x^k, define the exact and inexact energy error norms by

$$\epsilon_k = \|F'(x^k)^{1/2}\Delta x^k\|^2\,, \qquad \epsilon_k^\delta = \|F'(x^k)^{1/2}\delta x^k\|^2 = \frac{\epsilon_k}{1+\delta_k^2}$$

and the associated Kantorovich quantities as

$$h_k = \omega\,\|F'(x^k)^{1/2}\Delta x^k\|\,, \qquad h_k^\delta = \omega\,\|F'(x^k)^{1/2}\delta x^k\| = \frac{h_k}{\sqrt{1+\delta_k^2}}\,.$$

For given initial guess $x^0 \in D$ assume that the level set

$$\mathcal{L}_0 := \{x \in D \mid f(x) \le f(x^0)\}$$

is nonempty, closed, and bounded. Then the following results hold:

I. Linear convergence mode. Assume that x^0 satisfies

$$h_0 \le 2\overline{\Theta} < 2 \tag{8.14}$$

for some $\overline{\Theta} < 1$. Let $\delta_{k+1} \ge \delta_k$ throughout the inexact Newton iteration. Moreover, let the Galerkin approximation be controlled such that

$$\vartheta(h_k^\delta, \delta_k) = \frac{h_k^\delta + \delta_k\left(h_k^\delta + \sqrt{4 + (h_k^\delta)^2}\right)}{2\sqrt{1+\delta_k^2}} \le \overline{\Theta}\,. \tag{8.15}$$

Then the iterates x^k remain in $\mathcal{L}_0$ and converge at least linearly to the minimum point $x^* \in \mathcal{L}_0$ such that

$$\|F'(x^{k+1})^{1/2}\Delta x^{k+1}\| \le \overline{\Theta}\,\|F'(x^k)^{1/2}\Delta x^k\| \tag{8.16}$$

and

$$\|F'(x^{k+1})^{1/2}\delta x^{k+1}\| \le \overline{\Theta}\,\|F'(x^k)^{1/2}\delta x^k\|\,. \tag{8.17}$$

II. Quadratic convergence mode. Let, for some $\rho > 0$, the initial guess x^0 satisfy

$$h_0 < \frac{2}{1+\rho} \tag{8.18}$$

and the Galerkin approximation be controlled such that

$$\delta_k \le \frac{\rho h_k^\delta}{h_k^\delta + \sqrt{4 + (h_k^\delta)^2}}\,. \tag{8.19}$$

Then the inexact Newton iterates x^k remain in $\mathcal{L}_0$ and converge quadratically to the minimum point $x^* \in \mathcal{L}_0$ such that

$$\|F'(x^{k+1})^{1/2}\Delta x^{k+1}\| \le (1+\rho)\frac{\omega}{2}\|F'(x^k)^{1/2}\Delta x^k\|^2 \tag{8.20}$$

and

$$\|F'(x^{k+1})^{1/2}\delta x^{k+1}\| \le (1+\rho)\frac{\omega}{2}\|F'(x^k)^{1/2}\delta x^k\|^2\,. \tag{8.21}$$

III. Functional descent. *The convergence in terms of the functional can be estimated by*

$$-\tfrac{1}{6}h_k^\delta \epsilon_k^\delta \leq f(x^k) - f(x^{k+1}) - \tfrac{1}{2}\epsilon_k^\delta \leq \tfrac{1}{6}h_k^\delta \epsilon_k^\delta \,.$$

The proof in the general Newton-Galerkin case is—mutatis mutandis—the same as the one for the more special Newton-PCG case in Theorem 2.20. In passing we mention that the above discussed boundedness of the local energy norms and, via the Cauchy-Schwarz inequality, also of the local energy products is actually guaranteed by (8.14), (8.16), and (8.17) in the linear convergence mode or by (8.18), (8.20), and (8.21) in the quadratic convergence mode.

For linear elliptic PDEs, we have *computational approximation error estimates* available, typically incorporated within adaptive multilevel FEM (Section 1.4.5), which are a special case of Galerkin methods. Hence, we may readily satisfy the above threshold criteria (8.15) or (8.19), respectively. Thus we are only left with the decision of whether to use the linear or the quadratic convergence mode in such a setting—an important algorithmic question that deserves special attention.

Computational complexity model. In order to get some insight, we study a rather simple, but nevertheless meaningful complexity model. It starts from the fact that at the final iterate, say x_q, we want to meet the prescribed energy error tolerance criterion (2.110), i.e.,

$$\epsilon_q \doteq \mathrm{ETOL}^2 \,.$$

If we replace the absolute error parameter $\mathrm{ETOL}^2 \ll \epsilon_0$ by a relative error parameter $\mathrm{EREL} \ll 1$ with $\mathrm{ETOL}^2 = \mathrm{EREL}^2 \cdot \epsilon_0$, then we may rewrite the above final accuracy requirement as

$$\Theta_0 \cdot \Theta_1 \cdots \Theta_{q-1} \doteq \mathrm{EREL} \,,$$

which is equivalent to

$$\sum_{k=0}^{q-1} \log \frac{1}{\Theta_k} \doteq \log \frac{1}{\mathrm{EREL}} \,. \tag{8.22}$$

The number q of Newton steps is unknown in advance. Let A_k denote the amount of work for step k. Then we will want to minimize the total amount of work, i.e.,

$$A_{total} = \sum_{k=0}^{q} A_k = \min$$

subject to the constraint (8.22). For the solution of this *discrete optimization* problem, there exists a quite efficient established heuristics, the popular *greedy algorithm*—see, e.g., Chapter 9.3 in the introductory textbook [2] by M. Aigner. From this, we obtain the prescription that, at Newton step k, the algorithm should maximize the *information gain per unit work*, i.e.,

$$I_k = \frac{1}{A_k} \log \frac{1}{\Theta_k} = \max . \tag{8.23}$$

In order to maximize this quantity with respect to the variable δ_k, the general relation (2.109) is applicable, which reads

$$\Theta_k \leq \vartheta(h_k^\delta, \delta_k) .$$

To simplify matters, we study the case $h_k \to 0$ here. Thus we arrive at the rough model

$$\Theta_k \doteq \vartheta(0, \delta_k) = \frac{\delta_k}{\sqrt{1 + \delta_k^2}} ,$$

which, in view of (8.23), is equivalent to

$$\log \frac{1}{\Theta_k} \sim \log \left(1 + \frac{1}{\delta_k^2}\right) .$$

Next we compare two variants of Newton-Galerkin methods, the finite dimensional case (PCG) and the infinite dimensional case (FEM), which differ in the amount of work A_k as a function of δ_k.

*Inexact Newton-*PCG *method for discrete PDEs.* Assume that we attack a nonlinear discrete elliptic PDE by some inexact Newton method with PCG as inner iteration—as in the algorithm GIANT-PCG. This is exactly the situation treated in Section 8.2.2. For system dimension n, we have to consider

- the evaluation of the Jacobian matrix $J = F'(x^k)$, which is typically sparse, so that an amount $O(n)$ needs to be counted,
- the work per PCG step (evaluation of inner products), which for the sparse Jacobian J is also $O(n)$,
- the number m_k of PCG iterations at Newton step k: with preconditioner B we have (compare, e.g., Corollary 8.18 in the textbook [77])

$$m_k \sim \sqrt{\kappa(BJ)} \log 2 \left(1 + 1/\delta_k^2\right) .$$

Summing up, we arrive at the rough estimate

$$A_k \sim \left(c_1 + c_2 \log \left(1 + 1/\delta_k^2\right)\right) n \sim \mathrm{const} + \log \left(1 + 1/\delta_k^2\right) ,$$

where 'const' represents some positive constant. So we finally end up with

$$I_k \sim \frac{\log(1 + 1/\delta_k^2)}{\mathrm{const} + \log(1 + 1/\delta_k^2)} = \max .$$

The right hand side is a monotone *decreasing* function of δ_k, which directs us towards the smallest possible value of δ_k, i.e., to the *quadratic convergence mode*. It may be worth noting that the above analysis would lead to the same decision, if PCG were replaced by some linear multigrid method.

Inexact Newton multilevel FEM for continuous PDEs. For the inner iteration we now take an adaptive multilevel method for linear elliptic PDEs (such as the multiplicative multigrid algorithm UG by G. Wittum, P. Bastian et al. [22] or the additive multigrid algorithm KASKADE by P. Deuflhard, H. Yserentant et al. [78, 36, 23]). An example of such an algorithm is implemented in our code Newton-KASKADE. As a consequence of the adaptivity, the dimension n of subproblems to be solved at step k depends on δ_k. Let d denote the underlying spatial dimension. At iteration step k on refinement level j of the multilevel discretization, let n_k^j be the number of nodes and ϵ_k^j the local energy. With $l = l_k$ we mean the final discretization level, at which the prescribed final accuracy δ_k is achieved. On energy equilibrated meshes for linear elliptic PDEs, we have the following asymptotic theoretical result (see I. Babuška et al. [13])

$$\left(\frac{n_k^0}{n_k^l}\right)^{2/d} \sim \frac{\epsilon_k^\infty - \epsilon_k^l}{\epsilon_k^\infty} \leq \frac{\delta_k^2}{1 + \delta_k^2} \,.$$

Any decent multigrid solver for linear elliptic PDEs will require an amount of work proportional to the number of nodes, i.e.

$$A_k \sim n_k^l \sim n_k^0 \left(1 + 1/\delta_k^2\right)^{d/2} \,.$$

Inserting this result into I_k, we arrive at the rough estimate

$$I_k \sim \left(1 + 1/\delta_k^2\right)^{-d/2} \log\left(1 + 1/\delta_k^2\right) = \max \,.$$

For variable space dimension d this scalar function has its maximum at

$$\delta_k = 1/\sqrt{\exp(2/d) - 1}\,,$$

which, with the help of (2.113), then leads to the choice

$$\overline{\Theta} = \exp(-1/d)\,.$$

We thus have the approximate values

$$d = 2: \quad \delta_k = 0.76, \ \overline{\Theta} = 0.61\,, \qquad d = 3: \quad \delta_k = 1.03, \ \overline{\Theta} = 0.72\,.$$

Even though our rough complexity model might not cover such large values of δ_k, these results may nevertheless be taken as a clear indication to favor the *linear* over the quadratic convergence mode in an *adaptive* multilevel setting. Empirical tests actually suggested to use $\delta_k \approx 1$ corresponding to $\overline{\Theta} \approx 0.7$ as default values.

Example 8.7 By modification of a problem given by R. Rannacher [175], we consider the convex functional in two space dimensions (with x, y Euclidean coordinates here)

$$f(u) = \int_\Omega \left(1 + |\nabla u|^2\right)^{p/2} - gu \, dx, \ p > 1, \ x \in \Omega \subset \mathbb{R}^2, \ u \in W^{1,p}(\Omega)$$

for the specification $p = 1.4$, $g \equiv 0$. The functional gives rise to the first and second order expressions

$$\langle F(u), v \rangle \ = \ \int_\Omega \left(p(1 + |\nabla u|^2)^{p/2-1}\langle \nabla u, \nabla v\rangle - gv\right) dx \ ,$$

$$\langle w, F'(u)v \rangle \ = \ \int_\Omega p\Big((p-2)(1 + |\nabla u|^2)^{p/2-2}\langle \nabla w, \nabla u\rangle\langle \nabla u, \nabla v\rangle$$
$$+ (1 + |\nabla u|^2)^{p/2-1}\langle \nabla w, \nabla v\rangle\Big) dx \ .$$

With $\langle \cdot, \cdot \rangle$ the Euclidean inner product in $\mathbb{R}^2$, the term $\langle v, F'(u)v \rangle$ is strictly positive for $p \geq 1$.

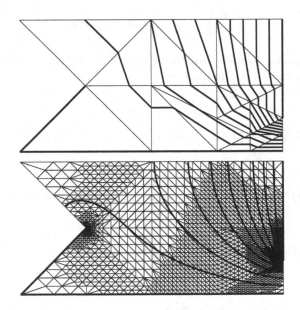

Fig. 8.4. Example 8.7. Newton-KASKADE iterates: *Top:* initial guess u^0 on initial coarse grid, *bottom:* iterate u^3 on automatically refined grid. *Thick lines:* homogeneous Dirichlet boundary conditions and level lines, *thin lines:* Neumann boundary conditions.

In order to solve this problem, we used the linear convergence mode in an adaptive Newton multilevel FEM with KASKADE to solve the arising linear elliptic PDEs. Figure 8.4 compares the starting guess u^0 on its coarse mesh ($j = 1$) with the inexact Newton iterate u^3 on its fine mesh ($j = 14$). The coarse mesh consists of $n_1 = 17$ nodes, the fine mesh of $n_{14} = 2054$ nodes; for comparison, a uniformly refined mesh at level $j = 14$ would have about $\bar{n}_{14} \approx 136000$ nodes. Note that—in the setting of multigrid methods, which require $O(n)$ operations—the total amount of work would be essentially blown up by the factor $\bar{n}_{14}/n_{14} \approx 65$. Apart from this clear computational saving, adaptivity also nicely models the two critical points on the boundary, the re-entrant corner and the discontinuity point.

8.3.2 Global Newton-Galerkin methods

In this section we study the inexact *global* Newton-Galerkin method

$$x^{k+1} = x^k + \lambda_k \delta x^k$$

in terms of iterates $x^k \in W^{1,p}$, inexact Newton corrections δx^k satisfying (8.13), and damping factors λ_k to be chosen appropriately. As the most prominent representatives of such methods we will take adaptive Newton multilevel FEMs, whenever it comes to numerical examples.

In Section 3.4.3 we had already discussed the finite dimensional analogue, the global inexact Newton-PCG method. With the theoretical considerations at the beginning of Section 8.3, we are prepared to modify the global convergence theorems for the Newton-PCG methods in such a way that they apply to the more general Newton-Galerkin case. In what follows, we just combine and modify our previous Theorems 3.23 and 3.26.

Theorem 8.7 *Notation as introduced above. Let $f : D \to \mathbb{R}^1$ be a strictly convex C^2-functional to be minimized over some open convex domain $D \subset W^{1,p}$ and $F'(x) = f''(x)$ be strictly positive. For $x, y \in D$, assume the affine conjugate Lipschitz condition*

$$\|F'(x)^{-1/2}(F'(y) - F'(x))(y - x)\| \leq \omega \|F'(x)^{1/2}(y - x)\|^2$$

with $0 \leq \omega < \infty$. Let Δx^k denote the exact and δx^k the inexact Newton correction. For each well-defined iterate $x^k \in D$, define the quantities

$$\epsilon_k = \|F'(x^k)^{1/2}\Delta x^k\|^2 \, , \qquad \epsilon_k^\delta = \|F'(x^k)^{1/2}\delta x^k\|^2 = \frac{\epsilon_k}{1 + \delta_k^2} \, ,$$

$$h_k = \omega \|F'(x^k)^{1/2}\Delta x^k\| \, , \qquad h_k^\delta = \omega \|F'(x^k)^{1/2}\delta x^k\| = \frac{h_k}{\sqrt{1 + \delta_k^2}} \, .$$

Moreover, let $x^k + \lambda \delta x^k \in D$ for $0 \leq \lambda \leq \lambda_{\max}^k$ with

$$\lambda_{\max}^k := \frac{4}{1 + \sqrt{1 + 8h_k^\delta/3}} \le 2 .$$

Then

$$f(x^k + \lambda \Delta x^k) \le f(x^k) - t_k(\lambda)\epsilon_k^\delta$$

where

$$t_k(\lambda) = \lambda - \tfrac{1}{2}\lambda^2 - \tfrac{1}{6}\lambda^3 h_k^\delta .$$

The optimal choice of damping factor is

$$\bar{\lambda}_k = \frac{2}{1 + \sqrt{1 + 2h_k^\delta}} \le 1 .$$

As in the local convergence case, h_k is the Kantorovich quantity and δ_k the relative Galerkin approximation error.

Adaptive damping and accuracy matching. Following our usual paradigm, the unknown theoretical quantities h_k and δ_k are replaced by computationally available estimates $[h_k]$ and $[\delta_k]$. For $[h_k]$ we just use the terms $E_{1,2,3}(\lambda)$ as given in Section 3.4.2 for the exact Newton method and in Section 3.4.3 for the inexact Newton-PCG method. On this basis we realize the correction strategy (3.88) with h_k replaced by the a-posteriori estimate $[h_k^\delta]$ and the prediction strategy (3.89) with h_{k+1} replaced by the a-priori estimate $[h_{k+1}^\delta]$. Unless stated otherwise, we choose the approximation error bound $\delta_k = 1$ as a default throughout the Newton-Galerkin iteration, thus eventually merging into the linear local convergence mode.

Example 8.8   Good versus bad initial coarse grid. We return to our previous Example 8.7, but this time for the critical value $p = 1$, which characterizes the (parametric) *minimal surface* problem. This value is critical, since then $u \in W^{1,1}$, a *nonreflexive* Banach space, which implies that the existence of a unique solution is no longer guaranteed. For special boundary conditions and inhomogeneities g, however, a unique solution can be shown to exist, even in $C^{0,1}$ (see, e.g., the textbooks by E. Zeidler [205]). Such a situation occurs, e.g., for

$$\Omega = \left[-\frac{\pi}{2}, 0\right] \times \left[-\frac{\pi}{2}, \frac{\pi}{2}\right] , \quad u|_{\partial\Omega} = s \cos x \cos y , \quad g \equiv 0 .$$

Taking the Z_2-symmetry along the x-axis into account, we may halve Ω and impose homogeneous Neumann boundary conditions at $y = 0$. The parameter s is set to $s = 3.5$. From a quick rough examination of the problem, we expect a boundary layer at $x = 0$. As initial guess u^0 we take the prescribed values on the Dirichlet boundary part and otherwise just zero.

Again we solve the problem by `Newton-KASKADE`. As good initial coarse grid we select the grid in Figure 8.5, left, which takes the expected boundary layer into account. As bad initial coarse grid we choose the one in Figure 8.5, right,

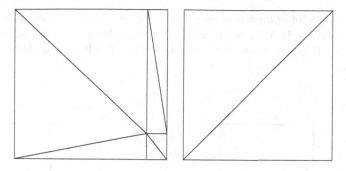

Fig. 8.5. Example 8.8. Good (left) and bad (right) initial coarse grid.

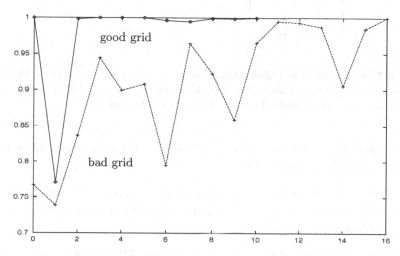

Fig. 8.6. Example 8.8. Comparative damping strategies for good and bad initial coarse grids.

which deliberately ignores any knowledge about the occurrence of a boundary layer.

In Figure 8.6, the comparative performance of our global Newton-KASKADE algorithm is documented in terms of the obtained damping factors for both initial grids. As expected from reports in the engineering literature, the bad coarse grid requires many more iterations to eventually capture the nonlinearity.

Example 8.9   Function space versus finite dimensional approach.
Once again, we return to Example 8.7, this time for the critical value $p = 1$. In Figure 8.7, we show two settings: On the left (Example 8.9a), a unique solution exists, which has been computed, but is not documented here; this

example serves for comparison only, see Table 8.11 below. Our main interest focuses on Example 8.9b, where *no (physical) solution exists.* For the initial guess u^0 we take the prescribed values on the Dirichlet boundary and zero otherwise.

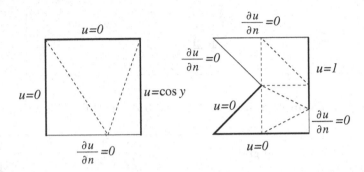

Fig. 8.7. Example 8.9. Domains and initial coarse grids. *Black lines:* Dirichlet boundary conditions, *grey lines:* Neumann boundary conditions. *Left:* Example 8.9a, unique solution exists. *Right:* Example 8.9b, no solution exists.

At Example 8.9b, we want to compare the algorithmic behavior of two different Newton-FEM approaches:

- our *function space* oriented approach, as presented in this section, and
- the *finite dimensional* approach, which is typically implemented in classical Newton-multigrid FEMs.

In the finite dimensional approach, the discrete FE problem is solved successively *on each of the mesh levels* so that there the damping factors will repeatedly run up to values $\lambda = 1$. In contrast to that behavior, our function space approach aims at directly solving the continuous problem by exploiting information available from the whole mesh refinement history. Consequently, if a unique solution exists, this approach will reach the local convergence phase in accordance with the mesh refinement process. Such a behavior has already been shown for our preceding Example 8.8 in Figure 8.6.

Figure 8.8 gives an account for Example 8.9b. In the finite dimensional option, damping factors $\lambda = 1$ arise repeatedly on each of the mesh refinement levels. After more than 60 Newton-FEM iterations, this approach gives the impression of a unique solvability of the problem—on the basis of the local convergence of the Newton-FEM algorithm on each of the successive meshes. Our function space option, however, terminates already after 20 Newton iterations for $\lambda < \lambda_{\min} = 0.01$.

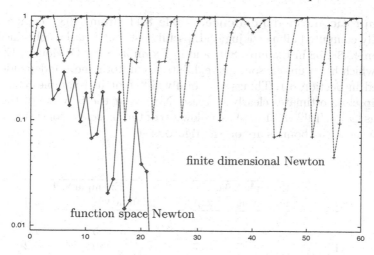

Fig. 8.8. Example 8.9b. Iterative damping factors for two Newton-FEM algorithms. To be compared with Figure 8.9.

To understand this discrepancy, we simultaneously look at the local energy norms ϵ_k, which measure the *exact* Newton corrections Δx^k, see Figure 8.9. The finite dimensional method ends up with 'sufficiently small' Newton corrections on each of the refinement levels, pretending some local *pseudoconvergence*. Our function space Newton method, however, stays with 'moderate size' corrections throughout the iteration.

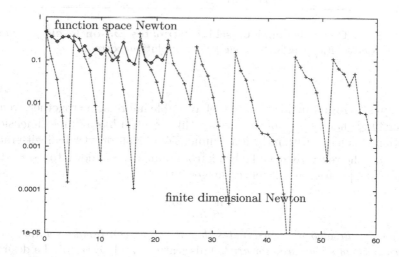

Fig. 8.9. Example 8.9b. Iterative energy norms $\epsilon_k^{1/2}$ for two Newton-FEM algorithms. To be compared with Figure 8.8.

Asymptotic mesh dependence. Table 8.11 (from [198]) compares the actually computed affine conjugate Lipschitz estimates $[\omega_j]$ as obtained from Newton-KASKADE in Example 8.9a and in Example 8.9b. Obviously, Example 8.9a, which has a unique solution, exhibits asymptotic mesh independence as studied in Section 8.1. Things are totally different in Example 8.9b, where the Lipschitz estimates clearly increase. Note that the blow-up of the lower bounds $[\omega_j]$ in Table 8.11 implies a blow-up of the Lipschitz constants ω_j—for this purpose the bounds are on the rigorous side.

j	Example 8.9a ♯ unknowns	$[\omega_j]$	Example 8.9b ♯ unknowns	$[\omega_j]$
0	4	1.32	5	7.5
1	7	1.17	10	4.2
2	18	4.55	17	7.3
3	50	6.11	26	9.6
4	123	5.25	51	22.5
5	158	20.19	87	50.3
6	278	19.97	105	1486.2
7	356	9.69	139	2715.6
8	487	8.47	196	5178.6
9	632	11.73	241	6837.2
10	787	44.21	421	12040.2
11	981	49.24	523	167636.0
12	1239	20.10	635	1405910.0
13	1610	32.93		
14	2054	37.22		

Table 8.11. Computational Lipschitz estimates $[\omega_j]$ on levels j. *Example 8.9a:* unique solution exists, *Example 8.9b:* no solution exists.

Interpretation. Putting all pieces of available information together, we now understand that on each of the levels j this problem has a finite dimensional solution x_j^*, unique within the finite dimensional Kantorovich ball with radius $\rho_j \sim 1/\omega_j$; however, these balls shrink from radius $\rho_1 \sim 1$ down to $\rho_{22} \sim 10^{-6}$. Frank extrapolation of this effect suggests that

$$\lim_{j\to\infty} \rho_j = 0 \,.$$

Obviously, *the algorithm insinuates that a unique continuous solution of the stated PDE problem does not exist.* This feature would certainly be desirable for any numerical PDE solver.

Exercises

Exercise 8.1 In Section 8.3.1, a rough computational complexity model of adaptive multilevel FEM for nonlinear elliptic PDEs leads to the problem (dropping the index k)

$$I \sim \left(1 + 1/\delta^2\right)^{-d/2} \log\left(1 + 1/\delta^2\right) = \max,$$

where d is the spatial dimension.

a) Calculate the maximum point δ and evaluate it for $d = 2$ and $d = 3$.

b) How would the rough model need to be changed, if the situation $h \neq 0$ were to be modeled?

Exercise 8.2 Consider the finite dimensional Newton sequence

$$x^{k+1} = x^k + \Delta x^k,$$

where x^0 is given and Δx^k is the solution of a linear system. In sufficiently large scale computations, rounding errors caused by direct elimination or truncation errors from iterative linear solvers will generate a different sequence

$$y^{k+1} = y^k + \Delta y^k + \epsilon_k,$$

where $y^0 = x^0$ is given, Δy^k is understood to be the exact Newton correction at y^k, and

$$\|\epsilon_k\| \leq \delta \|\Delta y^k\|,$$

Upon using analytical tools of Section 8.1, derive iterative error bounds for $\|y^k - x^k\|$ and $\|y^k - x^*\|$.

Exercise 8.3 Consider the nonlinear ODE boundary value problem

$$\dot{x} = f(x), \quad Ax(a) + Bx(b) = 0$$

with linear separable boundary conditions. We want to study asymptotic mesh independence for Gauss collocation methods of order $s \geq 1$ (compare Section 7.4). For the approximating space we select $X = W^{1,\infty}$ and impose the assumptions from Section 8.1. Let $X_j \subset X$ denote a finite dimensional subspace characterizing the collocation discretization with maximum mesh size τ_j. Assume that f is sufficiently smooth and the BVP is well-conditioned for all required arguments.

a) In view of (8.5), derive upper bounds δ_j such that

$$\|\Delta x_j - \Delta x\|_{W^{1,\infty}} \leq \delta_j$$

with the asymptotic property

$$\delta_j \to 0.$$

Hint: Compare the exact solution w of

$$\dot{w} + f_x(x_j)w = -f_x(x_j)x_j + f(x_j), \quad Aw(a) + Bw(b) = 0$$

and its approximation w_j using the error estimate (as given in [180, 49])

$$\|w - w_j\|_{W^{1,\infty}} \leq C\tau_j\|\ddot{w}\|_\infty,$$

where C is a bounded generic constant, which is independent of j.

b) Under the assumptions of Theorem 2.2 derive some bound

$$\|w - w_j\| \leq \sigma_j\|v_j\|^2$$

with the asymptotic property

$$\sigma_j \to 0.$$

Exercise 8.4   We consider linear finite element approximations on quasi-uniform triangulations for semilinear elliptic boundary value problems

$$F(x) = -\text{div}\nabla x - f(x) = 0, \quad x \in H_0^1(\Omega)$$

on convex polygonal domains $\Omega \subset \mathbb{R}^d$, $d \leq 3$. For this setting, we want to study asymptotic mesh independence. The notation is as in Section 8.1. In view of (8.5) and (8.9), derive upper bounds δ_j such that

$$\|\Delta x_j - \Delta x\| \leq \delta_j$$

and σ_j such that

$$\|w - w_j\| \leq \sigma_j\|v_j\|^2,$$

Assume that the above right hand term $f : \mathbb{R} \to \mathbb{R}$ is globally Lipschitz continuously differentiable. In particular, show that for the process of successive refinement the asymptotic properties

$$\lim_{j \to \infty} \delta_j = 0, \quad \lim_{j \to \infty} \sigma_j = 0$$

hold.

Hint: Exploit the H^2-regularity of $x_j + \Delta x$ and use the embedding $H^1 \hookrightarrow L_4$.

References

1. A. Abdulle and G. Wanner. 200 years of least squares method. *Elemente der Mathematik*, pages 45–60, 2002.
2. M. Aigner. *Diskrete Mathematik*. Vieweg, Braunschweig, Wiesbaden, 4. edition, 2001.
3. E.L. Allgower and K. Böhmer. Application of the mesh independence principle to mesh refinement strategies. *SIAM J. Numer. Anal.*, 24:1335–1351, 1987.
4. E.L. Allgower, K. Böhmer, F.A. Potra, and W.C. Rheinboldt. A mesh-independence principle for operator equations and their discretizations. *SIAM J. Numer. Anal.*, 23:160–169, 1986.
5. E. Anderson, Z. Bai, C. Bischof, J. Demmel, J. Dongarra, J. DuCroz, A. Greenbaum, S. Hammarling, A. McKenney, S. Ostruchov, and D. Sorensen. *LAPACK Users' Guide*. SIAM, Philadelphia, 1999.
6. D. Armbruster. Bifurcation Theory and Computer Algebra: An Initial Approach. In B.F. Caviness, editor, *Proceedings European Conference on Computer Algebra, Linz, Austria*, volume 204 of *Lecture Notes in Computer Science*, pages 126–136, Berlin, Heidelberg, New York, 1985. Springer.
7. L. Armijo. Minimization of functions having Lipschitz-continuous first partial derivatives. *Pacific J. Math.*, 204:126–136, 1966.
8. U.M. Ascher, J. Christiansen, and R.D. Russell. A collocation solver for mixed order systems of boundary value problems. *Math. Comp.*, 33:659–679, 1979.
9. U.M. Ascher, R.M.M. Mattheij, and R.D. Russell. *Numerical Solution of Boundary Value Problems for Ordinary Differential Equations*, volume 14 of *Classics in Applied Mathematics, Vol.13*. SIAM Publications, Philadelphia, PA, 2nd edition, 1995.
10. U.M. Ascher and M.R. Osborne. A note on solving nonlinear equations and the 'natural' criterion function. *J. Opt. Theory & Applications (JOTA)*, 55:147–152, 1987.
11. O. Axelsson. *Iterative Solution Methods*. Cambridge University Press, Cambridge, 1994.
12. I. Babuška and W.C. Rheinboldt. Error Estimates for Adaptive Finite Element Computations. *SIAM J. Numer. Anal.*, 15:736–754, 1978.
13. I. Babuška, R.B. Kellog, and J. Pitkäranta. Direct and Inverse Estimates for Finite Elements with Mesh Refinements. *Numer. Math.*, 15:447–471, 1979.
14. I. Babuška and W.C. Rheinboldt. Estimates for adaptive finite element computations. *SIAM J. Numer. Anal.*, 15:736–754, 1978.
15. G. Bader. *Numerische Behandlung von Randwertproblemen für Funktionaldifferentialgleichungen*. PhD thesis, Universität Heidelberg, Inst. Angew. Math., 1983.

16. G. Bader and U.M. Ascher. A new basis implementation for a mixed order boundary value ODE solver. *SIAM J. Sci. Stat. Comput.*, 8:483–500, 1987.

17. G. Bader and P. Kunkel. Continuation and collocation for parameter-dependent boundary value problems. *SIAM J. Sci. Stat. Comput.*, 10:72–88, 1989.

18. R.E. Bank. *PLTMG: A Software Package for Solving Elliptic Partial Differential Equations. Users' Guide 8.0.* Frontiers in Applied Mathematics. SIAM, 1998.

19. R.E. Bank and D.J. Rose. Global approximate Newton methods. *Numer. Math.*, 37:279–295, 1981.

20. R.E. Bank, A.H. Sherman, and A. Weiser. Refinement algorithms and data structures for regular local mesh refinement. In *Scientific Computing*, pages 3–17. North-Holland, 1983.

21. P. Bastian, K. Birken, K. Johannsen, S. Lang, N. Neuss, H. Rentz-Reichert, and C. Wieners. UG—A flexible software toolbox for solving partial differential equations. *Comp. Vis. Sci.*, 1:27–40, 1997.

22. P. Bastian and G. Wittum. Adaptive multigrid methods: The UG concept. In W. Hackbusch and G. Wittum, editors, *Adaptive Methods—Algorithms, Theory and Applications, Series Notes on Numerical Fluid Mechanics*, volume 46, pages 17–37. Vieweg, Braunschweig, 1994.

23. R. Beck, B. Erdmann, and R. Roitzsch. *KASKADE 3.0 — User's Guide*, 1996.

24. A. Ben-Israel and T.N.E. Greville. *Generalized Inverses: Theory and Applications.* Wiley & Sons, New York, London, Sydney, Toronto, 1974.

25. Å. Bjørck. Iterative refinement of linear least squares solutions I. *BIT*, 7:257–278, 1967.

26. Å. Bjørck. Least Squares Methods. In P. G. Ciarlet and J. L. Lions, editors, *Handbook of Numerical Analysis I*, pages 466–652. Elsevier Science Publishers (North-Holland), Amsterdam, New York, 1990.

27. Å. Bjørck and I.S. Duff. A Discrete Method for the Solution of Sparse Linear Least Squares Problems. Technical Report CSS 79, Harwell, AERE, 1980.

28. J. Blue. Robust Methods for Solving Systems of Nonlinear Equations. *SIAM J. Sci. Stat. Comput.*, 1:22–33, 1980.

29. H.G. Bock. Numerical treatment of inverse problems in chemical reaction kinetics. In K.H. Ebert, P. Deuflhard, and W. Jäger, editors, *Modelling of Chemical Reaction Systems*, pages 102–125. Springer-Verlag, Berlin, Heidelberg, New York, 1981.

30. H.G. Bock. Numerische Behandlung von zustandsbeschränkten und Chebychev-Steuerungsproblemen. Technical report, Carl-Cranz-Gesellschaft, DR 3.10, 1981.

31. H.G. Bock. Recent Advances in Parameter Identification Techniques for ODE's. In P. Deuflhard and E. Hairer, editors, *Numerical Treatment of Inverse Problems in Differential and Integral Equations*, volume 2 of *Progress in Scientific Computing*, pages 95–121. Birkhäuser, Boston, Basel, Stuttgart, 1983.

32. H.G. Bock. *Randwertproblemmethoden zur Parameteridentifizierung in Systemen nichtlinearer Differentialgleichungen.* PhD thesis, Universität Bonn, 1985.

33. H.G. Bock, E.A. Kostina, and J.P. Schlöder. On the role of natural level functions to achieve global convergence for damped Newton methods. In M. D. Powell and S. Scholtes, editors, *System Modelling and Optimization. Methods, Theory, and Applications*, pages 51–74. Kluwer, Amsterdam, 2000.

34. P.T. Boggs and J.E. Dennis jr. A continuous analogue analysis of nonlinear iterative methods. Technical report, Cornell University, Ithaca, Oct. 1974.

35. F. Bornemann and P. Deuflhard. The cascadic multigrid method for elliptic problems. *Numer. Math.*, 75:135–152, 1996.

36. F. Bornemann, B. Erdmann, and R. Kornhuber. Adaptive multilevel methods in three space dimensions. *Int. J. Num. Meth. in Eng.*, 36:3187–3203, 1993.

37. F.A. Bornemann, B. Erdmann, and R. Kornhuber. A posteriori error estimates for elliptic problems in two and three space dimensions. *SIAM J. Numer. Anal.*, 33:1188–1204, 1996.

38. D. Braess. Eine Möglichkeit zur Konvergenzbeschleunigung bei Iterationsverfahren für bestimmte nichtlineare Probleme. *Numer. Math.*, 14:468–475, 1970.

39. J. Bramble, J. Pasciak, and J. Xu. Parallel multilevel preconditioners. *Math. Comp.*, 55:1–22, (1990).

40. C.G. Broyden. A class of methods for solving nonlinear simultaneous equations. *Math. Comp.*, 19:577–583, 1965.

41. C.G. Broyden, J.E. Dennis jr., and J.J. Moré. On the local and superlinear convergence of quasi-Newton methods. *IMA J. Appl. Math.*, 12:223–246, 1973.

42. R. Bulirsch. Die Mehrzielmethode zur numerischen Lösung von nichtlinearen Randwertproblemen und Aufgaben der optimalen Steuerung. Technical report, Carl-Cranz-Gesellschaft, 1971.

43. P. Businger and G.H. Golub. Linear least squares solutions by Householder transformations. *Num. Math.*, 7:269–276, 1965.

44. A. Cauchy. Méthode générale pour la résolution des systèmes d'équations simultanées. *C.R. Acad. Sci. Paris*, 25:536–538, 1847.

45. A.R. Conn, N.I.M. Gould, and P.L. Toint. *Trust-Region Methods*. SIAM, MPS, Philadelphia, 2000.

46. M.G. Crandall and P.H. Rabinowitz. Bifurcation from simple eigenvalues. *J. Funct. Anal.*, 8:321–340, 1971.

47. G. Dahlquist. A special stability problem for linear multistep methods. *BIT*, 3:27–43, 1963.

48. D. Davidenko. On a new method of numerically integrating a system of nonlinear equations. *Dokl. Akad. Nauk SSSR*, 88:601–604, 1953.

49. C. de Boor and B. Swartz. Collocation at Gaussian Points. *SIAM J. Numer. Anal.*, 10:582–606, 1973.

50. M. Dellnitz and B. Werner. Computational methods for bifurcation problems with symmetries—with special attention to steady state and Hopf bifurcation points. *J. Comp. Appl. Math.*, 26:97–123, 1989.

51. R.S. Dembo, S.C. Eisenstat, and T. Steihaug. Inexact Newton Methods. *SIAM J. Numer. Anal.*, 18:400–408, 1982.

52. J.E. Dennis jr. A stationary Newton method for nonlinear functional equations. *SIAM J. Numer. Anal.*, 4:222–232, 1967.

53. J.E. Dennis jr. On the Kantorivich hypotheses for Newton's method. *SIAM J. Numer. Anal.*, 6:493–507, 1969.

54. J.E. Dennis jr, D. Gay, and R. Welsch. An adaptive nonlinear least-squares algorithm. *ACM Trans. Math. Software*, 7:369–383, 1981.

55. J.E. Dennis jr and J.J. Moré. A characterization of superlinear convergence and its application to quasi-newton methods. *Math. Comp.*, 28:549–560, 1974.

56. J.E. Dennis jr and R.B. Schnabel. Least Change Secant Updates for Quasi-Newton Methods. *SIAM Rev.*, 21:443–459, 1979.

57. J.E. Dennis jr and R.B. Schnabel. *Numerical Methods for Unconstrained Optimization and Nonlinear Equations.* Prentice Hall, 1983.

58. J.E. Dennis jr and H.F. Walker. Convergence Theorems for Least-Change Secant Update Methods. *SIAM J. Numer. Anal.*, 18:949–987, 1979.

59. P. Deuflhard. *Ein Newton-Verfahren bei fast-singulärer Funktionalmatrix zur Lösung von nichtlinearen Randwertaufgaben mit der Mehrzielmethode.* PhD thesis, Universität zu Köln, 1972.

60. P. Deuflhard. A modified Newton method for the solution of ill-conditioned systems of nonlinear equations with applications to multiple shooting. *Numer. Math.*, 22:289–315, 1974.

61. P. Deuflhard. A stepsize control for continuation methods and its special application to multiple shooting techniques. *Numer. Math.*, 33:115–146, 1979.

62. P. Deuflhard. Nonlinear Equation Solvers in Boundary Value Problem Codes. In I. Gladwell and D.K. Sayers, editors, *Computational Techniques for Ordinary Differential Equations*, pages 217–272. Academic Press, New York, NY, 1980.

63. P. Deuflhard. A relaxation strategy for the modified Newton method. In R. Bulirsch, W. Oettli, and J. Stoer, editors, *Optimization and Optimal Control*, Springer Lecture Notes in Math. 447, pages 38–55. Springer-Verlag, Berlin, Heidelberg, New York, 1981.

64. P. Deuflhard. Computation of periodic solutions of nonlinear ODE's. *BIT*, 24:456–466, 1984.

65. P. Deuflhard. Recent progress in extrapolation methods for ordinary differential equations. *SIAM Review*, 27:505–535, 1985.

66. P. Deuflhard. Uniqueness Theorems for Stiff ODE Initial Value Problems. In D.F. Griffiths and G.A. Watson, editors, *Proceedings 13th Biennial Conference on Numerical Analysis 1989*, pages 74–205, Harlow, Essex, UK, 1990. Longman Scientific & Technical.

67. P. Deuflhard. Global Inexact Newton Methods for Very Large Scale Nonlinear Problems. *IMPACT Comp. Sci. Eng.*, 3:366–393, 1991.

68. P. Deuflhard. Cascadic Conjugate Gradient Methods for Elliptic Partial Differential Equations. Algorithm and Numerical Results. In D.E. Keyes and J. Xu, editors, *Domain Decomposition Methods in Scientific and Engineering Computing*, volume 180 of *AMS Series Contemporary Mathematics*, pages 29–42, 1994.

69. P. Deuflhard and V. Apostolescu. An Underrelaxed Gauss-Newton Method for Equality Constrained Nonlinear Least Squares Problems. In J. Stoer, editor, *Optimization Techniques, Proceedings of the 8th IFIP Conference, Würzburg 1977, part 2*, volume 7 of *Lecture Notes Control Inf. Sci.*, pages 22–32. Wiley-Interscience-Europe, Berlin, Heidelberg, New York, 1978.

70. P. Deuflhard and G. Bader. Multiple-shooting techniques revisited. In P. Deuflhard and E. Hairer, editors, *Numerical treatment of inverse problems in differential and Integral Equation*, volume 2, pages 13–26. Birkhäuser, Boston, MA, 1983.

71. P. Deuflhard and F. Bornemann. *Scientific Computing with Ordinary Differential Equations*, volume 42 of *Texts in Applied Mathematics*. Springer, New York, 2002.

72. P. Deuflhard, B. Fiedler, and P. Kunkel. Efficient numerical pathfollowing beyond critical points. *SIAM J. Numer. Anal.*, 24:912–927, 1987.

73. P. Deuflhard, B. Fiedler, and P. Kunkel. Numerical pathfollowing beyond critical points in ODE models. In P. Deuflhard and B. Engquist, editors, *Large Scale Scientific Computing*, volume 7 of *Progress in Scientific Computing*, pages 37–50. Birkhäuser, Boston, MA, 1987.

74. P. Deuflhard, R. Freund, and A. Walter. Fast Secant Methods for the Iterative Solution of Large Nonsymmetric Linear Systems. *IMPACT Comp. Sci. Eng.*, 2:244–276, 1990.

75. P. Deuflhard, E. Hairer, and J. Zugck. One-step and extrapolation methods for differential-algebraic systems. *Numer. Math.*, 51:501–516, 1987.

76. P. Deuflhard and G. Heindl. Affine Invariant Convergence theorems for Newton's Method and Extensions to related Methods. *SIAM J. Numer. Anal.*, 16:1–10, 1979.

77. P. Deuflhard and A. Hohmann. *Numerical Analysis in Modern Scientific Computing: An Introduction*, volume 43 of *Texts in Applied Mathematics*. Springer, Berlin, Heidelberg, New York, 2nd edition, 2003.

78. P. Deuflhard, P. Leinen, and H. Yserentant. Concepts of an adaptive hierarchical finite element code. *IMPACT Comp. Sci. Eng.*, 1:3–35, 1989.

79. P. Deuflhard and U. Nowak. Extrapolation integrators for quasilinear implicit ODEs. In P. Deuflhard and B. Engquist, editors, *Large Scale Scientific Computing*, pages 37–50. Birkhäuser, Boston, Basel, Stuttgart, 1987.

80. P. Deuflhard, U. Nowak, and M. Weiser. Affine Invariant Adaptive Newton Codes for Discretized PDEs. Technical Report ZIB-Report 02–33, Zuse Institute Berlin (ZIB), 2002.

81. P. Deuflhard, J. Pesch, and P. Rentrop. A Modified Continuation Method for the Numerical Solution of Nonlinear Two-Point Boundary Value Problems by Shooting Techniques. *Numer. Math.*, 26:327–343, 1976.

82. P. Deuflhard and F.A. Potra. Asymptotic Mesh Independence of Newton-Galerkin Methods via a Refined Mysovskii Theorem. *SIAM J. Numer. Anal.*, 29:1395–1412, 1992.

83. P. Deuflhard and W. Sautter. On rank-deficient pseudoinverses. *Lin. Alg. Appl.*, 29:91–111, 1980.

84. P. Deuflhard and M. Weiser. Local Inexact Newton Multilevel FEM for Nonlinear Elliptic Problems. In M.-O. Bristeau, G. Etgen, W. Fitzgibbon, J.-L. Lions, J. Periaux, and M. Wheeler, editors, *Computational Science for the 21st Century*, pages 129–138. Wiley-Interscience-Europe, Tours, France, 1997.

85. P. Deuflhard and M. Weiser. Global Inexact Newton Multilevel FEM for Nonlinear Elliptic Problems. In W. Hackbusch and G. Wittum, editors, *Multigrid Methods*, volume 3 of *Lecture Notes in Computational Science and Engineering*, pages 71–89. Springer, Berlin, Heidelberg, New York, 1998.

86. E.D. Dickmanns. Optimal control for synergetic plane change. In *Proc. XXth Int. Astronautical Congress*, pages 597–631, 1969.

87. E.D. Dickmanns. Maximum range three-dimensional lifting planetary entry. Technical report, NASA, TR R-387, 1972.

88. E.D. Dickmanns and H.-J. Pesch. Influence of a reradiative heating constraint on lifting entry trajectories for maximum lateral range. Technical report, 11th International Symposium on Space Technology and Science: Tokyo, 1975.

89. E. Doedel, T. Champneys, T. Fairgrieve, Y. Kuznetsov, B. Sandstede, and X.-J. Wang. AUTO97 Continuation and bifurcation software for ordinary differential equations (with HomCont). Technical report, Concordia Univ., Montreal, 1997.

90. S.C. Eisenstat and H.F. Walker. Globally convergent inexact Newton methods. *SIAM J. Optimization*, 4:393–422, 1994.

91. S.C. Eisenstat and H.F. Walker. Choosing the forcing terms in an inexact Newton method. *SIAM J. Sci. Comput.*, 17:16–32, 1996.

92. B. Fischer. *Polynomial Based Iteration Methods for Symmetric Linear Systems*. Wiley and Teubner, Chichester, New York, Stuttgart, Leipzig, 1996.

93. R. Fletcher and C.M. Reeves. Function Minimization by Conjugate Gradients. *Comput. J.*, 7:149–154, 1964.

94. K. Gatermann and A. Hohmann. Hexagonal Lattice Dome—Illustration of a Nontrivial Bifurcation Problem. Technical Report TR 91–08, Zuse Institute Berlin (ZIB), Berlin, 1991.

95. K. Gatermann and A. Hohmann. Symbolic Exploitation of Symmetry in Numerical Pathfollowing. *IMPACT Comp. Sci. Eng.*, 3:330–365, 1991.

96. K.F. Gauss. Theoria Motus Corporum Coelestium. *Opera*, 7:225–245, 1809.

97. D.M. Gay and R.B. Schnabel. Solving systems of nonlinear equations by Broyden's method with projected updates. In O.L. Mangasarian, R.R. Meyer, and S.M. Robinson, editors, *Nonlinear Programming*, volume 3, pages 245–281. Academic Press, 1978.

98. C.W. Gear. *Numerical Initial Value Problems in Ordinary Differential Equations*. Prentice-Hall, Englewood Cliffs, NJ, 1971.

99. K. Georg. On tracing an implicitly defined curve by quasi-Newton steps and calculating bifurcation by local perturbations. *SIAM J. Sci. Stat. Comput.*, 2:35–50, 1981.

100. A. George, J.W. Liu, and E. Ng. User's guide for sparspak: Waterloo sparse linear equation package. Technical Report CS–78–30, Department of Computer Science, University of Waterloo, 1980.

101. P.E. Gill and W. Murray. Nonlinear Least Squares and Nonlinearly Constrained Optimization. In *Proc. Dundee Conf. Numer. Anal. 1975*, volume 506 of *Lecture Notes in Math.*, pages 134–147. Springer, Berlin, Heidelberg, New York, 1976.

102. R. Glowinski. *Lectures on Numerical Methods for nonlinear Variational Problems*. Tata Institute of Fundamental Research, 1980.

103. G.Moore. The Numerical Treatment of Non-Trivial Bifurcation Points. *Numer. Func. Anal. & Optimiz.*, 6:441–472, 1980.

104. G.H. Golub and G. Meurant. *Résolution Numérique des Grands Systèmes Linéaires*, volume 49 of *Collection de la Direction des Etudes et Recherches de l'Electricité de France*. Eyolles, Paris, France, 1983.

105. G.H. Golub and V. Pereyra. The differentiation of pseudoinverses and nonlinear least squares problems whose variables separate. *SIAM J. Numer. Anal.*, 10:413–432, 1973.

106. G.H. Golub and C. Reinsch. Singular value decomposition and least squares solutions. *Num. Math.*, 14:403–420, 1970.

107. G.H. Golub and C.F. van Loan. *Matrix Computations*. The Johns Hopkins University Press, 2 edition, 1989.

108. G.H. Golub and J.H. Wilkinson. Note on the Iterative Refinement of Least Squares Solutions. *Num. Math.*, 9:139–148, 1966.

109. M. Golubitsky and D. Schaeffer. A theory for imperfect bifurcation via singularity theory. *Comm. Pure Appl. Math.*, 32:21–98, 1979.

110. M. Golubitsky and D. Schaeffer. *Singularities and Groups in Bifurcation Theory. Vol. I.* Springer-Verlag, New York, 1984.

111. M. Golubitsky, I. Stewart, and D. Schaeffer. *Singularities and Groups in Bifurcation Theory. Vol. II.* Springer-Verlag, New York, 1988.

112. A. Griewank and G. F. Corliss, editors. *Automatic Differentiation of Algorithms: Theory, Implementation, and Application.* SIAM, Philadelphia, 1991.

113. W. Hackbusch. *Multi-Grid Methods and Applications.* Springer Verlag, Berlin, Heidelberg, New York, Tokyo, 1985.

114. E. Hairer, S.P. Nørsett, and G. Wanner. *Solving Ordinary Differential Equations I. Nonstiff Problems.* Springer-Verlag, Berlin, Heidelberg, New York, 2nd edition, 1993.

115. E. Hairer and G. Wanner. *Solving Ordinary Differential Equations II. Stiff and Differential-Algebraic Problems.* Springer-Verlag, Berlin, Heidelberg, New York, 2nd edition, 1996.

116. S.B. Hazra, V. Schulz, J. Brezillon, and N. Gauger. Aerodynamic shape optimization using simultaneous pseudo-timestepping. *J. Comp. Phys.*, 204:46–64, 2005.

117. T.J. Healey. A group-theoretic approach to computational bifurcation problems with symmetry. *Comp. Meth. Appl. Mech. Eng.*, 67:257–295, 1988.

118. M.D. Hebden. An algorithm for minimization using excact second derivatives. Technical Report TP–515, Harwell, AERE, 1973.

119. M.W. Hirsch and S. Smale. On algorithms for solving $f(x) = 0$. *Communications on Pure and Applied Mathematics*, 55:12, 1980.

120. A. Hohmann. *Inexact Gauss Newton Methods for Parameter Dependent Nonlinear Problems.* PhD thesis, Freie Universität Berlin, 1994.

121. A.S. Householder. *Principles of Numerical Analysis.* McGraw-Hill, New York, 1953.

122. I. Jankowski and H. Wozniakowski. Iterative refinement implies numerical stability. *BIT*, 17:303–311, 1977.

123. R.L. Jennrich. Asymptotic properties of nonlinear least squares estimators. *Ann. Math. Stat.*, 40:633–643, 1969.

124. A.D. Jepson. *Numerical Hopf Bifurcation.* PhD thesis, California Institute of Techonology, Pasadena, 1981.

125. L. Kantorovich. The method of successive approximations for functional equations. *Acta Math.*, 71:63–97, 1939.

126. L. Kantorovich. On Newton's Method for Functional Equations. (Russian). *Dokl. Akad. Nauk SSSR*, 59:1237–1249, 1948.

127. L. Kantorovich and G. Akhilov. *Functional analysis in normed spaces.* Fizmatgiz, Moscow, 1959. German translation: Berlin, Akademie-Verlag, 1964.

128. L. Kaufman. A variable projection method for solving separable nonlinear least squares problems. *BIT*, 15:49–57, 1975.

129. H.B. Keller. Newton's Method under Mild Differentiability Conditions. *J. Comp. Syst. Sci.*, 4:15–28, 1970.

130. H.B. Keller. Numerical solution of bifurcation and nonlinear eigenvalue problems. In P. H. Rabinowitz, editor, *Applications in Bifurcation Theory*, pages 359–384. Academic Press, New York, San Francisco, London, 1977.

131. H.B. Keller. *Numerical Methods for Two-Point Boundary Value Problems*. Dover Publ., New York, rev. and exp. ed. edition, 1992.

132. C.T. Kelley. *Iterative methods for linear and nonlinear equations*. SIAM Publications, Philadelphia, 1995.

133. C.T. Kelley and D.E. Keyes. Convergence analysis of pseudo-transient continuation. *SIAM J. Numer. Anal.*, 35:508–523, 1998.

134. N. Kollerstrom. Thomas Simpson and 'Newton's Method of Approximation': an enduring myth. *British Journal for History of Science*, 25:347–354, 1992.

135. R. Kornhuber. Nonlinear multigrid techniques. In J. F. Blowey, J. P. Coleman, and A. W. Craig, editors, *Theory and Numerics of Differential Equations*, pages 179–229. Springer Universitext, Heidelberg, New York, 2001.

136. J. Kowalik and M.R. Osborne. *Methods for Unconstrained Optimization Problems*. American Elsevier Publ. Comp., New York, 1968.

137. M. Kubiček. Algorithm 502. Dependence of solutions of nonlinear systems on a parameter. *ACM Trans. Math. Software*, 2:98–107, 1976.

138. P. Kunkel. *Quadratisch konvergente Verfahren zur Berechnung von entfalteten Singularitäten*. PhD thesis, Department of Mathematics, University of Heidelberg, 1986.

139. P. Kunkel. Quadratically convergent methods for the computation of unfolded singularities. *SIAM J. Numer. Anal.*, 25:1392–1408, 1988.

140. P. Kunkel. Efficient computation of singular points. *IMA J. Numer. Anal.*, 9:421–433, 1989.

141. P. Kunkel. A tree–based analysis of a family of augmented systems for the computation of singular points. *IMA J. Numer. Anal.*, 16:501–527, 1996.

142. E. Lahaye. Une méthode de résolution d'une catégorie d'équations transcendentes. *C.R. Acad. Sci. Paris*, 198:1840–1842, 1934.

143. K.A. Levenberg. A method for the solution of certain nonlinear problems least squares. *Quart. Appl. Math.*, 2:164–168, 1992.

144. P. Lindström and P.-Å. Wedin. A new linesearch algorithm for unconstrained nonlinear least squares problems. *Math. Progr.*, 29:268–296, 1984.

145. W. Liniger and R.A. Willoughby. Efficient integration methods for stiff systems of ordinary differential equations. *SIAM J. Numer. Anal.*, 7:47–66, 1970.

146. G.I. Marchuk and Y.A. Kuznetsov. On Optimal Iteration Processes. *Dokl. Akad. Nauk SSSR*, 181:1041–1945, 1968.

147. D.W. Marquardt. An algorithm for least-squares-estimation of nonlinear parameters. *ACM Trans. Math. Software*, 7:17–41, 1981.

148. S.F. McCormick. A revised mesh refinement strategy for Newton's method applied to nonlinear two-point boundary value problems. *Lecture Notes Math.*, 679:15–23, 1978.

149. R.G. Melhem and W.C. Rheinboldt. A comparison of Methods for Determining Turning Points of Nonlinear Equations. *Computing*, 29:201–226, 1982.

150. R. Menzel and H. Schwetlick. Über einen Ordnungsbegriff bei Einbettungsalgorithmen zur Lösung nichtlinearer Gleichungen. *Computing*, 16:187–199, 1976.

151. C.D. Meyer. *Matrix Analysis and Applied Linear Algebra*. SIAM Publications, Philadelphia, 2000.

152. J.J. Moré. The Levenberg-Marquardt algorithm: Implementation and theory. In G. Watson, editor, *Numerical Analysis*, volume 630 of *Lecture Notes in Mathematics*, pages 105–116. Springer Verlag, New York, 1978.

153. J.J. Moré, B.S. Garbow, and K.E. Hillstrom. Testing Unconstrained Optimization Software. *SIAM J. Appl. Math.*, 11:431–441, 1992.

154. A.P. Morgan, A.J. Sommese, and L.T. Watson. Finding All Isolated Solutions to Polynomial Systems Using HOMPACK. *ACM Trans. Math. Software*, 15:93–122, 1989.

155. I. Mysovskikh. On convergence of Newton's method. (Russian). *Trudy Mat. Inst. Steklov*, 28:145–147, 1949.

156. M.S. Nakhla and F.H. Branin. Determining the periodic response of nonlinear systems by a gradient method. *Circ. Th. Appl.*, 5:255–273, 1977.

157. M.Z. Nashed. *Generalized Inverses and Applications*. Academic Press, New York, 1976.

158. U. Nowak and P. Deuflhard. Towards parameter identification for large chemical reaction systems. In P. Deuflhard and E. Hairer, editors, *Numerical Treatment of Inverse Problems in Differential and Integral Equations*. Birkhäuser, Boston, Basel, Stuttgart, 1983.

159. U. Nowak and P. Deuflhard. Numerical identification of selected rate constants in large chemical reaction systems. *Appl. Num. Math.*, 1:59–75, 1985.

160. U. Nowak and L. Weimann. GIANT—A Software Package for the Numerical Solution of Very Large Systems of Highly Nonlinear Equations. Technical Report TR 90–11, Zuse Institute Berlin (ZIB), 1990.

161. U. Nowak and L. Weimann. A Family of Newton Codes for Systems of Highly Nonlinear Equations. Technical Report TR 91–10, Zuse Institute Berlin (ZIB), 1991.

162. H.J. Oberle. *BOUNDSCO, Hinweise zur Benutzung des Mehrzielverfahrens für die numerische Lösung von Randwertproblemen mit Schaltbedingungen*, volume 6 of *Hamburger Beiträge zur angewandten Mathematik, Reihe B*. Universität Hamburg, 1987.

163. J.M. Ortega and W.C. Rheinboldt. *Iterative Solution of Nonlinear Equations in Several Variables*. Classics in Appl. Math. SIAM Publications, Philadelphia, 2nd edition, 2000.

164. M.R. Osborne. Fisher's method of scoring. *Int. Stat. Rev.*, 60:99–117, 1992.

165. V. Pereyra. Iterative Methods for Solving Nonlinear Least Squares Problems. *SIAM J. Numer. Anal.*, 4:27–36, 1967.

166. M. Pernice and H.F. Walker. NITSOL: a Newton iterative solver for nonlinear systems. *SIAM J. Sci. Comp.*, 5:275–297, 1998.

167. L.R. Petzold. A description of DASSL: A differential/algebraic system system solver. In *Scientific Computing*, pages 65–68. North-Holland, Amsterdam, New York, London, 1982.

168. H. Poincaré. *Les Méthodes Nouvelles de la Mécanique Céleste*. Gauthier-Villars, Paris, 1892.

169. E. Polak. *Computational Methods in Optimization*. Academic Press, New York, 1971.

170. F.A. Potra. Monotone Iterative Methods for Nonlinear Operator Equations. *Numer. Funct. Anal. Optim.*, 9:809–843, 1987.

171. F.A. Potra and V. Pták. *Nondiscrete Induction and Iterative Processes*. Pittman, London, 1984.

172. F.A. Potra and W.C. Rheinboldt. Newton-like Methods with Monotone Convergence for Solving Nonlinaer Operator Equations. *Nonlinear Analysis. Theory, Methods, and Applications*, 11:697–717, 1987.

173. P.H. Rabinowitz. *Applications of Bifurcation Theory*. Academic Press, New York, San Francisco, London, 1977.

174. L.B. Rall. Note on the Convergence of Newton's Method. *SIAM J. Numer. Anal.*, 11:34–36, 1974.

175. R. Rannacher. On the Convergence of the Newton-Raphson Method for Strongly Nonlinear Finite Element Equations. In P. Wriggers and W. Wagner, editors, *Nonlinear Computational Mechanics—State of the Art*, pages 11–111. Springer, 1991.

176. W.C. Rheinboldt. An adaptive continuation process for solving systems of nonlinear equations. *Polish Academy of Science, Stefan Banach Center Publ.*, 3:129–142, 1977.

177. W.C. Rheinboldt. *Methods for Solving Systems of Nonlinear Equations*. SIAM, Philadelphia, 2nd edition, 1998.

178. W.C. Rheinboldt and J.V. Burkardt. A locally parametrized continuation process. *ACM Trans. Math. Software*, 9:215–235, 1983.

179. A. Ruhe and P.-Å. Wedin. Algorithms fo Separable Nonlinear Least Squares problems. *SIAM Rev.*, 22:318–337, 1980.

180. R.D. Russell and L.F. Shampine. A Collocation Method for Boundary Value Problems. *Numer. Math.*, 19:1–28, 1972.

181. Y. Saad. *Iterative methods for sparse linear systems*. SIAM, Philadelphia, 2nd edition, 2003.

182. Y. Saad and M.H. Schultz. GMRES: A generalized minimal residual method for solving nonsymmetric systems. *SIAM J. Sci. Statist. Comput.*, 7:856–869, 1986.

183. H. Schwetlick. Effective methods for computing turning points of curves implicitly defined by nonlinear equations. Preprint Sect. Math. 46, Universität Halle, Montreal, 1997.

184. M. Seager. A SLAP for the Masses. Technical Report UCRL-100267, Lawrence Livermore National Laboratory, livermore, California, USA, 1988.

185. V.V. Shaidurov. Some estimates of the rate of convergence for the cascadic conjugate-gradient method. Technical report, Otto-von-Guericke-Universität Magdeburg, 1994.

186. J. Stoer. On the Numerical Solution of Constrained Least-Squares Problems. *SIAM J. Numer. Anal.*, 8:382–411, 1971.

187. J. Stoer and R. Bulirsch. *Introduction to Numerical Analysis*. Springer-Verlag, Berlin, Heidelberg, New York, 1980.

188. M. Urabe. Galerkin's procedure for nonlinear periodic systems. *Arch. Rat. Mech. Anal.*, 20:120–152, 1965.

189. H.A. van der Vorst. Bi-CGSTAB: A fast and smoothly converging variant of Bi-CG for the solution of non-symmetric linear systems. *SIAM J. Sci. Stat. Comput.*, 12:631–644, 1992.

190. C. van Loan. On the Method of Weighting for Equality-Constrained Least-Squares Problems. *SIAM J. Numer. Anal.*, 22:851–864, 1985.

191. J. M. Varah. Alternate row and column elimination for solving linear systems. *SIAM J. Numer. Math.*, 13:71–75, 1976.

192. R.S. Varga. *Matrix Iterative Analysis*. Prentice Hall, Eaglewood Cliffs, N. J., 1962.

193. S. Volkwein and M. Weiser. Affine Invariant Convergence Analysis for Inexact Augmented Lagrangian SQP Methods. *SIAM J. Control Optim.*, 41:875–899, 2002.

194. W. Walter. *Differential and integral inequalities*. Springer, Berlin, Heidelberg, New York, 1970.

195. W. Walter. Private communication, 1987.

196. P.-Å. Wedin. On the Gauss-Newton method for the non-linear least squares problem. Technical Report 24, Swedisch Institute of Applied Mathematics, 1974.

197. M. Weiser and P. Deuflhard. The Central Path towards the Numerical Solution of Optimal Control Problems. Technical Report 01–12, Zuse Institute Berlin (ZIB), 2001.

198. M. Weiser, A. Schiela, and P. Deuflhard. Asymptotic Mesh Independence of Newton's Method Revisited. *SIAM J. Numer. Anal.*, 42:1830–1845, 2005.

199. C. Wulff, A. Hohmann, and P. Deuflhard. Numerical continuation of periodic orbits with symmetry. Technical Report SC 94–12, Zuse Institute Berlin (ZIB), 1994.

200. J. Xu. *Theory of Multilevel Methods*. PhD thesis, Department of Mathematics, Pennsylvania State University, 1989.

201. J. Xu. Iterative methods by space decomposition and subspace correction. *SIAM Review*, 34:581–613, 1992.

202. T. Yamamoto. A unified derivation of several error bounds for Newton's process. *J. Comp. Appl. Math.*, 12/13:179–191, 1985.

203. T.J. Ypma. Local Convergence of Inexact Newton Methods. *SIAM J. Numer. Anal.*, 21:583–590, 1984.

204. H. Yserentant. On the multilevel splitting of finite element spaces. *Numer. Math.*, 58:163–184, 1986.

205. E. Zeidler. *Nonlinear Functional Analysis and its Applications I–III*. Springer, Berlin, Heidelberg, New York, Tokyo, 1985–1990.

Software

This monograph presents a scheme to construct adaptive Newton-type algorithms in close connection with an associated affine invariant convergence analysis. Part of these algorithms are presented as informal programs in the text. Some, but not all of the described algorithms have been worked out in detail. Below follows a list of codes mentioned by name in the book, which can be downloaded via the web address

http://www.zib.de/Numerics/NewtonLib/

All of the there available programs (not only by the author and his group) are free as long as they are exclusively used for research or teaching purposes.

Iterative methods for large systems of linear equations:

- PCG – adaptive preconditioned conjugate gradient method for linear systems with symmetric positive definite matrix; *energy error* norm based truncation criterion (Section 1.4.2)
- GBIT – adaptive Broyden's 'good' rank-1 update method specialized for linear equations; *error* oriented truncation criterion (Section 1.4.4)

Exact global Newton methods for systems of nonlinear equations:

- NLEQ1 – popular production code; global Newton method with *error* oriented convergence criterion; arbitrary selection of direct linear equation solver; adaptive damping strategies slightly different from Section 3.3.3; no rank strategy
- NLEQ2 – production code; global Newton method with *error* oriented convergence criterion; QR-decomposition with subcondition number estimate; adaptive damping and rank strategy slightly different from Section 3.3.3
- NLEQ-RES – global Newton method with *residual* based convergence criterion and adaptive trust region strategy (Section 3.2.2)
- NLEQ-ERR – global Newton method with *error* oriented convergence criterion and adaptive trust region strategy (Section 3.3.3)
- NLEQ-OPT – global Newton method for gradient systems originating from *convex optimization*; energy error norm oriented or objective function based convergence criteria and adaptive trust region strategy (Section 3.4.2)

Local quasi-Newton methods for systems of nonlinear equations:

- QNERR – recursive implementation of Broyden's 'good' rank-1 update method; *error* oriented convergence criterion (Section 2.1.4)
- QNRES – recursive implementation of Broyden's 'bad' rank-1 update method; *residual* based convergence criterion (Section 2.2.3)

Continuation methods for parameter dependent systems of nonlinear equations:

- CONTI1 – global Newton continuation method (classical and tangent continuation); no path-following beyond turning points (Section 5.1.3)
- ALCON1 – global quasi-Gauss-Newton continuation method; adaptive path-following beyond turning points (Section 5.2.3)
- ALCON2 – global quasi-Gauss-Newton continuation method; adaptive path-following beyond turning points; computation of bifurcation diagrams including simple bifurcations (Sections 5.2.3, 5.3.2, and 5.3.3)

Global Gauss-Newton methods for nonlinear least squares problems:

- NLSCON – (older) global *constrained* (or unconstrained) Gauss-Newton method with *error* oriented convergence criterion; adaptive trust region strategies slightly different from Sections 4.3.4 and 4.1.2
- NLSQ-RES – global *unconstrained* Gauss-Newton method with projected *residual* based convergence criterion and adaptive trust region strategy (Section 4.2.3)
- NLSQ-ERR – global *unconstrained* Gauss-Newton method with *error* oriented convergence criterion and adaptive trust region strategies (Sections 4.3.4 and 4.3.5)

Inexact global Newton methods for large systems of nonlinear equations:

- GIANT – (older) global inexact Newton method with *error* oriented convergence criterion; adaptive trust region strategy slightly different from Sections 2.1.5 and 3.3.4; earlier version of GBIT for inner iteration
- GIANT-GMRES – global inexact Newton method with *residual* based convergence criterion and adaptive trust region strategy; GMRES for inner iteration (Sections 2.2.4 and 3.2.3)
- GIANT-GBIT – global inexact Newton method with *error* oriented convergence criterion and adaptive trust region strategy; GBIT for inner iteration (Sections 2.1.5 and 3.3.4)
- GIANT-PCG – global inexact Newton method for gradient systems originating from *convex function optimization*; energy error norm oriented or function based convergence criteria and adaptive trust region strategy; PCG for inner iteration (Sections 2.3.3 and 3.4.3)

Multiple shooting methods for ODE boundary value problems:

- BVPSOL – Multiple shooting method for two-point boundary value problems; exact global Newton method with *error* oriented convergence criterion and adaptive trust region strategies (Section 7.1.2)
- MULCON – Multiple shooting method for two-point boundary value problems; adaptive Gauss-Newton continuation method (Section 7.1.3)
- PERIOD – multiple shooting method for periodic solutions of ODEs; global underdetermined Gauss-Newton method with *error* oriented convergence criterion and adaptive trust region strategies (Section 7.3.1)
- PERHOM – multiple shooting method for periodic solutions of parameter dependent ODEs; adaptive *error* oriented underdetermined Gauss-Newton orbit continuation method (Section 7.3.2)
- PARKIN – single shooting method for parameter identification in large reaction kinetic ODE networks; global Gauss-Newton method with *error* oriented convergence measure and adaptive trust region strategies (Section 7.2)

Adaptive multilevel finite element methods for elliptic PDEs:

- KASKADE – function space oriented additive multilevel FEM for linear elliptic PDEs; hierarchical basis preconditioning in 2D; BPX preconditioning in 3D (Section 1.4.5)
- Newton-KASKADE – function space oriented global inexact Newton multilevel FEM for nonlinear elliptic PDEs originating from *convex optimization*; energy error norm oriented or objective functional based convergence criteria and adaptive trust region strategy; adaptive multilevel code KASKADE for inner iteration (Section 8.3); this code is still in the form of a research code which is not appropriate for public distribution (see above web address where possible cooperation is discussed)

Index